Fabrice Guillet, Howard J. Hamilton (Eds.)

Quality Measures in Data Mining

Studies in Computational Intelligence, Volume 43

Editor-in-chief
Prof. Janusz Kacprzyk
Systems Research Institute
Polish Academy of Sciences
ul. Newelska 6
01-447 Warsaw
Poland
E-mail: kacprzyk@ibspan.waw.pl

Further volumes of this series
can be found on our homepage:
springer.com

Vol. 26. Nadia Nedjah, Luiza de Macedo Mourelle (Eds.)
Swarm Intelligent Systems, 2006
ISBN 3-540-33868-3

Vol. 27. Vassilis G. Kaburlasos
Towards a Unified Modeling and Knowledge-Representation based on Lattice Theory, 2006
ISBN 3-540-34169-2

Vol. 28. Brahim Chaib-draa, Jörg P. Müller (Eds.)
Multiagent based Supply Chain Management, 2006
ISBN 3-540-33875-6

Vol. 29. Sai Sumathi, S.N. Sivanandam
Introduction to Data Mining and its Application, 2006
ISBN 3-540-34689-9

Vol. 30. Yukio Ohsawa, Shusaku Tsumoto (Eds.)
Chance Discoveries in Real World Decision Making, 2006
ISBN 3-540-34352-0

Vol. 31. Ajith Abraham, Crina Grosan, Vitorino Ramos (Eds.)
Stigmergic Optimization, 2006
ISBN 3-540-34689-9

Vol. 32. Akira Hirose
Complex-Valued Neural Networks, 2006
ISBN 3-540-33456-4

Vol. 33. Martin Pelikan, Kumara Sastry, Erick Cantú-Paz (Eds.)
Scalable Optimization via Probabilistic Modeling, 2006
ISBN 3-540-34953-7

Vol. 34. Ajith Abraham, Crina Grosan, Vitorino Ramos (Eds.)
Swarm Intelligence in Data Mining, 2006
ISBN 3-540-34955-3

Vol. 35. Ke Chen, Lipo Wang (Eds.)
Trends in Neural Computation, 2007
ISBN 3-540-36121-9

Vol. 36. Ildar Batyrshin, Janusz Kacprzyk, Leonid Sheremetor, Lotfi A. Zadeh (Eds.)
Preception-based Data Mining and Decision Making in Economics and Finance, 2006
ISBN 3-540-36244-4

Vol. 37. Jie Lu, Da Ruan, Guangquan Zhang (Eds.)
E-Service Intelligence, 2007
ISBN 3-540-37015-3

Vol. 38. Art Lew, Holger Mauch
Dynamic Programming, 2007
ISBN 3-540-37013-7

Vol. 39. Gregory Levitin (Ed.)
Computational Intelligence in Reliability Engineering, 2007
ISBN 3-540-37367-5

Vol. 40. Gregory Levitin (Ed.)
Computational Intelligence in Reliability Engineering, 2007
ISBN 3-540-37371-3

Vol. 41. Mukesh Khare, S.M. Shiva Nagendra (Eds.)
Artificial Neural Networks in Vehicular Pollution Modelling, 2007
ISBN 3-540-37417-5

Vol. 42. Bernd J. Krämer, Wolfgang A. Halang (Eds.)
Contributions to Ubiquitous Computing, 2007
ISBN 3-540-44909-4

Vol. 43. Fabrice Guillet, Howard J. Hamilton (Eds.)
Quality Measures in Data Mining, 2007
ISBN 3-540-44911-6

Fabrice Guillet
Howard J. Hamilton
(Eds.)

Quality Measures in Data Mining

With 51 Figures and 78 Tables

Springer

Fabrice Guillet
LINA-CNRS FRE 2729 - Ecole polytechnique
de l'université de Nantes
Rue Christian-Pauc-La Chantrerie - BP 60601
44306 NANTES Cedex 3-France
E-mail: Fabrice.Guillet@polytech.univ-nantes.fr

Howard J. Hamilton
Department of Computer Science
University of Regina
Regina, SK S4S 0A2-Canada
E-mail: hamilton@cs.uregina.ca

ISSN print edition: 1860-949X
ISSN electronic edition: 1860-9503
ISBN 978-3-642-07952-8
e-ISBN 978-3-540-44918-8

Springer is a part of Springer Science+Business Media
springer.com

Softcover reprint of the hardcover 1st edition 2007

Cover design: deblik, Berlin

Preface

Data Mining has been identified as one of the ten emergent technologies of the 21st century (MIT Technology Review, 2001). This discipline aims at discovering knowledge relevant to decision making from large amounts of data. After some knowledge has been discovered, the final user (a decision-maker or a data-analyst) is unfortunately confronted with a major difficulty in the validation stage: he/she must cope with the typically numerous extracted pieces of knowledge in order to select the most interesting ones according to his/her preferences. For this reason, during the last decade, the designing of quality measures (or interestingness measures) has become an important challenge in Data Mining.

The purpose of this book is to present the state of the art concerning quality/interestingness measures for data mining. The book summarizes recent developments and presents original research on this topic. The chapters include reviews, comparative studies of existing measures, proposals of new measures, simulations, and case studies. Both theoretical and applied chapters are included.

Structure of the book

The book is structured in three parts. The first part gathers four overviews of quality measures. The second part contains four chapters dealing with data quality, data linkage, contrast sets and association rule clustering. Lastly, in the third part, four chapters describe new quality measures and rule validation.

Part I: Overviews of Quality Measures

- **Chapter 1: Choosing the Right Lens: Finding What is Interesting in Data Mining**, by Geng and Hamilton, gives a broad overview of the use of interestingness measures in data mining. This survey reviews interestingness measures for rules and summaries, classifies them from several perspectives, compares their properties, identifies their roles in the data mining process, describes methods of analyzing the measures, reviews principles for selecting appropriate measures for applications, and predicts trends for research in this area.

- **Chapter 2: A Graph-based Clustering Approach to Evaluate Interestingness Measures: A Tool and a Comparative Study**, by Hiep *et al.*, is concerned with the study of interestingness measures. As interestingness depends both on the the structure of the data and on the decision-maker's goals, this chapter introduces a new contextual approach implemented in ARQAT, an exploratory data analysis tool, in order to help the decision-maker select the most suitable interestingness measures. The tool, which embeds a graph-based clustering approach, is used to compare and contrast the behavior of thirty-six interestingness measures on two typical but quite different datasets. This experiment leads to the discovery of five stable clusters of measures.

- **Chapter 3: Association Rule Interestingness Measures: Experimental and Theoretical Studies**, by Lenca *et al.*, discusses the selection of the most appropriate interestingness measures, according to a variety of criteria. It presents a formal and an experimental study of 20 measures. The experimental studies carried out on 10 data sets lead to an experimental classification of the measures. This studies leads to the design of a multi-criteria decision analysis in order to select the measures that best take into account the user's needs.

- **Chapter 4: On the Discovery of Exception Rules: A Survey**, by Duval *et al.*, presents a survey of approaches developed for mining exception rules. They distinguish two approaches to using an expert's knowledge: using it as syntactic constraints and using it to form as commonsense rules. Works that rely on either of these approaches, along with their particular quality evaluation, are presented in this survey. Moreover, this chapter also gives ideas on how numerical criteria can be intertwined with user-centered approaches.

PART II: FROM DATA TO RULE QUALITY

- **Chapter 5: Measuring and Modelling Data Quality for Quality-Awareness in Data Mining**, by Berti-Équille. This chapter offers an overview of data quality management, data linkage and data cleaning techniques that can be advantageously employed for improving quality awareness during the knowledge discovery process. It also details the steps of a pragmatic framework for data quality awareness and enhancement. Each step may use, combine and exploit the data quality characterization, measurement and management methods, and the related techniques proposed in the literature.

- **Chapter 6: Quality and Complexity Measures for Data Linkage and Deduplication**, by Christen and Goiser, proposes a survey of different measures that have been used to characterize the quality and complexity of data linkage algorithms. It is shown that measures in the space of record pair comparisons can produce deceptive quality results. Various measures are discussed and recommendations are given on how to assess data linkage and deduplication quality and complexity.

- **Chapter 7: Statistical Methodologies for Mining Potentially Interesting Contrast Sets**, by Hilderman and Peckham, focuses on contrast sets that aim at identifying the significant differences between classes or groups. They compare two contrast set mining methodologies, STUCCO and CIGAR, and discuss the underlying statistical measures. Experimental results show that both methodologies are statistically sound, and thus represent valid alternative solutions to the problem of identifying potentially interesting contrast sets.

- **Chapter 8: Understandability of Association Rules: A Heuristic Measure to Enhance Rule Quality**, by Natarajan and Shekar, deals with the clustering of association rules in order to facilitate easy exploration of connections between rules, and introduces the *Weakness* measure dedicated to this goal. The average linkage method is used to cluster rules obtained from a small artificial data set. Clusters are compared with those obtained by applying a commonly used method.

PART III: RULE QUALITY AND VALIDATION

- **Chapter 9: A New Probabilistic Measure of Interestingness for Association Rules, Based on the Likelihood of the Link**, by Lerman and Azé, presents the foundations and the construction of a probabilistic interestingness measure called the likelihood of the link index. They discuss two facets, symmetrical and asymmetrical, of this measure and the two stages needed to build this index. Finally, they report the results of experiments to estimate the relevance of their statistical approach.

- **Chapter 10: Towards a Unifying Probabilistic Implicative Normalized Quality Measure for Association Rules**, by Diatta *et al.*, defines the so-called normalized probabilistic quality measures (PQM) for association rules. Then, they consider a normalized and implicative PQM called M_{GK}, and discuss its properties.

- **Chapter 11: Association Rule Interestingness: Measure and Statistical Validation**, by Lallich *et al.*, is concerned with association rule validation. After reviewing well-known measures and criteria, the statistical validity of selecting the most interesting rules by performing a large number of tests is investigated. An original, bootstrap-based validation method is proposed that controls, for a given level, the number of false discoveries. The potential value of this method is illustrated by several examples.

- **Chapter 12: Comparing Classification Results between N-ary and Binary Problems**, by Felkin, deals with supervised learning and the quality of classifiers. This chapter presents a practical tool that will enable the data-analyst to apply quality measures to a classification task. More specifically, the tool can be used during the pre-processing step, when the analyst is considering different formulations of the task at hand. This tool is well suited for illustrating the choices for the number of possible class values to be used to define a classification problem and the relative difficulties of the problems that result from these choices.

Topics

The topics of the book include:

- Measures for data quality
- Objective vs subjective measures
- Interestingness measures for rules, patterns, and summaries
- Quality measures for classification, clustering, pattern discovery, etc.
- Theoretical properties of quality measures
- Human-centered quality measures for knowledge validation
- Aggregation of measures
- Quality measures for different stages of the data mining process,
- Evaluation of measure properties via simulation
- Application of quality measures and case studies

Review Committee

All published chapters have been reviewed by at least 2 referees.

- Henri Briand (LINA, University of Nantes, France)
- Rgis Gras (LINA, University of Nantes, France)
- Yves Kodratoff (LRI, University of Paris-Sud, France)
- Vipin Kumar (University of Minnesota, USA)
- Pascale Kuntz (LINA, University of Nantes, France)
- Robert Hilderman (University of Regina, Canada)
- Ludovic Lebart (ENST, Paris, France)
- Philippe Lenca (ENST-Bretagne, Brest, France)
- Bing Liu (University of Illinois at Chicago, USA)
- Amdo Napoli (LORIA, University of Nancy, France)
- Gregory Piatetsky-Shapiro (KDNuggets, USA)
- Gilbert Ritschard (Geneve University, Switzerland)
- Sigal Sahar (Intel, USA)
- Gilbert Saporta (CNAM, Paris, France)
- Dan Simovici (University of Massachusetts Boston, USA)
- Jaideep Srivastava (University of Minnesota, USA)
- Einoshin Suzuki (Yokohama National University, Japan)
- Pang-Ning Tan (Michigan State University, USA)
- Alexander Tuzhilin (Stern School of Business, USA)
- Djamel Zighed (ERIC, University of Lyon 2, France)

Associated Reviewers

Jérôme Azé,
Laure Berti-Equille,
Libei Chen,
Peter Christen,
Béatrice Duval,
Mary Felkin,
Liqiang Geng,
Karl Goiser,
Stéphane Lallich,
Rajesh Natajaran,
Ansaf Salleb,
Benoît Vaillant

Acknowledgments

The editors would like to thank the chapter authors for their insights and contributions to this book.

The editors would also like to acknowledge the member of the review committee and the associated referees for their involvement in the review process of the book. Without their support the book would not have been satisfactorily completed.

A special thank goes to D. Zighed and H. Briand for their kind support and encouragement.

Finally, we thank Springer and the publishing team, and especially T. Ditzinger and J. Kacprzyk, for their confidence in our project.

Regina, Canada and Nantes, France,
May 2006

Fabrice Guillet
Howard Hamilton

Contents

List of Contributors

Jérôme Azé
LRI, University of Paris-Sud, Orsay, France
aze@lri.fr

Laure Berti-Équille
IRISA, University of Rennes I, France
Laure.Berti-Equille@irisa.fr

Julien Blanchard
LINA CNRS 2729, Polytechnic School of Nantes University, France
Julien.Blanchard@univ-nantes.fr

Henri Briand
LINA CNRS 2729, Polytechnic School of Nantes University, France
Julien.Blanchard@univ-nantes.fr

Peter Christen
The Australian National University, Canberra, Australia
peter.christen@anu.edu.au

Jean Diatta
IREMIA, University of La Réunion, Saint-Denis, France
jdiatta@univ-reunion.fr

Béatrice Duval
LERIA, University of Angers, France
Beatrice.Duval@univ-angers.fr

Mary Felkin
LRI, University of Paris-Sud, Orsay, France
felkin@lri.fr

Liqiang Geng
Department of Computer Science, University of Regina, Canada
gengl@cs.uregina.ca

Karl Goiser
The Australian National University, Canberra, Australia
karl.goiser@anu.edu.au

Régis Gras
LINA CNRS 2729, Polytechnic School of Nantes University, France
Julien.Blanchard@univ-nantes.fr

Fabrice Guillet
LINA CNRS 2729, Polytechnic School of Nantes University, France
Fabrice.Guillet@univ-nantes.fr

Howard J. Hamilton
Department of Computer Science, University of Regina, Canada
hamilton@cs.uregina.ca

Robert J. Hilderman
Department of Computer Science,
University of Regina, Canada
hilder@cs.uregina.ca

Xuan-Hiep Huynh
LINA CNRS 2729, Polytechnic
School of Nantes University, France
Xuan-Hiep.Huynh@univ-nantes.fr

Pascale Kuntz
LINA CNRS 2729, Polytechnic
School of Nantes University, France
Julien.Blanchard@univ-nantes.fr

Stéphane Lallich
ERIC, University of Lyon 2, France
stephane.lallich@univ-lyon2.fr

Philippe Lenca
TAMCIC CNRS 2872, GET/ENST
Bretagne, France
philippe.lenca@enst-bretagne.fr

Israël-César Lerman
IRISA, University of Rennes I,
France
lerman@irisa.fr

Patrick Meyer
University of Luxembourg,
Luxembourg
patrick.meyer@uni.lu

Rajesh Natarajan
Cognizant Technology Solutions,
Chennai, India
rajesh.natarajan@cognizant.com

Terry Peckham
Department of Computer Science,
University of Regina, Canada
peckham@cs.uregina.ca

Elie Prudhomme
ERIC, University of Lyon 2, France
eprudhomme@eric.univ-lyon2.fr

Henri Ralambondrainy
IREMIA, University of La Réunion,
Saint-Denis, France
ralambon@univ-reunion.fr

Ansaf Salleb
UCCLS, Columbia University,
New York, U.S.A
Ansaf@ccls.columbia.edu

B. Shekar
QMIS, Indian Institute
of Management(IIMB),
Bangalore, India
shek@iimb.ernet.in

Olivier Teytaud
TAO-Inria, LRI,
University of Paris-Sud,
Orsay, France
teytaud@lri.fr

André Totohasina
ENSET, University of Antsiranana,
Madagascar
totohasina@yahoo.fr

Benoît Vaillant
TAMCIC CNRS 2872, GET/ENST
Bretagne, France
benoit.vaillant@enst-bretagne.fr

Christel Vrain
LIFO, University of Orléans, France
Christel.Vrain@univ-orleans.fr

Part I

Overviews on rule quality

Choosing the Right Lens: Finding What is Interesting in Data Mining*

Liqiang Geng and Howard J. Hamilton

Department of Computer Science, University of Regina, Regina, Saskatchewan, Canada {gengl, hamilton}@cs.uregina.ca

Summary. Interestingness measures play an important role in data mining regardless of the kind of patterns being mined. Good measures should select and rank patterns according to their potential interest to the user. Good measures should also reduce the time and space cost of the mining process. This survey reviews the interestingness measures for rules and summaries, classifies them from several perspectives, compares their properties, identifies their roles in the data mining process, and reviews the analysis methods and selection principles for appropriate measures for applications.

Key words: Data Mining, Knowledge Discovery, Interestingness Measures, Association Rules, Summaries.

1 Introduction

Measuring the interestingness of discovered patterns is an active and important area of data mining research. Although much work has been conducted in this area, so far there is no widespread agreement on a formal definition of interestingness in this context. Based on the diversity of definitions presented to date, interestingness is perhaps best treated as a very broad concept, which emphasizes conciseness, coverage, reliability, peculiarity, diversity, novelty, surprisingness, utility, and actionability. These nine more specific criteria are used to determine whether or not a pattern is interesting as shown in Table 1.

These factors are sometimes correlated with rather than independent of one another. For example, actionability may be a good approximation for surprisingness and vice versa [41], conciseness often coincides with generality,

* This article is a shortened and adapted version of Geng, L. and Hamilton, H. J., Interestingness Measures for Data Mining: A Survey, *ACM Computing Surveys*, 38, 2006, in press, which is copyright © 2006 by the Association for Computing Machinery, Inc. Used by permission of the ACM.

Liqiang Geng and Howard J. Hamilton: *Choosing the Right Lens: Finding What is Interesting in Data Mining*, Studies in Computational Intelligence (SCI) **43**, 3–24 (2007)
www.springerlink.com

Criteria	Description	References
Conciseness	A pattern contains few attribute-value pairs. A pattern set contains few patterns	Bastide et al., 2000 [5] Padmanabhan and Tuzhilin, 2000 [33]
Generality/Coverage	A pattern covers a relatively large subset of a dataset.	Agrawal and Srikant, 1994 [2]; Webb and Brain, 2002 [47]
Reliability	The relationship described by a pattern occurs in a high percentage of applicable cases	Ohsaki et al., 2004 [31]; Tan et al., 2002 [43]
Peculiarity	A pattern is far away from the other discovered patterns according to some distance measure	Barnett and Lewis, 1994 [4]; Knorr et al., 2000 [22]; Zhong et al., 2003 [51]
Diversity	The elements of a the pattern differ significantly from each other or the patterns in a set differ significantly from each other	Hilderman and Hamilton, 2001 [19]
Novelty	A person did not know the pattern before and is not able to infer it from other known patterns	Sahar, 1999 [36]
Surprisingness	A pattern contradicts a person's existing knowledge or expectations	Bay and Pazzani, 1999 [6]; Carvalho and Freitas, 2000 [8] Liu et al., 1997 [27]; Liu et al., 1999 [28]; Silberschatz and Tuzhilin, 1995 [40]; Silberschatz and Tuzhilin, 1996 [41];
Utility	A pattern is of use to a person in reaching a goal.	Chan et al., 2003 [9]; Lu et al., 2001 [29]; Shen et al., 2002 [39]; Yao et al., 2004 [48]
Actionability/ Applicability	A pattern enables decision making about future actions in the domain	Ling et al., 2002 [26]; Wang et al., 2002 [46]

Table 1. Criteria for interestingness.

and generality conflicts with peculiarity. The conciseness, generality, reliability, peculiarity, and diversity factors depend only on the data and the patterns, and thus can be considered as objective. The novelty, surprisingness, utility, and actionability factors depend on the person who uses the patterns as well as the data and patterns themselves, and thus can be considered as subjective. The comparison of what is classified as interesting by both the objective and the subjective interestingness measures (IMs) to what human subjects choose has rarely been tackled. Two recent studies have compared the ranking of rules by human experts to the ranking of rules by various IMs and suggested choosing the measure that produces the most similar ranking to that of the experts [31, 43]. These studies were based on specific data sets and experts, and their results cannot be taken as general conclusions. It must be noted that research on IMs has not been restricted to the field of data mining. There is widespread activity in many research areas, including information retrieval [3], pattern recognition [11], statistics [18], and machine learning [30]. Within these fields, many techniques, including discriminant analysis, outlier detection, genetic algorithms, and validation techniques involving bootstrap or Monte Carlo methods, have a prominent role in data mining and provide foundations for IMs. So far, researchers have proposed IMs for various kinds of patterns, analyzed their theoretical properties and empirical evaluations, and proposed strategies for selecting appropriate measures for different domains and different requirements. The most common patterns that can be evaluated by IMs include association rules, classification rules, and summaries. In Section 2, we review the IMs for association rules and classification rules. In Section 3, we review the IMs for summaries. In Section 4, we present the conclusions.

2 IMs for association rules and classification rules

In data mining research during the past decade, most IMs have been proposed for evaluating association rules and classification rules. An association rule is defined in the following way [2]. Let $I = \{i_1, i_2, \ldots, i_m\}$ be a set of items. Let D a set of transactions, where each transaction T is a set of items such that $T \subseteq I$. An *association rule* is an implication of the form $X \rightarrow Y$, where $X \subset I$, $Y \subset I$, and $X \cap Y = \phi$. The rule holds for the data set D with confidence c and support s if c% of transactions in D that contain X also contain Y and s% of transactions in D contain $X \cup Y$. In this paper, we assume that the *support* and *confidence* measures yield fractions from $[0, 1]$ rather than percentages. A *classification rule* is an implication of the form $X_1 = x_1, X_2 = x_2, \ldots, X_n = x_n \rightarrow Y = y$, where X_i is a conditional attribute, x_i is a value that belongs to the domain of X_i, Y is the class attribute and y is class value. The rule $X_1 = x_1, X_2 = x_2, \ldots, X_n = x_n \rightarrow Y = y$ specifies that if an object satisfies the condition $X_1 = x_1, X_2 = x_2, \ldots, X_n = x_n$, it can be classified into the category of y. Although association rules and classification rules have different purposes, mining techniques, and evaluation methods, IMs are often applied to them in the same way. In this survey, we will distinguish between them only as necessary.

Based on the levels of involvement of the user and the amount of additional information required beside the raw data set, IMs for association rules can be classified into objective measures and subjective measures.

An *objective measure* is based on only the raw data. No additional knowledge about the data is required. Most objective measures are based on theories in probability, statistics, or information theory. They represent the correlation or distribution of the data.

A *subjective measure* is a one that takes into account both the data and the user who uses the data. To define a subjective measure, the user's domain or background knowledge about the data, which are usually represented as beliefs and expectations, is needed. The data mining methods involving subjective measures are usually interactive. The key issue for mining patterns based on subjective measures is the representation of user's knowledge, which has been addressed by various frameworks and procedures for data mining [27, 28, 40, 41, 36].

2.1 Objective measures for association rules or classification rules

Objective measures have been thoroughly studied by many researchers. Tables 2 and 3 list 38 common measures [43, 24, 31]. In these tables, A and B represent the antecedent and the consequent of a rule, respectively. $P(A)$ denotes the probability of A, $P(B|A)$ denotes the conditional probability of B given A. Association measures based on probability are usually functions of contingency tables. These measures originate from various areas, such as statistics (correlation coefficient, Odds ratio, Yule's Q, and Yule's Y [43, 49, 31]), information theory (J-measure and mutual information [43, 49, 31]), and information retrieval (accuracy and sensitivity/recall [23, 49, 31]).

Given an association rule $A \rightarrow B$, the two main interestingness factors for the rule are generality and reliability. *Support* $P(AB)$ or *coverage* $P(A)$ is used to represent the generality of the rule. *Confidence* $P(B|A)$ or correlation is used to represent the reliability of the rule. Some researchers have suggested that a good IM should include both generality and reliability. For example, Tan et al [42], proposed the IS measure $IS = \sqrt{I \times support}$, where $I = \frac{P(AB)}{P(A)P(B)}$ is the ratio between the joint probability of two variables with respect to their expected probability under the independence assumption. Since $IS = \sqrt{I \times support} = \sqrt{\frac{P(AB)^2}{P(A)P(B)}} = \frac{P(AB)}{\sqrt{P(A)P(B)}}$, we can see that this measure also represents the cosine angle between A and B. Lavrac et al. proposed a weighted relative accuracy $WRAcc = P(A)(P(B|A) - P(B))$ [23]. This measure combines the coverage $P(A)$ and the correlation between B and A. This measure is identical to the Piatetsky-Shapiro's measure $P(AB) - P(A)P(B)$ [15]. Other measures involving these two factors include Yao and Liu's two way support [49], Jaccard [43], Gray and Orlowska's interestingness weighting dependency [16], and Klosgen's measure [21]. All these measure combine support $P(AB)$ (or coverage $P(A)$) and a correlation factor, $(P(B|A) - P(B)$ or $\frac{P(B|A)}{P(B)})$.

Tan et al. refer to a measure that includes both support and correlation as an *appropriate measure*. They argue that any appropriate measure can be

Measure	Formula
Support	$P(AB)$
Confidence/Precision	$P(B\|A)$
Coverage	$P(A)$
Prevalence	$P(B)$
Recall	$P(A\|B)$
Specificity	$P(\neg B\|\neg A)$
Accuracy	$P(AB) + P(\neg A \neg B)$
Lift or Interest	$\frac{P(B\|A)}{P(B)}$ or equivalently $\frac{P(AB)}{P(A)P(B)}$
Leverage	$P(B\|A) - P(A)P(B)$
Added Value/Change of Support	$P(B\|A) - P(B)$
Relative Risk	$\frac{P(B\|A)}{P(B\|\neg A)}$
Jaccard	$\frac{P(AB)}{P(A)+P(B)-P(AB)}$
Certainty Factor	$\frac{P(B\|A)-P(B)}{1-P(B)}$
Odds Ratio	$\frac{P(AB)P(\neg A \neg B)}{P(A \neg B)P(\neg B A)}$
Yule's Q	$\frac{P(AB)P(\neg A \neg B)-P(A \neg B)P(\neg AB)}{P(AB)P(\neg A \neg B)+P(A \neg B)P(\neg AB)}$
Yule's Y	$\frac{\sqrt{P(AB)P(\neg A \neg B)}-\sqrt{P(A \neg B)P(\neg AB)}}{\sqrt{P(AB)P(\neg A \neg B)}+\sqrt{P(A \neg B)P(\neg AB)}}$
Klosgen	$\sqrt{P(AB)} \times (P(B\|A) - P(B))$
Brin's Conviction	$\frac{P(A)P(\neg B)}{P(A \neg B)}$
Gray and Orlowska's Interestingness weighting Dependency (GOI-D)	$((\frac{P(AB)}{P(A)P(B)})^k - 1) \times P(AB)^m$, k, m: Coefficients of dependency and generality, weighting the relative importance of the two factors.

Table 2. Definitions of objective IMs for rules, part I.

used to rank discovered patterns [42]. They also show that the behaviors of such measures, especially where support is low, are similar.

Bayardo and Agrawal studied the relationship between support, confidence, and other measures from another angle [7]. They define a partial ordered relation based on support and confidence as follows. For rules r_1 and r_2, if $support(r_1) \leq support(r_2)$ and $confidence(r_1) \leq confidence(r_2)$, we have $r_1 \leq_{sc} r_2$. Any rule r in the upper border, for which there is no r' such that $r \leq_{sc} r'$, is called a sc-optimal rule. For those measures that are monotone in both support and confidence, the most interesting rules are sc-optimal rules. For example, the Laplace measure $\frac{N(AB)+1}{N(A)+2}$, where $N(AB)$ and $N(A)$ denote the number of records that include AB and A, respectively, can be transformed to $\frac{|D|support(A \to B)+1}{\frac{|D|support(A \to B)}{confidence(A \to B)}+2}$, where $|D|$ denotes the number of records in the data set. Since $|D|$ is a constant, the Laplace measure can be considered as a function of *support* and *confidence* of the rule $A \to B$. It is easy to show

Measure	Formula
Collective Strength	$\frac{P(AB)+P(\neg B\|\neg A)}{P(A)P(B)+P(\neg A)P(\neg B)} \times \frac{1-P(A)P(B)-P(\neg A)P(\neg B)}{1-P(AB)-P(\neg B\|\neg A)}$
Laplace Correction	$\frac{N(AB)+1}{N(A)+2}$
Gini Index	$P(A) \times (P(B\|A)^2 + P(\neg B\|A)^2) + P(\neg A) \times (P(B\|\neg A)^2 + P(\neg B\|\neg A)^2) - P(B)^2 - P(\neg B)^2$
Goodman and Kruskal's Predictive Association	$\frac{\sum_i max_j P(A_iB_j)+\sum_j max_i P(A_iB_j)-max_i P(A_i)-max_i P(B_j)}{2-max_i P(A_i)-max_i P(B_j)}$
Normalized Mutual Information	$\frac{\sum_i \sum_j P(A_iB_j)\times log_2 \frac{P(A_iB_j)}{P(A_i)P(B_j)}}{-\sum_i P(A_i)\times log_2 P(A_i)}$
J-Measure	$P(AB)log(\frac{P(B\|A)}{P(B)}) + P(A\neg B)log(\frac{P(\neg B\|A)}{P(\neg B)})$
Yao and Liu's One Way Support	$P(B\|A) \times log_2 \frac{P(AB)}{P(A)P(B)}$
Yao and Liu's Two Way Support	$P(AB) \times log_2 \frac{P(AB)}{P(A)P(B)}$
Yao and Liu's Two Way Support Variation	$P(AB)\times log_2 \frac{P(AB)}{P(A)P(B)} + P(A\neg B)\times log_2 \frac{P(A\neg B)}{P(A)P(\neg B)} + P(\neg AB)\times log_2 \frac{P(\neg AB)}{P(\neg A)P(B)} + P(\neg A\neg B) \times log_2 \frac{P(\neg A\neg B)}{P(\neg A)P(\neg B)}$
ϕ-Coefficient (Linear Correlation Coefficient)	$\frac{P(AB)-P(A)P(B)}{\sqrt{P(A)P(B)P(\neg A)P(\neg B)}}$
Piatetsky-Shapiro's	$P(AB) - P(A)P(B)$
Cosine	$\frac{P(AB)}{\sqrt{P(A)P(B)}}$
Loevinger	$1 - \frac{P(A)P(\neg B)}{P(A\neg B)}$
Information Gain	$log \frac{P(AB)}{P(A)P(B)}$
Sebag-Schoenauer	$\frac{P(AB)}{P(A\neg B)}$
Least Contradiction	$\frac{P(AB)-P(A\neg B)}{P(B)}$
Odd Multiplier	$\frac{P(AB)P(\neg B)}{P(B)P(A\neg B)}$
Example and Counter Example Rate	$1 - \frac{P(A\neg B)}{P(AB)}$
Zhang	$\frac{P(AB)-P(A)P(B)}{max(P(AB)P(\neg B),P(B)P(A\neg B))}$

Table 3. Definitions of objective IMs for rules, part II.

that the Laplace measure is monotone in both support and confidence. This property is useful when the user is only interested in the single most interesting rule, since we only need to check the *sc*-optimal rule set, which contains fewer rules than the entire rule set.

Yao et al. identified a fundamental relationship between preference relations and IMs for association rules: there exists a real valued IM if and only if the preference relation is a weak order [50]. A weak order is a relation that is *asymmetric* and *negative transitive.* It is a special type of partial order and more general than a total order. Other researchers studied more general forms of IMs. Jaroszewicz and Simovici proposed a general measure based on distribution divergence [20]. The chi-square, Gini measures, and entropy gain measures can be obtained from this measure by setting different parameters. In this survey, we do not elaborate on the individual measures. Instead, we emphasize the properties of these measures and how to analyze them and choose from among them for data mining applications.

Properties of the measures: Many objective measures have been proposed for different applications. To analyze these measures, some properties for the measures have been proposed. We consider three sets of properties that have been proposed in the literature. Piatetsky-Shapiro [15] proposed three principles that should be obeyed by any objective measure F.

P1. $F = 0$ if A and B are statistically independent; i.e., $P(AB) = P(A)P(B)$.

P2. F monotonically increases with $P(AB)$ when $P(A)$ and $P(B)$ remain the same.

P3. F monotonically decreases with $P(A)$ (or $P(B)$) when $P(AB)$ and $P(B)$ (or $P(A)$) remain the same.

The first principle states that an association rule that occurs by chance has zero interest value, i.e., it is not interesting. In practice, this principle may seem too rigid and some researchers propose a constant value for the independent situations [43]. The second principle states that the greater the support for $A \to B$ is, the greater the interestingness value is when the support for A and B is fixed, i.e., the more positive correlation A and B have, the more interesting the rule is. The third principle states that if the supports for $A \to B$ and B (or A) are fixed, the smaller the support for A (or B) is, the more interesting the pattern is. According to these two principles, when the cover of A and B are identical or the cover of A contains the cover of B (or vice versa), the IM should attain its maximum value.

Tan et al. proposed five properties based on operations on 2×2 contingency tables [43].

O1. F should be symmetric under variable permutation.

O2. F should be the same when we scale either any row or any column by a positive factor.

O3. F should become -F under row/column permutation, i.e., swapping the rows or columns in the contingency table makes interestingness values change sign.

O4. F should remain the same under both row and column permutation.

O5. F should have no relationship with the count of the records that do not contain A and B.

Property O1 states that rule $A \to B$ and $B \to A$ should have the same interestingness values. To provide additional symmetric measures, Tan et al. transformed each asymmetric measure F to a symmetric one by taking the maximum value of $F(A \rightarrow B)$ and $F(B \rightarrow A)$. For example, they define a symmetric confidence measure as $max(P(B|A), P(A|B))$ [43]. Property O2 requires invariance with the scaling the rows or columns. Property O3 states that $F(A \rightarrow B) = -F(A \rightarrow \neg B) = -F(\neg A \rightarrow B)$. This property means that the measure can identify both positive and negative correlations. Property O4 states that $F(A \rightarrow B) = F(\neg A \rightarrow \neg B)$. Property O3 is a special case of property O4. Property O5 states that the measure should only take into account the number of the records containing A or B or both. Support does not satisfy this property, while confidence does.

Lenca et al. proposed five properties to evaluate association measures [24].

Q1. F is constant if there is no counterexample to the rule.

Q2. $P(A\neg B)$ is linear concave or shows a convex decrease around 0+.

Q3. F increases as the total number of records increases.

Q4. The threshold is easy to fix.

Q5. The semantics of the measure are easy to express.

Lenca et al. claimed that properties Q1, Q4, and Q5 are desirable for measures, but that properties Q2 and Q3 may or may not be desired by users. Property Q1 states that rules with confidence of 1 should have the same interestingness value regardless of the support, which contradicts Tan et al. who suggested that a measure should combine support and association aspects. Property Q2 was initially proposed by Gras et al. [15]. It describes the manner in which the interestingness value decreases as a few counterexamples are added. If the user can tolerate a few counterexamples, a concave decrease is desirable. If the system strictly requires confidence of 1, a convex decrease is desirable. Property Q3 contradicts property O2, which implies that if the total number of the records is increased, while all probabilities are fixed, the interestingness measure value will not change. In contrast, Q3 states that if all probabilities are fixed, the interestingness measure value increases with the size of the data set. This property reflects the idea that rules obtained from large data sets tend to be more reliable.

In Tables 4 and 5, we indicate which of the properties hold for each of the measures listed in Tables 2 and 3, respectively. For property Q2, we assume the total number of the records is fixed and when the number of the records for $A\neg B$ increases, the number of the records for AB decreases correspondingly. In Tables 4 and 5, we use 0, 1, 2, 3, 4, 5, and 6 to represent convex decrease, linear decrease, concave decrease, not sensitive, increase, not applicable, and depending on parameters, respectively.

P1: $F = 0$ if A and B are statistically independent.

P2: F monotonically increases with $P(AB)$.

P3: F monotonically decreases with $P(A)$ and $P(B)$.

O1: F is symmetric under variable permutation.

O2: F is the same when we scale any row or column by a positive factor.

O3: F becomes $-F$ under row/column permutation.

O4: F remains the same under both row and column permutation.

Measure	P1	P2	P3	O1	O2	O3	O4	O5	Q1	Q2	Q3
Support	N	Y	N	Y	N	N	N	N	N	1	N
Confidence/Precision	N	Y	N	N	N	N	N	N	Y	1	N
Coverage	N	N	N	N	N	N	N	N	N	3	N
Prevalence	N	N	N	N	N	N	N	N	N	1	N
Recall	N	Y	N	N	N	N	N	Y	N	2	N
Specificity	N	N	N	N	N	N	N	N	N	3	N
Accuracy	N	Y	Y	Y	N	N	Y	N	N	1	N
Lift or Interest	N	Y	Y	Y	N	N	N	N	N	2	N
Leverage	N	Y	Y	N	N	N	N	Y	N	1	N
Added Value	Y	Y	Y	N	N	N	N	N	N	1	N
Relative Risk	N	Y	Y	N	N	N	N	N	N	1	N
Jaccard	N	Y	Y	Y	N	N	N	Y	N	1	N
Certainty Factor	Y	Y	Y	N	N	N	Y	N	N	0	N
Odds ratio	N	Y	Y	Y	Y	Y	Y	N	Y	0	N
Yule's Q	Y	Y	Y	Y	Y	Y	Y	N	Y	0	N
Yule's Y	Y	Y	Y	Y	Y	Y	Y	N	Y	0	N
Klosgen	Y	Y	Y	N	N	N	N	N	N	0	N
Brin's Conviction	N	Y	N	N	N	N	Y	N	Y	0	N
Gray and Orlowska's Interestingness weighting Dependency (GOI-D)	N	Y	N	N	N	N	N	Y	N	6	N

Table 4. Properties of objective IMs for rules, part I.

O5: F has no relationship with the count of the records that do not contain A and B.

Q1: F is constant if there is no counterexample to the rule.

Q2: $P(A \neg B)$ is linear concave or shows a convex decrease around 0+.

Q3: F increases as the total number of records increases.

Selection strategies for measures: Due to the overwhelming number of IMs, a means of selecting an appropriate measure for an application is an important issue. So far, two methods have been proposed for comparing and analyzing the measures, namely ranking and clustering. Analysis can be conducted based on either the properties of the measures or empirical evaluations on data sets. Table 6 classifies the studies that are summarized here.

Tan et al. [43] proposed a method to rank the measures based on a specific data set. In this method, the user is first required to rank a set of mined patterns, and then the measure that has the most similar ranking results for these patterns is selected for further use. This method is not directly applicable if the number of the patterns is overwhelming. Instead, the method selects the patterns that have the greatest standard deviations in their rankings by the IMs. Since these patterns cause the greatest conflict among the measures, they should be presented to the user for ranking. The method then selects the measure that gives rankings that are most consistent with the manual

Measure	P1	P2	P3	O1	O2	O3	O4	O5	Q1	Q2	Q3
Collective Strength	N	Y	Y	Y	N	Y	Y	N	N	0	N
Laplace Correction	N	Y	N	N	N	N	N	N	N	1	N
Gini Index	Y	N	N	N	N	N	Y	N	N	0	N
Goodman and Kruskal's	Y	N	N	Y	N	N	Y	N	N	5	N
Normalized Mutual Information	Y	Y	Y	N	N	N	Y	N	N	5	N
J-Measure	Y	N	N	N	N	N	N	N	Y	0	N
Yao and Liu's based on one way support	Y	Y	Y	N	N	N	N	Y	N	0	N
Yao and Liu's based on two way support	Y	Y	Y	Y	N	N	N	Y	N	0	N
Yao and Liu's two way support variation	Y	N	N	Y	N	N	Y	N	N	0	N
ϕ-Coefficient (Linear Correlation Coefficient)	Y	Y	Y	Y	N	Y	Y	N	N	0	N
Piatetsky-Shapiro's	Y	Y	Y	Y	N	Y	Y	N	N	1	N
Cosine	N	Y	Y	Y	N	N	N	Y	N	2	N
Loevinger	Y	Y	N	N	N	N	N	N	Y	4	N
Information gain	Y	Y	Y	Y	N	N	N	Y	N	2	N
Sebag-Schoenauer	N	Y	Y	N	N	N	N	Y	Y	0	N
Least Contradiction	N	Y	Y	N	N	N	N	Y	N	2	N
Odd Multiplier	N	Y	Y	N	N	N	N	N	Y	0	N
Example and Counter Example Rate	N	Y	Y	N	N	N	N	Y	Y	2	N
Zhang	Y	N	N	N	N	N	N	N	N	0	N

Table 5. Properties of objective IMs for rules, part II.

Analysis method	Based on properties	Based on data sets
Ranking	Lenca et al., 2004 [24]	Tan et al., 2002 [43]
Clustering	Vaillant et al., 2004 [44]	Vaillant et al., 2004 [44]

Table 6. Analysis methods for objective association rule IMs.

ranking. This method is based on the specific data set and needs the user's involvement.

Another method to select the appropriate measure is based on the multi-criteria decision aid Lenca et al. [24]. In this method, marks and weights are assigned to each property that the user thinks is of importance. For example, if a symmetric property is desired, a measure is assigned 1 if it is symmetric, and 0 if it is asymmetric. With each row representing a measure and each column representing a property, a decision matrix is created. An entry in the matrix represents the mark for the measure according to the property. By applying the multi-criteria decision process to the table, one can obtain a ranking of results. With this method, the user is not required to rank the mined patterns. Instead, he or she needs to identify the desired properties and specify their significance for a particular application.

Another method for analyzing measures is to cluster the IMs into groups [44]. As with the ranking method, the clustering method can be based on either the properties of the measures or the results of experiments on data sets. *Property based clustering*, which groups the measures based on the similarity of their properties, works on a decision matrix, with each row representing a measure, and each column representing a property. *Experiment based clustering* works on a matrix with each row representing a measure and each column representing a measure applied to a data set. Each entry represents a similarity value between the two measures on the specified data set. Similarity is calculated on the rankings of the two measures on the data set. Vaillant et al. showed consistent results using the two clustering methods with twenty data sets.

Form-dependent objective measures: A *form dependent measure* is an objective measure based on the form of the rules. We consider form dependent measures based on peculiarity, surprisingness, and conciseness.

The *neighborhood-based unexpectedness* measure [10] is based on peculiarity. The intuition for this method is that if a rule has a different consequent from neighboring rules, it is interesting. The distance $Dist(R_1, R_2)$ between two rules $R_1 : X_1 \rightarrow Y_1$ and $R_2 : X_2 \rightarrow Y_2$ is defined as $Dist(R_1, R_2) = \delta_1|X_1Y_1 - X_2Y_2| + \delta_2|X_1 - X_2| + \delta_3|Y_1 - Y_2|$, where $X - Y$ denotes the symmetric difference between X and Y, $|X|$ denotes the cardinality of X, and δ_1, δ_2, δ_3 are the weights determined by the user. Based on this distance, the $r-$neighborhood of the rule R_0, denoted as $N(R_0, r)$, is defined as $\{R : Dist(R, R_0) \leq r,\ R \text{ is a potential rule}\}$, where $r > 0$ is the *radius* of the neighborhood. Then the authors proposed two IMs. The first one is called *unexpected confidence*: if the confidence of a rule r_0 is far from the average confidence of rules in its neighborhood, this rule is interesting. Another measure is based on the sparsity of neighborhood, i.e., the if the number of mined rules in the neighborhood is far fewer than the number of all potential rules in the neighborhood, it is considered to be interesting.

Another form-dependent measure is called *surprisingness*. Most of the literature uses subjective IMs to represent the surprisingness of classification rules. Taking a different perspective, [14] defined two objective IMs for this purpose, based on the form of the rules.

The first measure is based on the generalization of the rule. Suppose there is a classification rule $A_1, A_2, \ldots, A_n \rightarrow C$. When we remove one of the conditions, say A_1, we get a more general rule $A_2, \ldots, A_n \rightarrow C_1$. If $C_1 = C$, we count one, otherwise we count zero. Then we do the same for $A_2, \ldots$ and A_n and count the sum of the times C_i differs from C. The result, an integer in the interval $[0, n]$, is defined as the *raw surprisingness* of the rule, denoted as $Surp_{raw}$. *Normalized surprisingness* $Surp_{norm}$, defined as $Surp_{raw}/n$, takes on real values in the interval $[0, 1]$. If all classes that the generalized rules predict are different from the original class C, $Surp_{norm}$ takes on value 1, which means that the rule is most interesting. If all classes that the generalized rules predict are the same as C, $Surp_{norm}$ takes on value 0, which means that the rule is not interesting at all, since all its generalized forms make the same prediction. This method can be regarded as a neighborhood-based method, where the neighborhood of a rule R is the set of the rules with one condition removed from R.

Freitas also proposed another measure based on information gain [14]. It is defined as the reciprocal of the average information gain for all the condition attributes in a rule. It is based on the assumption that a larger information gain indicates a better attribute for classification, and thus the user may be more aware of it and consequently the rules containing these attributes may be of less interest. This measure is biased towards the rules that have less than average information gain for all their condition attributes.

Conciseness, a form-dependent measure, is often used for rule sets rather than single rules. We consider two methods for evaluating the conciseness of rules. The first method is based on logical redundancy [33, 5, 25]. In this method, no measure is defined for the conciseness; instead, algorithms are designed to find non-redundant rules. For example, Li and Hamilton propose both an algorithm to find a minimum rule set and an inference system. The set of association rules discovered by the algorithm is minimum in that no redundant rules are present. All other association rules that satisfy confidence and support constraints can be derived from this rule set using the inference system.

The second method to evaluate the conciseness of a rule set is called the Minimum Description Length (MDL) principle. It takes into account both the complexity of the theory (rule set in this context) and the accuracy of the theory. The first part of the MDL measure, $L(H)$, is called the *theory cost*, which measures the theory complexity, where H is theory. The second part, $L(D|H)$, measures the degree to which the theory fails to account for the data, where D denotes the data. For a group of theories (rule sets), a more complex theory tends to fit the data better than a simpler one, and therefore, it has a higher $L(H)$ value and a smaller $L(D|H)$ value. The theory with the shortest description length has the best balance between these two factors and is preferred. Detailed MDL measures for classification rules and decision trees can be found in [13, 45]

Objective IMs indicate the support and degree of correlation of a pattern for a given data set. However, they do not take into account the knowledge of the user who uses the data.

2.2 Subjective IMs

In many applications, the user may have background knowledge and the patterns that have the highest rankings according to the objective measures may not be novel.

Subjective IMs are measures that take into account both the data and the user's knowledge. They are appropriate when (1) the background knowledge of the users varies, (2) the interests of the users vary, and (3) the background knowledge of the user evolves. The advantage of subjective IMs is that they can reduce the number of patterns mined an tailor the mining results for a specific user. However, at the same time, subjective IMs pose difficulties for the user. First, unlike the objective measures considered in the previous section, subjective measures may not be representable by simple mathematical formulas. Mining systems with different subjective interesting measures have various knowledge representation forms [27, 28]. The user must learn the appropriate specifications for these measures. Secondly, subjective measures usually require more human interaction during the mining process [36].

Three kinds of subjective measures are based on unexpectedness [41], novelty, and actionability [35]. An *unexpectedness measure* leads to finding patterns that are surprising to the user, while an *actionability measure* tries to identify patterns with which the user can take actions to his or her advantage. While the former tries to find patterns that contradict the user's knowledge, the latter tries to find patterns that conform to situations identified by the user as actionable [28].

Unexpectedness and Novelty: To find unexpected or novel patterns in data, three approaches can be distinguished based on the roles of the unexpectedness measures in the mining process: (1) the user provides a formal specification of his/her knowledge, and after obtaining the mining results, the system chooses the unexpected patterns to present to the user [27, 28, 40, 41]; (2) according to the user's interactive feedback, the system removes the uninteresting patterns [36]; and (3) the system applies the user's specification as constraints during the mining process to narrow down the search space and provide fewer results [32]. Let us consider each of these approaches in turn.

The first approach is to use IMs to filter interesting patterns from mined results. Liu et al. proposed a technique to rank the discovered rules according to their interestingness [28]. They classify the interestingness of discovered rules into three categories, finding unexpected patterns, confirming the user's knowledge, and finding actionable patterns. According to Liu et al., an *unexpected pattern* is a pattern that is unexpected or previously unknown to the user [28], which corresponds to our terms surprising and novel. Three types of unexpectedness of a rule are identified, unexpected consequent, unexpected condition, and totally unexpected patterns. A *confirming pattern* is intended to validate a user's knowledge by the mined patterns. An *actionable pattern* can help the user do something to his or her advantage. In this case, the user should provide the situations under which he or she may take actions. For all three categories, the user must provide some patterns reflecting his or her knowledge about potential actions. This knowledge is in the form of fuzzy rules. The system matches each discovered pattern against the fuzzy rules. The discovered patterns are then ranked according to their degrees of matching. The authors proposed different IMs for the three categories. All these IMs are based on functions of fuzzy values that represent the match between the user's knowledge and the discovered patterns.

Liu et al. proposed two specifications for defining a user's vague knowledge, T1 and T2 [27]. T1 can express the positive and negative relations between a condition variable and a class, the relation between a range (or a subset) of values of condition variables and class, and even more vague impressions that there is a relation between a condition variable and a class. T2 extends T1 by separating the user's knowledge into core and supplement. The *core* represents the user's certain knowledge and the *supplement* represents the user's uncertain knowledge. The core and a subset of supplement have a relation with a class. Then the user proposes match algorithms for obtaining IMs for these two kinds of specifications for conforming rules, unexpected conclusion rules, and unexpected condition rules.

Silberschatz and Tuzhilin relate unexpectedness to a belief system [41]. To define beliefs, they use arbitrary predicate formulae in first order logic rather than if-then rules. They also classify beliefs into hard beliefs and soft beliefs. A *hard belief* is a constraint that cannot be changed with new evidence. If the evidence (rules mined from data) contradicts hard beliefs, a mistake is assumed to have been made in acquiring the evidence. A *soft belief* is a belief that the user is willing to change as new patterns are discovered. They adopt Bayesian approach and assume that the degree of belief is measured with the conditional probability. Given evidence E (discovered rules), the degree of belief in a is updated with the following Bayes rule:

$$P(\alpha|E,\xi) = \frac{P(E|\alpha,\xi)P(\alpha|\xi)}{P(E|\alpha,\xi)P(\alpha|\xi) + P(E|\neg\alpha,\xi)P(\neg\alpha|\xi)}, \tag{1}$$

where ξ is the context. Then, the IM for pattern p relative to a soft belief system B is defined as the relative difference of the prior and posterior probabilities:

$$I(p,B) = \sum_{\alpha \in B} \frac{|P(\alpha|p,\xi) - P(\alpha|\xi)|}{P(\alpha|\xi)}.$$

The second approach to finding unexpectedness and novelty is to eliminate uninteresting patterns. To reduce the amount of computation and interactions with the user in filtering interesting association rules, Sahar proposed a method that removes uninteresting rules rather than selecting interesting rules [36]. The method consists of three steps. (1) The system selects the *best candidate rule* as the rule with one condition attribute and one consequence attribute that has the largest cover list. The *cover list* of a rule R is all mined rules that contain the condition and consequence of R. (2) The best candidate rule is presented to the user to be classified into four categories, not-true-not-interesting, not-true-interesting, true-not-interesting, and true-and-interesting. Sahar describes a rule as being not-interesting if it is "common knowledge," i.e., "not novel" in our terminology. If the best candidate rule R is *not-true-not-interesting* or *true-not-interesting*, the system removes it and its cover list. If the rule is *not-true-interesting*, the system removes this rule and all rules in its cover list that have the same condition. Finally, if the rule is *true-interesting*, the system keeps it. This process iterates until the rule set is empty or the user halts the process. The remaining patterns are true and interesting to the user.

The advantage of this method is that the users are not required to provide specifications; instead, they work with the system interactively. They only need to classify simple rules into true or false, interesting or uninteresting, and then the system can eliminate a significant number of uninteresting rules. The drawback of this method is that although it makes the rule set smaller, it does not rank the interestingness of the remaining rules.

The third approach to finding unexpectedness and novelty is to constrain the search space. Instead of filtering uninteresting rules after the mining process, Padmanabhan and Tuzhilin proposed a method to narrow down the mining space based on the user's expectations [32]. Here, a user's beliefs are represented in the same format as mined rules. Only surprising rules, i.e., rules that contradict existing beliefs are mined. The algorithm to find surprising rules consists of two parts, ZoominUR and ZoomoutUR. For a given belief $X \rightarrow Y$, ZoominUR finds all rules of the form $X, A \rightarrow \neg Y$, i.e., more specific rules that have the contradictory consequence to the belief. Then ZoomoutUR generalizes the rules found by ZoomoutUR. For rule $X, A \rightarrow \neg Y$, ZoomoutUR finds all X' the rules $X', A \rightarrow \neg Y$, where X' is a subset of X.

Actionability: Ling et al. proposed a measure to find optimal actions for profitable Customer Relationship Management [26]. In this method, a decision tree is mined from the data. The non-leaf nodes correspond to the customer's conditions. The leaf nodes relate to profit that can be obtained from the customer. A cost for changing a customer's condition is assigned. Based on the cost information and the profit gains, the system finds the *optimal action*, i.e., the action that maximizes *profit_gain* $- \sum$ *cost*.

Wang et al. proposed an integrated method to mine association rules and recommend the best one with respect to the profit to the user [46]. In addition to support and confidence, the system incorporates two other measures, *rule profit* and recommendation profit. The rule profit is defined as the total profit obtained in the transactions that match this rule. The *recommendation profit* is the average profit for each transaction that matches the rule. The recommendation system chooses the rules in the order of recommendation profit, rule profit, and conciseness.

3 IMs for Summaries

Summarization is considered as one of the major tasks in knowledge discovery and is the key issue in online analytical processing (OLAP) systems. The essence of summarization is the formation of interesting and compact descriptions of raw data at different concept levels, which are called *summaries*. For example, sales information in a company may be summarized to the levels of area, *City*, *Province*, and *Country*. It can also be summarized to the levels of time, *Week*, *Month*, and *Year*. The combination of all possible levels for all attributes produces many summaries. Finding interesting summaries with IMs is accordingly an important issue.

Diversity has been widely used as an indicator of the interestingness of a summary. Although diversity is difficult to define, it is widely accepted that it is determined by two factors, the proportional distribution of classes in the population and the number of classes [19]. Table 7 lists several measures for diversity [19]. In this definition, p_i denotes the probability for class i, and $\overline{q}$ denotes the average probability for all classes.

Measure	Definitions
Variance	$\frac{\sum_{i=1} (p_i - \overline{q})^2}{m-1}$
Simpson	$\sum_{i=1} p_i^2$
Shannon	$-\sum_{i=1} p_i log_2 p_i$
Berger	$max(p_i)$
Schutz	$\frac{\sum_{i=1} \lvert p_i - \overline{q} \rvert}{2m\overline{q}}$
Theil	$\frac{\sum_{i=1} \lvert p_i log_2 p_i - \overline{q} log_2 \overline{q} \rvert}{m\overline{q}}$
Atkinson	$1 - \prod_{i=1} \frac{p_i}{\overline{q}}$

Table 7. IMs for diversity.

Based on a wide variety of previous work in statistics, information theory, economics, ecology, and data mining, Hilderman and Hamilton proposed some general principles that a good measure should satisfy (see [19] for many citations; other research in statistics [1], economics [12], and data mining [52] is also relevant).

1. *Minimum Value Principle.* Given a vector $(n_1, \ldots, n_m)$, where $n_i = n_j$ for all i, j, measure $f(n_1, \ldots, n_m)$ attains its minimum value.

This property indicates that the uniform distribution is the most uninteresting.

2. *Maximum Value Principle.* Given a vector $(n_1, \ldots, n_m)$, where $n_1 = N - m + 1$, $n_i = 1$, $i = 2, \ldots, m$, and $N > m$, $f(n_1, \ldots, n_m)$ attains its maximum value. This property shows that the most uneven distribution is the most interesting.

3. *Skewness principle.* Given a vector $(n_1, \ldots, n_m)$, where $n_1 = N - m + 1$, $n_i = 1, i = 2, \ldots, m$, and $N > m$, and a vector $(n_1 - c, n_2, \ldots, n_m, n_{m+1}, n_{m+c})$, where $n_1 - c > 1$, $n_i = 1$, $i = 2, \ldots, m + c$, $f(n_1, \ldots, n_m) > f(n_1 - c, n_2, \ldots, n_{m+c})$.

This property specifies that when the number of the total frequency remains the same, the IM for the most uneven distribution decreases when the number of the classes of tuples increases. This property has a bias for small number of classes.

4. *Permutation Invariance Principle.* Given a vector $(n_1, \ldots, n_m)$ and any permutation $(i_1, \ldots, i_m)$ of $(1, \ldots, m)$, $f(n_1, \ldots, n_m) = f(n_{i_1}, \ldots, n_{i_m})$.

This property specifies that interestingness for diversity has nothing to do with the order of the class; it is only determined by the distribution of the counts.

5. *Transfer principle.* Given a vector $(n_1, \ldots, n_m)$ and $0 < c < n_j < n_i$, $f(n_1, \ldots, n_i + c, \ldots, n_j - c, \ldots, n_m) > f(n_1, \ldots, n_i, \ldots, n_j, \ldots, n_m)$.

This property specifies that when a positive transfer is made from the count of one tuple to another tuple whose count is greater, the interestingness increases.

Properties 1 and 2 are special cases of Property 5. They represent boundary conditions of property 5.

These principles can be used to identify the interestingness of a summary according to its distributions.

In data cube systems, a cell in the summary instead of the summary itself might be interesting. *Discovery-driven exploration* guides the exploitation process by providing users with IM for cells in data cube based on statistical models [38]. Initially, the user specifies a starting summary and a starting cell, and the tool automatically calculates three interestingness values for each cell based on statistical models. The first value indicates the interestingness of this cell itself, the second value indicates the interestingness it would be if we drilled down from this cell to more detailed summaries, and the third value indicates which paths to drill down from this cell. The user can follow the guidance of the tool to navigate through the space of the data cube.

The process by automatically finding the underlying reasons for an exception can be simplified [37]. The user identifies an interesting difference between two cells. The system presents the most relevant data in more detailed cubes that account for the difference.

Subjective IMs for summaries can also be used to find unexpected summaries for user. Most of the objective IMs can be transformed to subjective measures by replacing the average probability by the expected probability. For example variance $\frac{\sum_{i=1}(p_i-\overline{q})^2}{m-1}$ becomes $\frac{\sum_{i=1}(p_i-e_i)^2}{m-1}$ where p_i is the observed probability for a cell i, $\overline{q}$ is the average probability, and e_i is the expected probability for cell i. It is difficult for a user to specify all expectations quickly and consistently. The user may prefer to specify expectations for one or a few summaries in the data cube. Therefore, a propagation method is needed to propagate the expectations to all other summaries. Hamilton et al. proposed a propagation method for this purpose [17].

4 Conclusions

To reduce the number of the mined results, many interestingness measures have been proposed for various kinds of the patterns. In this paper, we surveyed various IMs used in data mining. We summarized nine criteria to determine and define interestingness. Based on the form of the patterns, we distinguished measures for rules and those for summaries. Based on the degree of human interaction, we distinguished objective measures and subjective measures. Objective IMs are based on probability theory, statistics, and information theory. Therefore, they have strict principles and foundations and their properties can be formally analyzed and compared. We surveyed the properties of different kinds of objective measures, the analysis methods, and strategies for selecting such measures for applications. However, objective measures do not take into account the context of the domain of application and the goals and background knowledge of the user. Subjective measures incorporate the user's background knowledge and goals, respectively, and are suitable for more experienced users and interactive data mining. It is widely accepted that no single measure is superior to all others or suitable for all applications.

Since objective measures and subjective measures both have merits and limitation, in the future, the combination of objective and subjective measures is a possible research direction.

Of the nine criteria for interestingness, novelty (at least in the way we have defined it) has received the least attention. The prime difficulty is in modeling either the full extent of what the user knows or what the user does not know, in order to identify what is new. Nonetheless, novelty remains a crucial factor in the human appreciation for interesting results.

So far, the subjective measures have employed a variety of representations for the user's background knowledge, which lead to different measure definitions and inferences.

Choosing IMs that reflect real human interest remains an open issue. One promising approach is to use meta learning to automatically select or combine appropriate measures. Another possibility is to develop an interactive user interface based on visually interpreting the data according to the selected measure to assist the selection process. Extensive experiments comparing the results of IMs with actual human interest are required.

References

1. Aczel J. and Daroczy, Z. *On Measures of Information and Their Characterizations.* Academic Press, New York, 1975.
2. Agrawal, R. and Srikant, R. Fast algorithms for mining association rules. In *Proceedings of the 20th International Conference on Very Large Data Bases,* pages 487–499, Santiago, Chile, 1994.
3. Baeza-Yates, R. and Ribeiro-Neto, B. *Modern Information Retrieval.* Addison Wesley, Boston, 1999.

4. Barnett, V., and Lewis, T. *Outliers in Statistical Data.* John Wiley and Sons, New York, 1994.
5. Bastide, Y., Pasquier, N., Taouil, R., Stumme, G., and Lakhal, L. Mining minimal non-redundant association rules using frequent closed itemsets. In *Proceedings of the First International Conference on Computational Logic*, pages 972–986, London, UK, 2000.
6. Bay, S.D. and Pazzani, M.J. Detecting change in categorical data: mining contrast sets. In *Proceedings of the Fifth International Conference on Knowledge Discovery and Data Mining (KDD99)*, pages 302–306, San Diego, USA, 1999.
7. Bayardo, R.J. and Agrawal R. Mining the most interesting rules. In *Proceedings of the Fifth International Conference on Knowledge Discovery and Data Mining (KDD99)*, pages 145–154, San Diego, USA, 1999.
8. Carvalho, D.R. and Freitas, A.A. A genetic algorithm-based solution for the problem of small disjuncts. In *Proceedings of the Fourth European Conference on Principles of Data Mining and Knowledge Discovery (PKDD2000)*, pages 345–352, Lyon, France, 2000.
9. Chan, R., Yang, Q., and Shen, Y. Mining high utility itemsets. In *Proceedings of the Third IEEE International Conference on Data Mining (ICDM03)*, pages 19–26, Melbourne, FL, 2003.
10. Dong G. and Li, J. Interestingness of discovered association rules in terms of neighborhood-based unexpectedness. In *Proceedings of Second Pacific Asia Conference on Knowledge Discovery in Databases (PAKDD98)*, pages 72–86, Melbourne, 1998.
11. Duda, R.O., Hart, P.E., and Stork, D.G. *Pattern Classification.* Wiley-Interscience, 2001.
12. Encaoua, D. and Jacquemin, A. Indices de concentration et pouvoir de monopole. *Revue Economique*, 29(3):514–537, 1978.
13. Forsyth, R.S., Clarke, D.D., and Wright, R.L. Overfitting revisited: an information-theoretic approach to simplifying discrimination trees. *Journal of Experimental and Theoretical Artificial Intelligence*, 6:289–302, 1994.
14. Freitas, A.A. On objective measures of rule surprisingness. In *Proceedings of the Second European Symposium on Principles of Data Mining and Knowledge Discovery (PKDD1998)*, pages 1–9, Nantes, France, 1998.
15. Gras, R., Couturier, R., Blanchard, J., Briand, H., Kuntz, P., and Peter, P. Quelques critres pour une mesure de qualit des rgles d'association. *Revue des Nouvelles Technologies de l'Information*, pages 3–31, 2004.
16. Gray, B. and Orlowska, M.E. Ccaiia: clustering categorical attributes into interesting association rules. In *Proceedings of Second Pacific Asia Conference on Knowledge Discovery in Databases (PAKDD98)*, pages 132–143, Melbourne, 1998.
17. Hamilton, H.J., Geng, L., Findlater, L., and Randall, D.J. Spatio-temporal data mining with expected distribution domain generalization graphs. In *Proceedings 10th Symposium on Temporal Representation and Reasoning/International Conference on Temporal Logic (TIME-ICTL 2003)*, pages 181–191, Cairns, Australia, 2003.
18. Hastie, T., Tibshirani, R., and Friedman, J. *The Elements of Statistical Learning: Data Mining, Inference, and Prediction.* Springer, New York, 2001.
19. Hilderman, R.J. and Hamilton, H.J. *Knowledge Discovery and Measures of Interest.* Kluwer Academic Publishers, 2001.

20. Jaroszewicz, S. and Simovici, D.A. A general measure of rule interestingness. In *Proceedings of the Fifth European Conference on Principles of Data Mining and Knowledge Discovery (PKDD2001)*, pages 253–265, Freiburg, Germany, 2001.
21. Klosgen, W. Explora: A multipattern and multistrategy discovery assistant. In *Advances in Knowledge Discovery and Data Mining (Fayyad et al. eds)*, pages 249–271, California, 1996. AAAI Press/MIT Press.
22. Knorr, E.M., Ng, R.T., and Tucakov, V. Distance based outliers: Algorithms and applications. *International Journal on Very Large Data Bases*, 8:237–253, 2000.
23. Lavrac, N., Flach, P., and Zupan, B. Rule evaluation measures: A unifying view. In *Proceedings of the Ninth International Workshop on Inductive Logic Programming (ILP'99)), (Dzeroski and Flach eds)*, pages 174–185, Bled, Slovenia, 1999. Springer-Verlag.
24. Lenca P., Meyer P., Vaillant B., Lallich S. A. Multicriteria decision aid for im selection. Technical Report Technical Report LUSSI-TR-2004-01-EN, LUSSI Department, GET/ENST Bretagne, France, 2004.
25. Li, G. and Hamilton, H.J. Basic association rules. In *Proceedings of 2004 SIAM International Conference on Data Mining (SDM04)*, pages 166–177, Orlando, USA, 2004.
26. Ling C., Chen, T., Yang Q., and Chen J. Mining optimal actions for profitable crm. In *Proceedings of the 2002 IEEE International Conference on Data Mining (ICDM2002)*, pages 767–770, Maebashi City, Japan, 2002.
27. Liu, B., Hsu, W., and Chen, S. Using general impressions to analyze discovered classification rules. In *Proceedings of the Third International Conference on Knowledge Discovery and Data Mining (KDD97)*, pages 31–36, Newport Beach, California, USA, 1997.
28. Liu, B., Hsu, W., Mun, L., and Lee, H. Finding interesting patterns using user expectations. *IEEE Transactions on Knowledge and Data Engineering*, 11(6):817–832, 1999.
29. Lu, S., Hu, H., and Li, F. Mining weighted association rules. *Intelligent Data Analysis*, 5(3):211–225, 2001.
30. Mitchell, T.M. *Machine Learning.* McGraw-Hill, 1997.
31. Ohsaki, M., Kitaguchi, S., Okamoto, K., Yokoi, H., and Yamaguchi, T. Evaluation of rule ims with a clinical dataset on hepatitis. In *Proceedings of the Eighth European Conference on Principles of Data Mining and Knowledge Discovery (PKDD2004)*, pages 362–373, Pisa, Italy, 2004.
32. Padmanabhan, B. and Tuzhilin A. A belief-driven method for discovering unexpected patterns. In *Proceedings of the Fourth International Conference on Knowledge Discovery and Data Mining (KDD 98)*, pages 94–100, New York City, 1998.
33. Padmanabhan, B. and Tuzhilin A. Small is beautiful: Discovering the minimal set of unexpected patterns. In *Proceedings of Proceedings of the Sixth International Conference on Knowledge Discovery and Data Mining (KDD 2000)*, pages 54–63, Boston, USA, 2000.
34. Piatetsky-Shapiro, G. Discovery, analysis and presentation of strong rules. In *Knowledge Discovery in Databases (Piatetsky-Shapiro and Frawley eds)*, pages 229–248, MIT Press, Cambridge, MA, 1991.
35. Piatetsky-Shapiro, G. and Matheus, C. The interestingness of deviations. In *Proceedings of KDD Workshop 1994 (KDD 94)*, pages 77–87, Seattle, USA, 1994.

36. Sahar, S. Interestingness via what is not interesting. In *Proceedings of the Fifth International Conference on Knowledge Discovery and Data Mining*, pages 332–336, San Diego, USA, 1999.
37. Sarawagi, S. Explaining differences in multidimensional aggregates. In *Proceedings of the 25th International Conference on Very Large Data Bases (VLDB)*, pages 42–53, Edinburgh, Scotland, 1999.
38. Sarawagi, S., Agrawal, R., and Megiddo, N. Discovery driven exploration of olap data cubes. In *Proceedinds of the Sixth International Conference of Extending Database Technology (EDBT'98)*, pages 168–182, Valencia, Spain, 1998.
39. Shen, Y.D., Zhang, Z., and Yang, Q. Objective-oriented utility-based association mining. In *Proceedings of the 2002 IEEE International Conference on Data Mining (ICDM02)*, pages 426–433, Maebashi City, Japan, 2002.
40. Silberschatz, A. and Tuzhilin, A. On subjective measures of interestingness in knowledge discovery. In *First International Conference on Knowledge Discovery and Data Mining*, pages 275–281, Montreal, Canada, 1995.
41. Silberschatz, A. and Tuzhilin, A. What makes patterns interesting in knowledge discovery systems. *IEEE Transactions on Knowledge and Data Engineering*, 8(6):970–974, 1996.
42. Tan, P and Kumar, V. Ims for association patterns: A perspective. Technical Report Technical Report 00-036, Department of Computer Science, University of Minnesota, 2000.
43. Tan, P., Kumar, V., and Srivastava, J. Selecting the right im for association patterns. In *Proceedings of Proceedings of the Eighth International Conference on Knowledge Discovery and Data Mining (KDD02)*, pages 32–41, Edmonton, Canada, 2002.
44. Vaillant, B., Lenca, P., and Lallich, S. A clustering of ims. In *Proceedings of the Seventh International Conference on Discovery Science, (DS'2004)*, pages 290–297, Padova, Italy, 2004.
45. Vitanyi, P.M.B. and Li, M. Minimum description length induction, bayesianism, and kolmogorov complexity. *IEEE Transactions on Information Theory*, 46(2):446–464, 2000.
46. Wang, K., Zhou, S. and Han, J. Profit mining: from patterns to actions. In *Proceedings of the Eighth Conference on Extending Database Technology (EDBT 2002)*, pages 70–87, Prague, 2002.
47. Webb, G.I., and Brain, D. Generality is predictive of prediction accuracy. In *Proceedings of the 2002 Pacific Rim Knowledge Acquisition Workshop (PKAW 2002)*, pages 117–130, Tokyo, Japan, 2002.
48. Yao, H., Hamilton, H.J., and Butz, C.J. A foundational approach for mining itemset utilities from databases. In *Proceedings of SIAM International Conference on Data Mining*, pages 482–486, Orlando, FL, 2004.
49. Yao, Y.Y. and Zhong, N. An analysis of quantitative measures associated with rules. In *Proceedings of the Third Pacific-Asia Conference on Knowledge Discovery and Data Mining*, pages 479–488, Beijing, China, 1999.
50. Yao, Y.Y., Chen, Y., and Yang, X.D. A measurement-theoretic foundation of rule interestingness evaluation. In *Foundations and New Directions in Data Mining, (Lin et al. eds)*, pages 221–227, Melbourne, Florida, 2003.

51. Zhong, N., Yao, Y.Y., and Ohshima, M. Peculiarity oriented multidatabase mining. *IEEE Transactions on Knowledge and Data Engineering*, 15(4):952–960, 2003.
52. Zighed, D., Auray, J.P., and Duru, G. Sipina: Methode et logiciel. In *Editions Lacassagne*, Lyon, France, 1992.

A Graph-based Clustering Approach to Evaluate Interestingness Measures: A Tool and a Comparative Study

Xuan-Hiep Huynh, Fabrice Guillet, Julien Blanchard, Pascale Kuntz, Henri Briand, and Régis Gras

LINA CNRS 2729 - Polytechnic School of Nantes University, La Chantrerie BP 50609 44306 Nantes cedex 3, France {1stname.name}@polytech.univ-nantes.fr

Summary. Finding interestingness measures to evaluate association rules has become an important knowledge quality issue in KDD. Many interestingness measures may be found in the literature, and many authors have discussed and compared interestingness properties in order to improve the choice of the most suitable measures for a given application. As interestingness depends both on the data structure and on the decision-maker's goals, some measures may be relevant in some context, but not in others. Therefore, it is necessary to design new contextual approaches in order to help the decision-maker select the most suitable interestingness measures. In this paper, we present a new approach implemented by a new tool, ARQAT, for making comparisons. The approach is based on the analysis of a correlation graph presenting the clustering of objective interestingness measures and reflecting the post-processing of association rules. This graph-based clustering approach is used to compare and discuss the behavior of thirty-six interestingness measures on two prototypical and opposite datasets: a highly correlated one and a lowly correlated one. We focus on the discovery of the stable clusters obtained from the data analyzed between these thirty-six measures.

1 Introduction

As the number of discovered rules increases, end-users, such as data analysts and decision makers, are frequently confronted with a major challenge: how to validate and select the most interesting of those rules. Over the last decade the Knowledge Discovery in Databases (KDD) community has recognized this challenge – often referred to as interestingness – as an important and difficult component of the KDD process (Klemettinen et al. [15], Tan et al. [30]). To tackle this problem, the most commonly used approach is based on the construction of Interestingness Measures (IM).

In defining association rules, Agrawal et al. [1] [2] [3], introduced two IMs: support and confidence. These are well adapted to Apriori algorithm

Xuan-Hiep Huynh et al.: *A Graph-based Clustering Approach to Evaluate Interestingness Measures: A Tool and a Comparative Study*, Studies in Computational Intelligence (SCI) **43**, 25–50 (2007)
www.springerlink.com

constraints, but are not sufficient to capture the whole aspects of the rule interestingness. To push back this limit, many complementary IMs have been then proposed in the literature (see [5] [14] [30] for a survey). They can be classified in two categories [10]: subjective and objective. Subjective measures explicitly depend on the user's goals and his/her knowledge or beliefs. They are combined with specific supervised algorithms in order to compare the extracted rules with the user's expectations [29] [24] [21]. Consequently, subjective measures allow the capture of rule novelty and unexpectedness in relation to the user's knowledge or beliefs. Objective measures are numerical indexes that only rely on the data distribution.

In this paper, we present a new approach and a dedicated tool ARQAT (Association Rule Quality Analysis Tool) to study the specific behavior of a set of 36 IMs in the context of a specific dataset and in an exploratory analysis perspective, reflecting the post-processing of association rules. More precisely, ARQAT is a toolbox designed to help a data-analyst to capture the most suitable IMs and consequently, the most interesting rules within a specific ruleset.

We focus our study on the objective IMs studied in surveys [5] [14] [30]. The list of IMs is added with four complementary IMs (Appendix A): Implication Intensity (II), Entropic Implication Intensity (EII), TIC (information ratio modulated by contra-positive), and IPEE (probabilistic index of deviation from equilibrium). Furthermore, we present a new approach based on the analysis of a correlation graph (CG) for clustering objective IMs.

This approach is applied to compare and discuss the behavior of 36 IMs on two prototypical and opposite datasets: a strongly correlated one (mushroom dataset [23]) and a lowly correlated one (synthetic dataset). Our objective is to discover the stable clusters and to better understand the differences between IMs.

The paper is structured as follows. In Section 2, we present related works on objective IMs for association rules. Section 3 presents a taxonomy of the IMs based on two criteria: the "subject" (deviation from independence or equilibrium) of the IMs and the "nature" of the IMs (descriptive or statistical). In Section 4, we introduce the new tool ARQAT for evaluating the behavior of IMs. In Section 5, we detail the correlation graph clustering approach. And, Section 6 is dedicated to a specific study on two prototypical and opposite datasets in order to extract the stable behaviors.

2 Related works on objective IMs

The surveys on the objective IMs mainly address two related research issues: *(1)* defining a set of principles or properties that lead to the design of a good IM, *(2)* comparing the IM behavior from a data-analysis point of view. The results yielded can be useful to help the user select the suitable ones.

Considering the principles of a good IM issue, Piatetsky-Shapiro [25] introduced the Rule-Interest, and proposed three underlying principles for a good IM on a rule $a \rightarrow b$ between two itemsets a and b: 0 value when a and b are independent, monotonically increasing with *a and b*, monotonically decreasing with a or b. Hilderman and Hamilton [14] proposed five principles: minimum value, maximum value, skewness, permutation invariance, transfer. Tan et al. [30] defined five interestingness principles: symmetry under variable permutation, row/column scaling invariance, anti-symmetry under row/column permutation, inversion invariance, null invariance. Freitas [10] proposed an "attribute surprisingness" principle. Bayardo and Agrawal [5] concluded that the most interesting rules according to some selected IMs must reside along a support/confidence border. The work allows for improved insight into the data and supports more user-interaction in the optimized rule-mining process. Kononenko [19] analyzed the biases of eleven IMs for estimating the quality of multi-valued attributes. The values of information gain, J-measure, Gini-index, and relevance tend to linearly increase with the number of values of an attribute. Zhao and Karypis [33] used seven different criterion functions with clustering algorithms to maximize or minimize a particular one. Gavrilov et al. [11] studied the similarity measures for the clustering of similar stocks. Gras et al. [12] discussed a set of ten criteria: increase, decrease with respect to certain expected semantics, constraints for semantics reasons, decrease with trivial observations, flexible and general analysis, discriminative residence with the increment of data volume, quasi-inclusion, analytical properties that must be countable, two characteristics of formulation and algorithms.

Some of these surveys also address the related issue of the IM comparison by adopting a data-analysis point of view. Hilderman and Hamilton [14] used the five proposed principles to rank summaries generated from databases and used sixteen diversity measures to show that: six measures matched five proposed principles, and nine remaining measures matched at least one proposed principle. By studying twenty-one IMs, Tan et al. [30] showed that an IM cannot be adapted to all cases and use both a support-based pruning and standardization methods to select the best IMs; they found that, in some cases many IMs are highly correlated with each other. Eventually, the decision-maker will select the most suitable measure by matching the five proposed properties. Vaillant et al. [31] evaluated twenty IMs to choose a user-adapted IM with eight properties: asymmetric processing of a and b for an association rule $a \rightarrow b$, decrease with n_b, independence, logical rule, linearity with $n_{a\bar{b}}$ around 0^+, sensitivity to n, easiness to fix a threshold, intelligibility. Finally, Huynh et al. [16] introduced the first result of a new clustering approach for classifying thirty-four IMs with a correlation analysis.

3 A taxonomy of objective IMs

In this section, we propose a taxonomy of the objective IMs (details in Appendixes A and B) according to two criteria: the "subject" (deviation from independence or equilibrium), and the "nature" (descriptive or statistical). The conjunction of these criteria seems to us essential to grasp the meaning of the IMs, and therefore to help the user choose the ones he/she wants to apply.

In the following, we consider a finite set T of transactions. We denote an association rule by $a \rightarrow b$ where a and b are two disjoint itemsets. The itemset a (respectively b) is associated with a transaction subset $A = T(a) = \{t \in T, a \subseteq t\}$ (respectively $B = T(b)$). The itemset $\overline{a}$ (respectively $\overline{b}$) is associated with $\overline{A} = T(\overline{a}) = T - T(a) = \{t \in T, a \not\subseteq t\}$ (respectively $\overline{B} = T(\overline{b})$). In order to accept or reject the general trend to have b when a is present, it is quite common to consider the number $n_{a\overline{b}}$ of negative examples (contra-examples, counter-examples) of the rule $a \rightarrow b$. However, to quantify the "surprisingness" of this rule, consider some definitions are functions of $n = |T|$, $n_a = |A|$, $n_b = |B|$, $n_{\overline{a}} = |\overline{A}|$, $n_{\overline{b}} = |\overline{B}|$.

Let us denote that, for clarity, we also keep the probabilistic notations $p(a)$ (respectively $p(b)$, $p(a \text{ and } b)$, $p(a \text{ and } \overline{b})$) as the probability of a (respectively b, $a \text{ and } b$, $a \text{ and } \overline{b}$). This probability is estimated by the frequency of a: $p(a) = \frac{n_a}{n}$ (respectively $p(b) = \frac{n_b}{n}$, $p(a \text{ and } b) = \frac{n_{ab}}{n}$, $p(a \text{ and } \overline{b}) = \frac{n_{a\overline{b}}}{n}$).

3.1 Subject of an IM

Generally speaking, an association rule is more interesting when it is supported by lots of examples and few negative examples. Thus, given n_a, n_b and n, the interestingness of $a \rightarrow b$ is minimal when $n_{a\overline{b}} = \min(n_a, n_{\overline{b}})$ and maximal when $n_{a\overline{b}} = \max(0, n_a - n_b)$. Between these extreme situations, there exist two significant configurations in which the rules appear non-directed relations and therefore can be considered as neutral or non-existing: the independence and the equilibrium. In these configurations, the rules are to be discarded.

Independence

Two itemsets a and b are independent if $p(a \text{ and } b) = p(a) \times p(b)$, i.e. $n.n_{a\overline{b}} = n_a n_{\overline{b}}$. In the independence case, each itemset gives no information about the other, since knowing the value taken by one of the itemsets does not alter the probability distribution of the other itemset: $p(b \backslash a) = p(b \backslash \overline{a}) = p(b)$ and $p(\overline{b} \backslash a) = p(\overline{b} \backslash \overline{a}) = p(\overline{b})$ (the same for the probabilities of a and $\overline{a}$ given b or $\overline{b}$). In other words, knowing the value taken by an itemset lets our uncertainty about the other itemset intact. There are two ways of deviating from the independent situation: either the itemsets a and b are positively correlated ($p(a \text{ and } b) > p(a) \times p(b)$), or they are negatively correlated ($p(a \text{ and } b) < p(a) \times p(b)$).

Equilibrium

We define the equilibrium of a rule $a \rightarrow b$ as the situation where examples and negative examples are equal in numbers: $n_{ab} = n_{a\overline{b}} = \frac{1}{2}n_a$ [7]. In this situation, the itemset a is as concomitant with b as with $\overline{b}$ in the data. So a rule $a \rightarrow b$ at equilibrium is as directed towards b as towards $\overline{b}$. There are two ways of deviating from this equilibrium situation: either a is more concomitant with b than with $\overline{b}$, or a is more concomitant with $\overline{b}$ than with b.

Deviation from independence and from equilibrium

As there exist two different notions of neutrality, the objective interestingness of association rules must be measured from (at least) two complementary points of view: the deviation from independence, and the deviation from equilibrium. These are what we call the two possible subjects for the rule IMs. These deviations are directed in favor of examples and in disfavor of negative examples.

Definition 1. An IM m evaluates a deviation from independence if the IM has a fixed value at the independence:

$$m(n, n_a, n_b, \frac{n_a n_{\overline{b}}}{n}) = constant$$

Definition 2. An IM m evaluates a deviation from equilibrium if the IM has a fixed value at the equilibrium:

$$m(n, n_a, n_b, \frac{n_a}{2}) = constant$$

Independence is a function of four parameters n, n_a, n_b and $n_{a\overline{b}}$[1], whereas equilibrium is a function of the two parameters n_a and $n_{a\overline{b}}$. Thus, all the IMs of deviation from independence depend on the four parameters, while the IMs of deviation from equilibrium do not depend on n_b and n generally. The only exceptions to this principle are IPEE [7] and the Least Contradiction [4]. IPEE (see the formula in Appendix A) measures the statistical significance of the deviation from equilibrium. It depends on n. The Least Contradiction depends on n_b (see the formula in Appendix B). This is a hybrid IM which has a fixed value at equilibrium – as the IMs of deviation from equilibrium – but decreases with n_b – as the IMs of deviation from independence.

Comparison of the filtering capacities

We aim at filtering the rules with a threshold on the IMs (by retaining only the high values of the IMs), and at comparing the numbers of rules that are

[1] Here we have chosen $n_{a\overline{b}}$ as a parameter, but we could have chosen another cardinality of the joint distribution of the itemsets a and b, such as n_{ab}.

rejected by the IMs of deviation from equilibrium and from independence. Let us consider a rule with the cardinalities n, n_a, n_b, and $n_{a\bar{b}}$. By varying $n_{a\bar{b}}$ with fixed n, n_a, and n_b, one can distinguish two different cases:

- Case 1: $n_b \geq \frac{n}{2}$. Then $\frac{n_a n_{\bar{b}}}{n} \leq \frac{n_a}{2}$, and the rule goes through the independence before going through the equilibrium when $n_{a\bar{b}}$ increases.
- Case 2: $n_b \leq \frac{n}{2}$. Then $\frac{n_a n_{\bar{b}}}{n} \geq \frac{n_a}{2}$, and the rule goes through the equilibrium before going through the independence when $n_{a\bar{b}}$ increases.

Let us now compare an IM of deviation from equilibrium m_{eql} and an IM of deviation from independence m_{idp} for these two cases. In order to have a fair comparison, we suppose that the two IMs have similar behaviors: same value for a logical rule, same value for equilibrium/independence, same decrease speed with regard to the negative examples. For example, m_{eql} and m_{idp} can be the Descriptive Confirmed-Confidence [18] and the Loevinger index respectively [22] (Appendix B). As shown in figure 1, m_{idp} is more filtering than m_{eql} in case 1, whereas m_{eql} is more filtering than m_{idp} in case 2. More precisely, in case 1, m_{idp} contributes to rejecting the bad rules, while in case 2 it is m_{eql}. This confirms that the IMs of deviation from equilibrium and the IMs of deviation from independence are complementary, the second ones not being systematically "better" than the first ones[2]. In particular, the IMs of deviation from equilibrium must not be neglected when itemsets are rare (low frequency). In this situation, case 2 is more frequent than case 1.

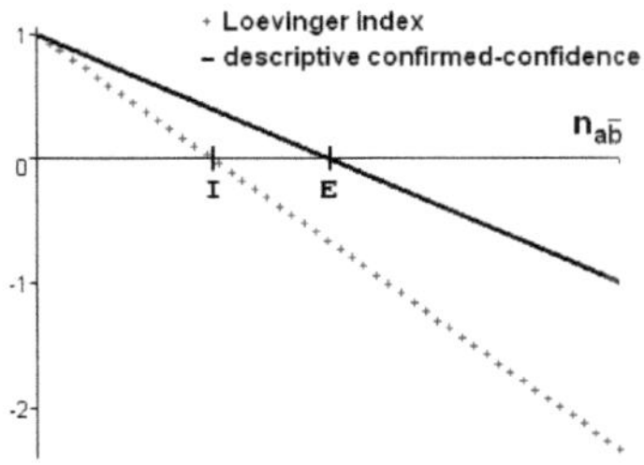

(a) case 1 ($n_b \geq \frac{n}{2}$)

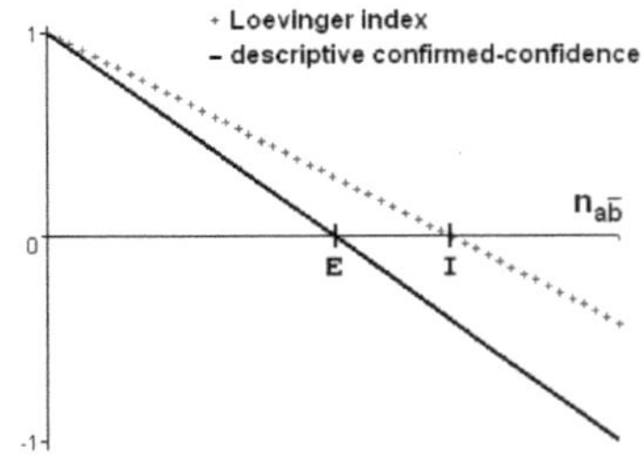

(b) case 2 ($n_b \leq \frac{n}{2}$)

Fig. 1. Comparison of Descriptive Confirmed-Confidence and Loevinger index (E: equilibrium, I: independence)

In our IM taxonomy, the subject of an IM could be the deviation from independence or the deviation from equilibrium. However, as some IMs do not assess any of the two deviation, a third cluster must be added ("other measures" in Tab. 1). The IMs of this cluster generally have a fixed value

[2] Numerous authors consider that a good IM must vanish at independence (principle originally proposed in [25]). This amounts to saying that IMs of deviation from independence are better than IMs of deviation from equilibrium.

only for the rules with no negative examples ($n_{a\bar{b}} = 0$) or for the rules with no examples ($n_{ab} = 0$). Most of them are similarity measures.

3.2 Nature of an IM

The objective IMs can also be classified according to their descriptive or statistical nature.

Descriptive IMs

The descriptive (or frequential) IMs do not vary with the cardinality expansion (when all the data cardinalities are increased or decreased in equal proportion). A descriptive IM m satisfies $m(n, n_a, n_b, n_{a\bar{b}}) = m(\alpha.n, \alpha.n_a, \alpha.n_b, \alpha.n_{a\bar{b}})$ for any strictly positive constant α. These IMs take the data cardinalities into account only in a relative way (by means of the frequencies $p(a)$, $p(b)$, $p(a \; and \; \bar{b})$) and not in an absolute way (by means of the cardinalities n_a, n_b, $n_{a\bar{b}}$).

Statistical IMs

The statistical IMs are those which vary with the cardinality expansion. They take into account the size of the phenomena studied. Indeed, a rule is statistically more valid when it is accessed on a large amount of data. Among the statistical IMs, one can find the probabilistic IMs, which compare the observed distribution to an expected distribution, such as the II measure presented in Appendix A.

3.3 IM taxonomy

A taxonomy according to the nature and subject of the objective IMs is given below (Tab. 1). On the column, we can see that most of the IMs are descriptive. Another observation shows that IPEE is the only one statistical IM computing the deviation from equilibrium.

4 ARQAT tool

ARQAT (Fig. 2) is an exploratory analysis tool that embeds thirty-six objective IMs studied in surveys (See Appendix B for a complete list of selected IMs).

It provides graphical views structured in five task-oriented groups: ruleset analysis, correlation and clustering analysis, interesting rules analysis, sensitivity analysis, and comparative analysis.

Subject \ Nature	Descriptive IMs	Statistical IMs
Measures of deviation from equilibrium	– Confidence (5), – Laplace (21), – Sebag & Schoenauer (31), – Example & Contra-Example (13), – Descriptive Confirm (9), – Descriptive Confirmed-Confidence (10), – Least Contradiction (22)	– IPEE (16)
Measures of deviation from independence	– Phi-Coefficient (28), – Lift (23), – Loevinger (25), – Conviction (6), – Dependency (8), – Pavillon (27), – J-measure (18), – Gini-index (14), – TIC (33), – Collective Strength (4), – Odds Ratio (26), – Yules's Q (34), – Yule's Y (35), – Klosgen (20), – Kappa (19)	– II (15), – EII$\alpha = 1$ (11), – EII$\alpha = 2$ (12), – Lerman (24), – Rule Interest (30)
Other measures	– Support (32), – Causal Support (3), – Jaccard (17), – Cosine (7), – Causal Confidence (0), – Causal Confirm (1), – Causal Confirmed-Confidence (2), – Putative Causal Dependency (29)	

Table 1. Taxonomy of the objective IMs

The ARQAT input is a set of association rules R where each association rule $a \rightarrow b$ must be associated with the four cardinalities n, n_a, n_b, and $n_{a\bar{b}}$.

In the first stage, the input ruleset is preprocessed in order to compute the IM values of each rule, and the correlations between all IM pairs. The results are stored in two tables: an IM table (R×I) where rows are rules and columns are IM values, and a correlation matrix (I×I) crossing IMs. At this stage, the ruleset may also be sampled (filtering box in Fig. 2) in order to focus the study on a more restricted subset of rules.

In the second stage, the data-analyst can drive the graphical exploration of results through a classical web-browser. ARQAT is structured in five groups of task-oriented views. The first group (1 in Fig. 2) is dedicated to ruleset and simple IM statistics to better understand the structure of the IM table (R×I). The second group (2) is oriented to the study of IM correlation in table (I×I) and IM clustering in order to select the most suitable IMs. The third one (3) focuses on rule ordering to select the most interesting rules. The fourth group (4) proposes to study the sensitivity of IMs. The last group (5) offers the possibility to compare the results obtained from different rulesets.

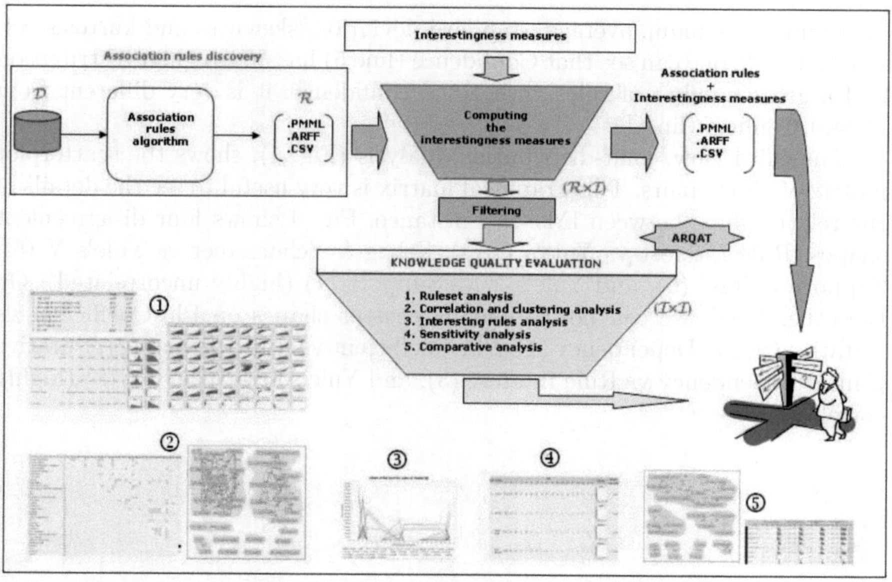

Fig. 2. ARQAT structure.

In this section, we focus on the description of the first three groups and we illustrate them with the same ruleset: 123228 association rules extracted by Apriori algorithm (support 12%) from the mushroom dataset [23].

4.1 Ruleset statistics

The basic statistics are summarized on three views of ARQAT. The first one, ruleset characteristics, shows the distributions underlying rule cardinalities, in order to detect "borderline cases". For instance, in Tab. 2, the first line gives the number of "logical" rules i.e. rules without negative examples. We can notice that the number of logical rules is here very high (≈13%).

N	Type	Count	Percent
1	$n_{a\bar{b}} = 0$	16158	13.11%
2	$n_{a\bar{b}} = 0$ & $na < nb$	15772	12.80%
3	$n_{a\bar{b}} = 0$ & $na < nb$ & $nb = n$	0	0.00%
4	$n_a > n_b$	61355	49.79%
5	$nb = n$	0	0.00%

Table 2. Some ruleset characteristics of the mushroom ruleset.

The second view, IM distribution (Fig. 3), draws the histograms for each IM. The distributions are also completed with classically statistical indexes:

minimum, maximum, average, standard deviation, skewness and kurtosis values. In Fig. 3, one can see that Confidence (line 5) has an irregular distribution and a great number of rules with 100% confidence; it is very different from Causal Confirm (line 1).

The third view, joint-distribution analysis (Fig. 4), shows the scatterplot matrix of all IM pairs. This graphical matrix is very useful to see the details of the relationships between IMs. For instance, Fig. 4 shows four disagreement shapes: Rule Interest vs Yule's Q (4), Sebag & Schoenauer vs Yule's Y (5), Support vs TIC (6), and Yule's Y vs Support (7) (highly uncorrelated). On the other hand, we can notice four agreement shapes on Phi-Coefficient vs Putative Causal Dependency (1), Phi-Coefficient vs Rule Interest (2), Putative Causal Dependency vs Rule Interest (3), and Yule's Q vs Yule's Y (8) (highly correlated).

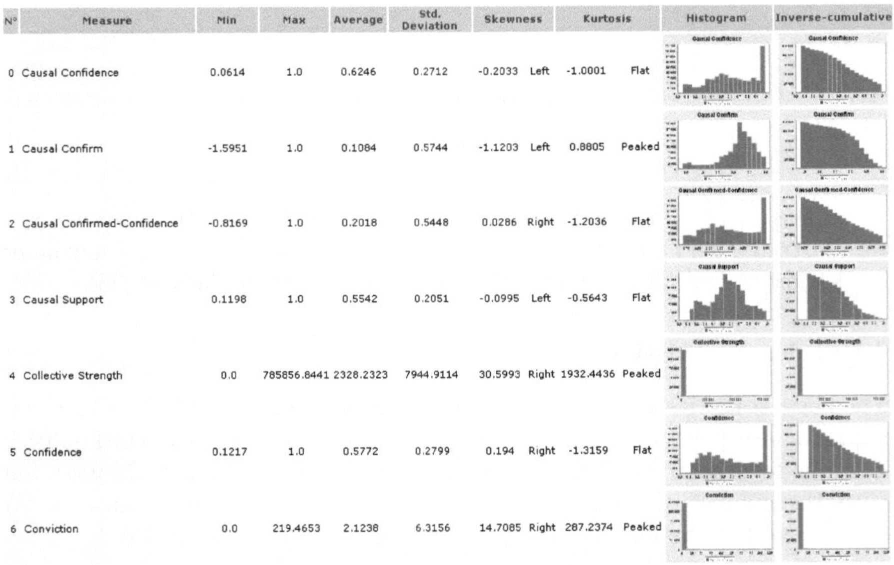

N°	Measure	Min	Max	Average	Std. Deviation	Skewness	Kurtosis	Histogram	Inverse-cumulative
0	Causal Confidence	0.0614	1.0	0.6246	0.2712	-0.2033 Left	-1.0001 Flat		
1	Causal Confirm	-1.5951	1.0	0.1084	0.5744	-1.1203 Left	0.8805 Peaked		
2	Causal Confirmed-Confidence	-0.8169	1.0	0.2018	0.5448	0.0286 Right	-1.2036 Flat		
3	Causal Support	0.1198	1.0	0.5542	0.2051	-0.0995 Left	-0.5643 Flat		
4	Collective Strength	0.0	785856.8441	2328.2323	7944.9114	30.5993 Right	1932.4436 Peaked		
5	Confidence	0.1217	1.0	0.5772	0.2799	0.194 Right	-1.3159 Flat		
6	Conviction	0.0	219.4653	2.1238	6.3156	14.7085 Right	287.2374 Peaked		

Fig. 3. Distribution of some IMs on the mushroom dataset.

4.2 Correlation analysis

This task aims at delivering IM clustering and facilitating the choice of a subset of IMs that is best-adapted to describe the ruleset. The correlation values between IM pairs are computed in the preprocessing stage by using the Pearson's correlation coefficient and stored in the correlation matrix (I×I). Two visual representations are proposed. The first one is a simple summary matrix in which each significant correlation value is visually associated with a

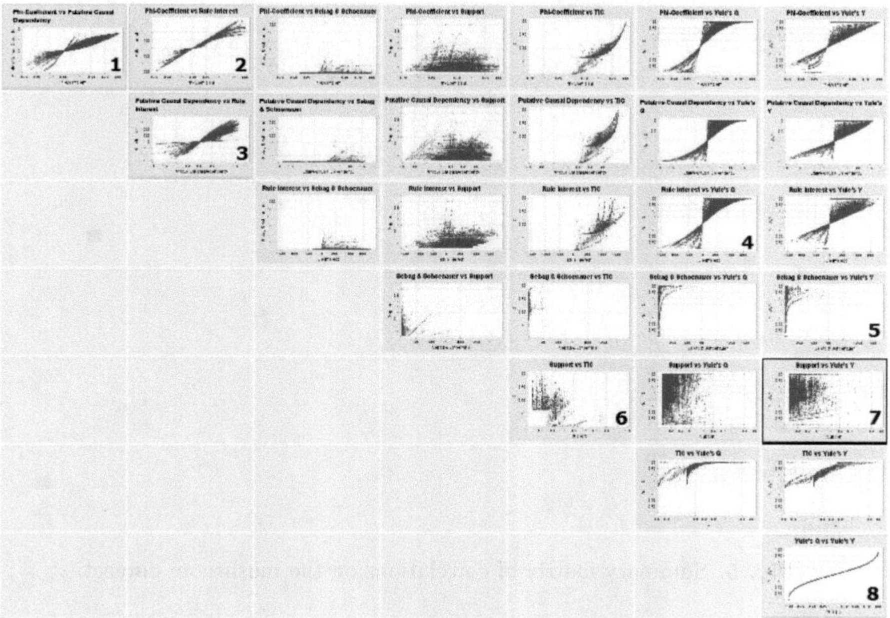

Fig. 4. Scatterplot matrix of joint-distributions on the mushroom dataset.

different color (a level of gray). For instance, the furthest right dark cell from Fig. 5 shows a low correlation value between Yule's Y and Support. The other seventy-nine gray cells correspond to high correlation values.

The second one (Fig. 6) is a graph-based view of the correlation matrix. As graphs are a good means to offer relevant visual insights on data structure, the correlation matrix is used as the relation of an undirected and valued graph, called "correlation graph". In a correlation graph, a vertex represents an IM and an edge value is the correlation value between two vertices/IMs. We also add The possibility to set a minimal threshold τ (maximal threshold θ respectively) to retain only the edges associated with a high correlation (respectively low correlation); the associated subgraphs are denoted by CG+ and CG0.

These two subgraphs can then be processed in order to extract clusters of IMs: each cluster is defined as a connected subgraph. In CG+, each cluster gathers correlated or anti-correlated IMs that may be interpreted similarly: they deliver a close point of view on data. Moreover, in CG0, each cluster contains uncorrelated IMs: i.e. IMs that deliver a different point of view.

Hence, as each graph depends on a specific ruleset, the user can use the graphs as data insight, which graphically help him/her select the minimal set of the IMs best adapted to his/her data. For instance in Fig. 6, CG+ graph contains twelve clusters on thirty-six IMs. The user can select the most representative IM in each cluster, and then retain it to validate the rules.

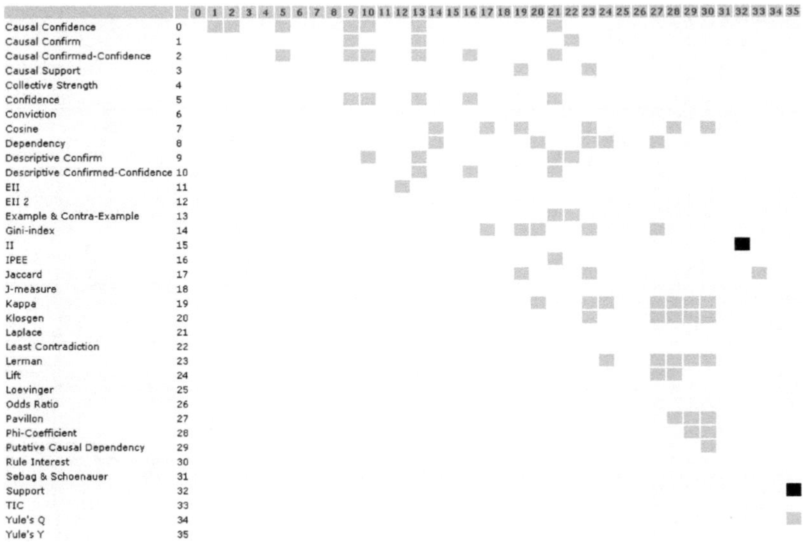

Fig. 5. Summary matrix of correlations on the mushroom dataset.

A close observation on the CG0 graph (Fig. 6) shows an uncorrelated cluster formed by II, Support and Yule's Y measures (also the two dark cells in Fig. 5). This observation is confirmed on Fig. 4 (7). CG+ graph shows a trivial cluster where Yule's Q and Yule's Y are strongly correlated. This is also confirmed in Fig. 4 (8), showing a functional dependency between the two IMs. These two examples show the interest of using the scatterplot matrix complementarily (Fig. 4) with the correlation graphs CG0, CG+ (Fig. 6) in order to evaluate the nature of the correlation links, and overcome the limits of the correlation coefficient.

4.3 Interesting rule analysis

In order to help a user select the most interesting rules, two specific views are implemented. The first view (Fig. 7) collects a set of a given number of interesting rules for each IM in one cluster, in order to answer the question: how interesting are the rules of this cluster?. The selected rules can alternatively be visualized with parallel coordinate drawing (Fig. 8). The main interest of such a drawing is to rapidly see the IM rankings of the rules.

These two views can be used with the IM values of a rule or alternatively with the rank of the value. For instance, Fig. 7 and Fig. 8 use the rank to evaluate the union of the ten interesting rules for each of the ten IMs in the C0 cluster (see Fig. 6). The Y-axis in Fig. 8 holds the rule rank for the corresponding IM. By observing the concentration lines on low rank values, one can obtain four IMs: Confidence(5), Descriptive Confirmed-Confidence(10),

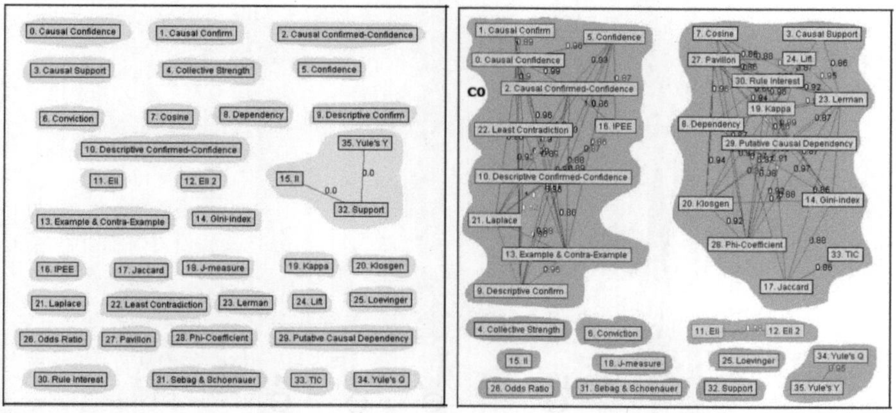

Fig. 6. CG0 and CG+ graphs on the mushroom dataset (clusters are highlighted with a gray background).

Example & Contra-Example(13), and IPEE (16) (on points 1, 2, 3, 4 respectively) that are good for a majority of interesting rules. This can also be retrieved from columns 5, 10, 13, 16 of Fig. 7. Among these four IMs, IPEE is the most suitable choice because of the lowest rule ranks obtained.

Measure Order		0	1	2	5	9	10	13	21	22	16	Rule's presentation
30	R107560	1	19121	1	1	41	1	1	8	5388	1	BROAD FREE ONE ==>veil_color=WHITE
31	R107562	1	18997	1	1	41	1	1	8	5361	1	BROAD ONE veil_color=WHITE ==>FREE
32	R107594	1	8972	1	1	18	1	1	3	2574	1	CLOSE FREE ONE ==>veil_color=WHITE
33	R107596	1	8914	1	1	18	1	1	3	2564	1	CLOSE ONE veil_color=WHITE ==>FREE
34	R122275	1	13800	1	1	32	1	1	5	3977	1	BROAD FREE ==>veil_color=WHITE
35	R122283	1	18299	1	1	38	1	1	6	5145	1	FREE stalk_surf_above=SMOOTH ==>veil_color=WHITE
36	R122285	1	18179	1	1	38	1	1	6	5134	1	stalk_surf_above=SMOOTH veil_color=WHITE ==>FREE
37	R122296	1	20903	1	1	55	1	1	10	6193	1	FREE stalk_surf_below=SMOOTH ==>veil_color=WHITE
38	R122308	65969	8772	40612	23743	10	23743	23743	23714	1013	1	FREE ==>ONE veil_color=WHITE

Fig. 7. Union of the ten interesting rules of the cluster $C0$ on the mushroom dataset (extract).

5 Focus on graph-based clustering approach

When considering a large set of IMs, the graph-based view of the correlation matrix may be quite complex. In order to highlight the more "natural" clusters, we propose to construct two types of subgraphs: the correlated ($CG+$) and the uncorrelated ($CG0$) partial subgraph. In this section we present the different filtering thresholds for their construction. We also extend the correlation graphs to graphs of stable clusters ($\overline{CG0}$ and $\overline{CG+}$) in order to compare several rulesets.

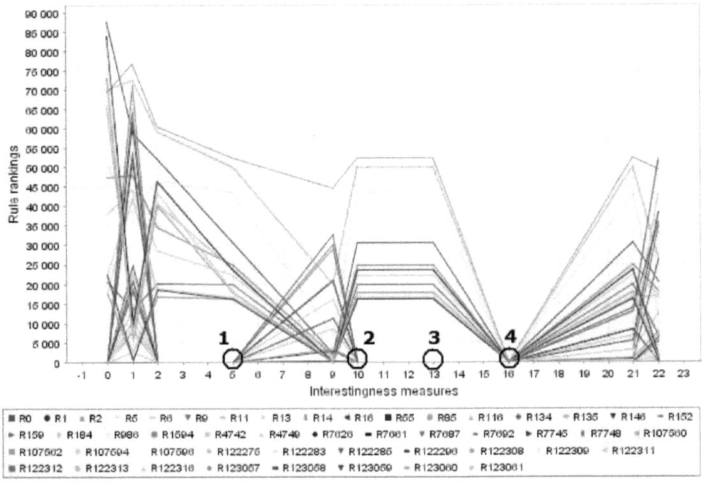

Fig. 8. Plot of the union of the ten interesting rules of the cluster $C0$ on the mushroom dataset.

5.1 Principles

Let $R(D) = \{r_1, r_2, ..., r_p\}$ denote a set of p association rules derived from a dataset D. Each rule $a \rightarrow b$ is described by its itemsets (a, b) and its cardinalities $(n, n_a, n_b, n_{a\bar{b}})$. Let M be the set of q available IMs for our analysis $M = \{m_1, m_2, ..., m_q\}$. Each IM is a numerical function on rule cardinalities: $m(a \rightarrow b) = f(n, n_a, n_b, n_{a\bar{b}})$. For each IM $m_i \in M$, we can construct a vector $m_i(R) = \{m_{i1}, m_{i2}, ..., m_{ip}\}, i = 1..q$, where m_{ij} corresponds to the calculated value of the IM m_i for a given rule r_j.

The correlation value between any two IMs $m_i, m_j\{i, j = 1..q\}$ on the set of rules R is calculated by using a Pearson's correlation coefficient $\rho(m_i, m_j)$ [27], where $\overline{m_i}, \overline{m_j}$ are the average values calculated of vector $m_i(R)$ and $m_j(R)$ respectively:

$$\rho(m_i, m_j) = \frac{\sum_{k=1}^{p}[(m_{ik} - \overline{m_i})(m_{jk} - \overline{m_j})]}{\sqrt{[\sum_{k=1}^{p}(m_{ik} - \overline{m_i})^2][\sum_{k=1}^{p}(m_{jk} - \overline{m_j})^2]}}$$

In order to make the interpretation of the large set of correlation values easier, we introduce the following definitions:

Definition 3. Two IMs m_i and m_j are τ-*correlated* with respect to the dataset D if their absolute correlation value is greater than or equal to a given threshold τ: $|\rho(m_i, m_j)| \geq \tau$. And, conversely, two IMs m_i and m_j are θ-*uncorrelated* with respect to the dataset D if the absolute value of their correlation value is lower than or equal to a threshold value θ: $|\rho(m_i, m_j)| \leq \theta$.

For θ-uncorrelated IMs, we use a statistical test of significance by choosing a level of significance of the test $\alpha = 0.05$ for hypothesis testing (common values for α are: $\alpha = 0.1, 0.05, 0.005$). The threshold θ is then calculated by the following formula: $\theta = 1.960/\sqrt{p}$ in a population of size p [27]. The assignment $\tau = 0.85$ of τ-correlated is used because this value is widely acceptable in the literature.

As the correlation coefficient is symmetrical, the $q(q-1)/2$ correlation values can be stored in one half of the table $q \times q$. This table $(I \times I)$ can also be viewed as the relation of an undirected and valued graph called correlation graph, in which a vertex value is an IM and an edge value is the correlation value between two vertices/IMs.

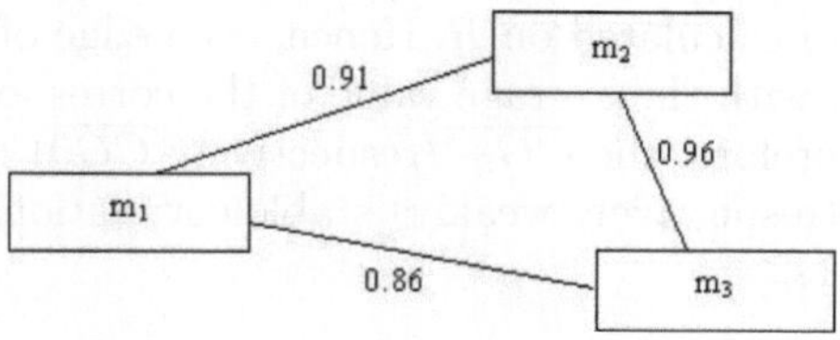

Fig. 9. An illustration of the correlation graph.

For instance, Fig. 9 can be the correlation graph obtained on five association rules $R(D)=\{r_1, r_2, r_3, r_4, r_5\}$ extracted from a dataset D and three IMs $M = \{m_1, m_2, m_3\}$ whose values and correlations are given in Tab. 3.

$R \times I$	m_1	m_2	m_3
r_1	0.84	0.89	0.91
r_2	0.86	0.90	0.93
r_3	0.88	0.94	0.97
r_4	0.94	0.95	0.99
r_5	0.83	0.87	0.84

$I \times I$	m_1	m_2	m_3
m_1		0.91	0.86
m_2			0.96
m_3			

Table 3. Correlation values for three IMs and five association rules.

5.2 Correlated versus uncorrelated graphs

Unfortunately, when the correlation graph is complete, it is not directly human-readable. We need to define two transformations in order to extract more limited and readable subgraphs. By using definition 3, we can extract the *correlated partial subgraph (CG+)*: the subgraph composed of edges associated with a τ-correlated. On the same way, the *uncorrelated partial subgraph (CG0)* where we only retain edges associated with correlation values close to 0 (θ-uncorrelated).

These two partial subgraphs can then be used as a visualization support in order to observe the correlative liaisons between IMs.

We can also observe the clusters of IMs corresponding with the connected parts of the graphs.

5.3 Extension to graph of stable clusters

In order to facilitate the comparison between several correlation matrices, we have introduced some extensions to define the stable clusters between IMs.

Definition 4. The $\overline{CG+}$ graph (respectively $\overline{CG0}$ graph) of a set of k rulesets $R = \{R(D_1), ..., R(D_k)\}$ is defined as the average graph of intersection of the k partially correlated (respectively uncorrelated) subgraphs $CG+_k$ (respectively $CG0_k$) calculated on R. Hence, each edge of $\overline{CG+}$ (respectively $\overline{CG0}$) is associated with the average value of the corresponding edge in the k $CG+_k$ graphs. Therefore, the $\overline{CG+}$ (respectively $\overline{CG0}$) graph allows visualizing the strongly (respectively weakly) stable correlations, as being common to k studied rulesets.

Definition 5. We call τ-stable (respectively θ-stable) clusters the connected part of the $\overline{CG+}$ (respectively $\overline{CG0}$) graph.

6 Study of IM behavior on two prototypical and opposite datasets

We have applied our method to two "opposite" datasets: D_1 and D_2, in order to compare correlation behavior and more precisely, to discover some stable clusters.

6.1 Data description

Our experiments are based on the categorical mushroom dataset (D_1) from Irvine machine-learning database repository and a synthetic dataset (D_2). The latter is obtained by simulating the transactions of customers in retail businesses, the dataset was generated using the IBM synthetic data generator [3]. D_2 has the typical characteristic of the Agrawal dataset T5.I2.D10k. We also generate the set of association rules (ruleset) R_1 (respectively R_2) from the dataset D_1 (respectively D_2) using the Apriori algorithm [2] [3]. For a closer evaluation of the IM behavior of the most interesting rules from these two rulesets, we have extracted $R_1^{'}$ (respectively $R_2^{'}$) from R_1 (respectively R_2) as the union of the first 1000 rules ($\approx 1\%$, ordered by decreasing IM values) issued from each IM (see Tab. 4).

In our experiment, we compared and analyzed the thirty-six IMs defined in Appendix B. We must notice that EII($\alpha = 1$) and EII($\alpha = 2$) are two entropic versions of the II measure.

Dataset	Items (Average length)	Transactions	Number of rules (support threshold)	$R(D)$	θ	τ	$\overline{R(D)}$
D_1	118 (22)	8416	123228 (12%)	R_1	0.005	0.85	$\overline{R_1}$
			10431 (12%)	$R_1^{'}$	0.020	0.85	$\overline{R_1^{'}}$
D_2	81 (5)	9650	102808 (0.093%)	R_2	0.003	0.85	$\overline{R_2}$
			7452 (0.093%)	$R_2^{'}$	0.012	0.85	$\overline{R_2^{'}}$

Table 4. Description of the datasets.

6.2 Discussion

The analysis aims at finding stable relations between the IMs studied over the four rulesets. We investigate in: (1) the $\overline{CG0}$ graphs in order to identify the IMs that do not agree for ranking the rules, (2) the $\overline{CG+}$ graph in order to find the IMs that do agree for ranking the rules.

Ruleset	Number of correlations		Number of clusters	
	τ-correlated	θ-uncorrelated	CG+	CG0
R_1	79	2	12	34
$R_1^{'}$	91	15	12	21
R_2	65	0	14	36
$R_2^{'}$	67	17	12	20

Table 5. Comparison of correlation.

CG+ and *CG0*

Fig. 10 shows four *CG+* graphs obtained from the four rulesets. As seen before, the sample rulesets and the original rulesets have close results so we can use the sample rulesets for representing the original rulesets. This observation is useful when we evaluate the CG+ graphs but not for CG0 graphs. For example, with the CG+ graph of R_1 (Fig. 10), one can choose the largest cluster containing the fourteen IMs (Causal Support, Pavillon, Lift, Lerman, Putative Causal Dependency, Rule Interest, Phi-Coefficient, Klosgen, Dependency, Kappa, Gini-index, Cosine, Jaccard, TIC) for his/her first choice. In this cluster one can also see the weak relation between TIC and the other IMs of the cluster. Tab. 5 also shows the two opposite tendencies obtained from the number of τ-correlated computed: $79(R_1) \rightarrow 91(R_1^{'}), 65(R_2) \rightarrow 67(R_2^{'})$.

With the four CG0 graphs (Fig. 11), one can easily see that the number of θ-uncorrelated increases when the most interesting rules are selected: $2(R_1) \rightarrow 15(R_1^{'}), 0(R_2) \rightarrow 17(R_2^{'})$ (Fig. 11, Tab. 5).

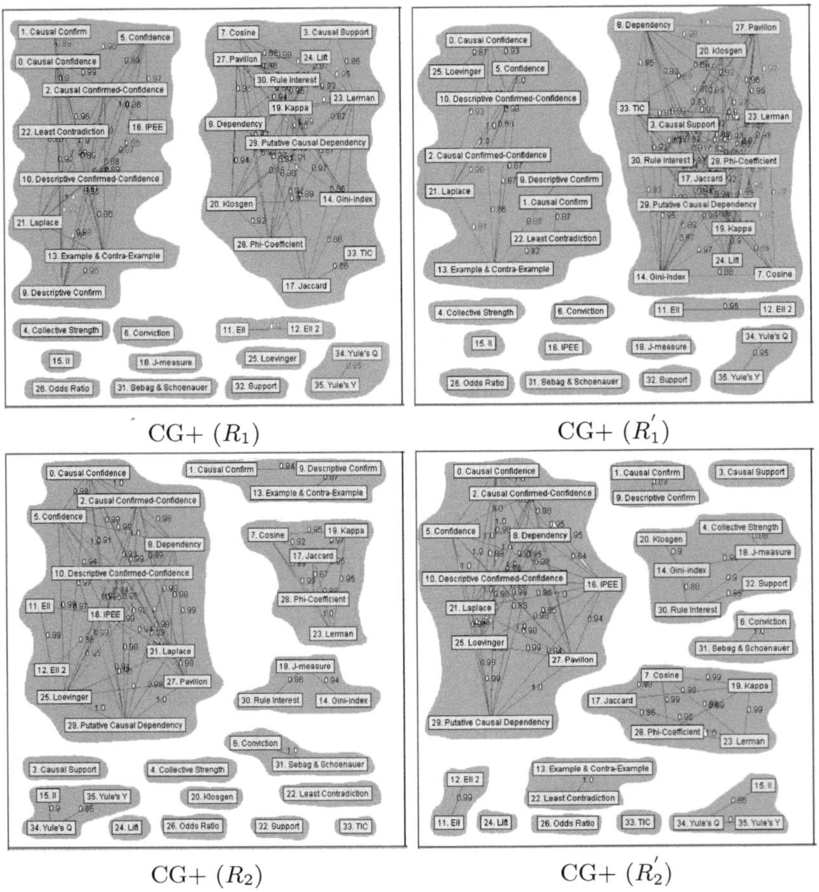

Fig. 10. CG+ graphs (clusters are highlighted in gray).

$\overline{CG0}$ graphs: uncorrelated stability

Uncorrelated graphs first show that there are no θ-stable clusters that appear on the four rulesets studied in Fig. 11. Secondly, there is no $\overline{CG0}$ graph from these datasets. A close observation of four CG0 graphs shows that at least one IM in each cluster will later be clustered around in a τ-stable cluster of $\overline{CG+}$ graph (Fig. 11, Fig. 12) like Yule's Y, Putative Causal Dependency, EII($\alpha = 2$), Cosine, Laplace so that the stronger the θ-uncorrelated, the more interesting the IM that participated in the θ-uncorrelated.

$\overline{CG+}$ graph: correlated stability

From Tab. 5, we can see that, R_1' is approximately twice as correlated as R_2'. As seen in Fig. 12, five τ-stable clusters found come from the datasets studied.

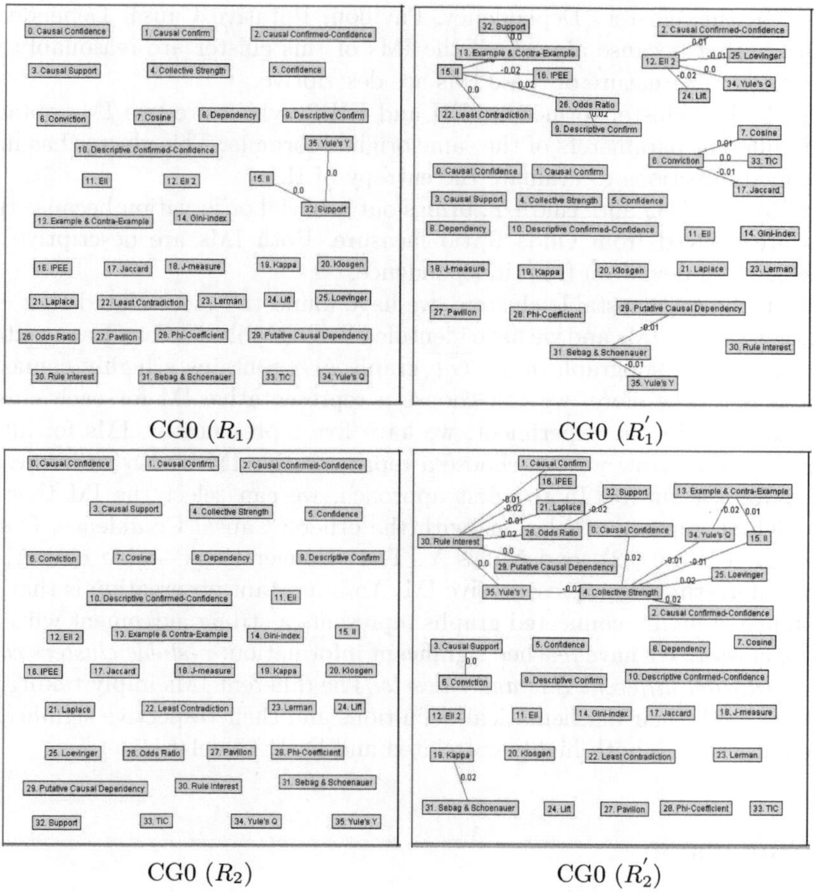

CG0 (R_1) CG0 (R_1')

CG0 (R_2) CG0 (R_2')

Fig. 11. CG0 graphs.

By briefly analyzing these τ-stable clusters, some interesting observations are drawn.

(C1), the largest cluster, (Confidence, Causal Confidence, Causal Confirmed-Confidence, Descriptive Confirmed-Confidence, Laplace) has most of its IMs extended from Confidence measure. From this cluster, we can easily see a highly connected component – each vertex must have an edge with the other vertices – indicating the strong agreement of the five IMs.

According to the taxonomy (Tab. 1), this cluster is associated with descriptive IMs that are sensible to equilibrium.

(C2), another cluster, has two highly connected components which are formed by Phi-Coefficient, Lerman, Kappa, Cosine and Jaccard. Most of these IMs are similarity measures. According to the taxonomy (Tab. 1) this cluster is to measure the deviation from independence.

(C3), this cluster (Dependency, Pavillon, Putative Causal Dependency) is interesting because almost all the IMs of this cluster are reasonably well correlated. The nature of these IMs are descriptive.

(C4), is a cluster formed by EII and EII 2, which are two IMs obtained with different parameters of the same original formula. This cluster has many extended directions to evaluate the entropy of II.

(C5), Yule's Q and Yule's Y, brings out a trivial observation because these IMs are derived from Odds Ratio measure. Both IMs are descriptive and measuring of deviation from independence.

In looking for τ-stable clusters, we have found the τ-correlated that exist between various IMs and we have identified five τ-stable clusters. Each τ-stable cluster forms a subgraph in a $\overline{CG+}$ graph, also contains a highly connected component. Therefore, we can choose a representative IM for each cluster. For example, in our experiment, we have five representative IMs for all the thirty-six IMs. How we can choose a representative IM is also an interesting study for the future. In the first approach, we can select the IM that has the highest number of relations with the others: Causal Confidence, Cosine, Klosgen, EII($\alpha = 2$), and Yule's Y. The stronger the τ-stable cluster, the more interesting the representative IM. An important observation is that, the existence of highly connected graphs represents a strong agreement with a τ-stable cluster. We have reached significant information: *τ-stable clusters can be obtained from different IMs and rulesets.* The different IMs imply taking into account both their mathematical definitions and their respective significance. The datasets are both highly correlated and lowly correlated.

7 Conclusion

We have studied and compared the various IMs described in the literature in order to help the decision-maker to better understand the behavior of the IMs in the stage of post-processing of association rules. A new approach called correlation graph implemented by a new tool, ARQAT, with two types: CG+ and CG0 is proposed to evaluate IMs by using graphs as a visual insight on the data.

With this approach, the decision-maker has a few IMs to decide and as a graphical representation to select the most interesting rules to examine. Another interesting result obtained from this work is that we have found some stable clusters between IMs, five such τ-stable clusters have been found with the $\overline{CG+}$ graph. Our approach is highly related to the real value of the dataset and the number of proposed IMs.

Our future research will investigate the two following directions: first, we will improve the correlation analysis by introducing a better measure than linear correlation whose limits are stressed in the literature; second, we will also improve the IM clustering analysis with IM aggregation techniques to facilitate the user's decision making from the most suitable IMs.

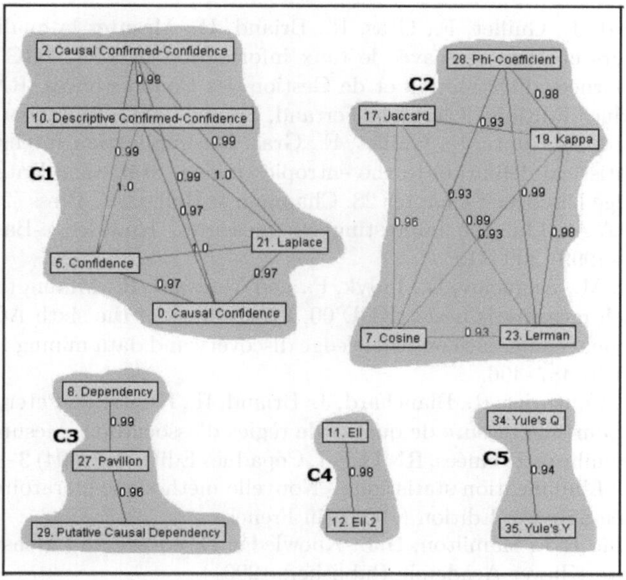

Fig. 12. $\overline{CG}+$ graph.

References

1. Agrawal, R., Imielinski, T., Swami, A.: Mining association rules between sets of items in large databases. Proceedings of the ACM-SIGMOD International Conference on Management of Data. Washington DC, USA (1993) 207–216
2. Agrawal, R., Srikant, R.: Fast algorithms for mining association rules. VLDB'94, Proceedings of the 20th International Conference on Very Large Data Bases. Santiago, Chile (1994) 487–499
3. Agrawal, R., Mannila, H., Srikant, R., Toivonen H., Verkano, A.I.: Fast discovery of association rules. Advances in Knowledge Discovery in Databases. (1996) 307–328
4. Azé, J., Kodratoff, Y.: A study of the Effect of Noisy Data in Rule Extraction Systems. EMCSR'02, Proceedings of the Sixteenth European Meeting on Cybernetics and Systems Research. (2002) 781–786
5. Bayardo, Jr.R.J., Agrawal, R.: Mining the most interesting rules. KDD'99, Proceedings of the Fifth ACM SIGKDD international conference on Knowledge discovery and data mining. San Diego, CA, USA (1999) 145–154
6. Blanchard, J., Guillet, F., Gras, R., and Briand, H.: Using information-theoretic measures to assess association rule interestingness. ICDM'05, Proceedings of the 5th IEEE International Conference on Data Mining, IEEE Computer Society Press, (2005) 66–73.
7. Blanchard, J., Guillet, F., Gras, R., Briand, H.: Assessing rule interestingness with a probabilistic measure of deviation from equilibrium. ASMDA'05, Proceedings of the 11th International Symposium on Applied Stochastic Models and Data Analysis. (2005) 191–200

8. Blanchard, J., Guillet, F., Gras, R., Briand, H.: Mesurer la qualité des règles et de leurs contraposées avec le taux informationnel TIC. EGC'04, Actes de 4èmes journées d'Extraction et de Gestion des Connaissances, RNTI-E-2, Vol. 1. Cépaduès Editions, Clermont Ferrand, France (2004) 287–298 (in French)
9. Blanchard, J., Kuntz, P., Guillet, F., Gras, R.: Implication Intensity: from the basic statistical definition to the entropic version. Statistical Data Mining and Knowledge Discovery, Chapter 28. Chapman & Hall, CRC Press (2003) 475–493
10. Freitas, A.A.: On rule interestingness measures. Knowledge-Based Systems, 12(5-6). (1999) 309–315
11. Gavrilov, M., Anguelov, D., Indyk, P., and Motwani, R.: Mining the stock market: which measure is best?. KDD'00, Proceedings of the sixth ACM SIGKDD international conference on Knowledge discovery and data mining . Boston, MA, USA (2000) 487–496.
12. Gras, R., Couturier, R., Blanchard, J., Briand, H., Kuntz, P., Peter, P.: Quelques critères pour une mesure de qualité de règles d'association. Mesures de Qualité pour la Fouille de Données, RNTI-E-1. Cépaduès Editions (2004) 3–31 (in French)
13. Gras, R.: L'implication statistique - Nouvelle méthode exploratoire de données. La Pensée Sauvage Édition (1996) (in French)
14. Hilderman, R.J., Hamilton, H.J.: Knowledge Discovery and Measures of Interestingness. Kluwer Academic Publishers (2001)
15. Klemettinen, M., Mannila, H., Ronkainen, P., Toivonen, H., and Verkano, A.I.: Finding interesting rules from larges sets of discovered association rules. ICIKM'94, Proceedings of the Third International Conference on Information and Knowledge Management. Ed. Nabil R. Adam, Bharat K. Bhargava and Yelena Yesha, Gaithersburg, Maryland. ACM Press, (1994) 401–407.
16. Huynh, X.-H., Guillet, F., Briand, H.: Clustering interestingness measures with positive correlation. ICEIS'05, Proceedings of the 7th International Conference on Enterprise Information Systems. (2005) 248–253
17. Kaufman, L., Rousseeuw, P.J.: Finding Groups in Data: An Introduction to Cluster Analysis. Wiley, New York, (1990)
18. Kodratoff, Y.: Comparing Machine Learning and Knowledge Discovery in DataBases: An Application to Knowledge Discovery in Texts. Machine Learning and Its Applications, LNCS 2049. Springer-Verlag, (2001) 1–21
19. Kononenco, I.: On biases in estimating multi-valued attributes. IJCAI'95. (1995) 1034–1040
20. Lenca, P., Lallich, S., Vaillant, B.: On the robustness of association rules. Proceedings of the IEEE International Conference on Cybernetics and Intelligent Systems. (2006) 596–601
21. Liu, B., Hsu, W., Mun, L., Lee, H.: Finding interestingness patterns using user expectations. IEEE Transactions on Knowledge and Data Mining (11). (1999) 817–832
22. Loevinger, J.: A systematic approach to the construction and evaluation of tests of ability. Psychological Monographs. (1947)
23. Newman, D.J., Hettich,S., Blake, C.L., Merz, C.J.: [UCI] Repository of machine learning databases, http://www.ics.uci.edu/~mlearn/MLRepository.html. University of California, Irvine, Department of Information and Computer Sciences, (1998).
24. Padmanabhan, B., Tuzhilin, A. : A belief-driven method for discovering unexpected patterns. KDD'98, Proceedings of the Fourth International Conference on Knowledge Discovery and Data Mining. (1998) 94–100

25. Piatetsky-Shapiro, G.: Discovery, analysis and presentation of strong rules. Knowledge Discovery in Databases, G. Piatetsky-Shapiro and W. Frawley editors. MIT Press, Cambridge, MA (1991) 229–248
26. Piatetsky-Shapiro, G., Steingold, S.: Measuring Lift Quality in Database Marketing. SIGKDD Explorations 2(2). (2000) 76–80
27. Ross, S.M.: Introduction to probability and statistics for engineers and scientists. Wiley, (1987)
28. Sebag, M., Schoenauer, M.: Generation of rules with certainty and confidence factors from incomplete and incoherent learning bases. EKAW'88, Proceedings of the European Knowledge Acquisition Workshop. Gesellschaft fr Mathematik und Datenverarbeitung mbH (1988) 28.1–28.20
29. Silberschatz, A., Tuzhilin, A.: What makes patterns interesting in knowledge discovery systems. IEEE Transactions on Knowledge Data Engineering 8(6). (1996) 970–974
30. Tan, P.N., Kumar, V., Srivastava, J.: Selecting the right objective measure for association analysis. Information Systems 29(4). (2004) 293–313
31. Vaillant, B., Lenca, P., Lallich, S.: A clustering of interestingness measures. DS'04, the 7th International Conference on Discovery Science LNAI 3245. (2004) 290–297
32. Vaillant, B., Lallich, S., Lenca, P.: Modeling of the counter-examples and association rules interestingness measures behavior. The 2006 International Conference on Data Mining. (2006)
33. Zhao, Y., Karypis, G.: Criterion functions for document clustering: experiments and analysis. Technical Report TR01-40, Deparment of Computer Science, University of Minnesota. (2001) 1–30

A Complementary IMs: II, EII, TIC, IPEE

A.1 Implication Intensity

Initially introduced by Gras [13], the implicative intensity II aims at quantifying the "surprisingness" of a rule.

Intuitively, it is more surprising to discover that a rule has a small number of negative examples when the dataset is large. Hence, the objective of the implicative intensity is to express the unlikelihood of $n_{a\bar{b}}$ in T.

More precisely, we compare the observed number of negative examples $n_{a\bar{b}}$ with the number $N_{a\bar{b}}$ of expected negative examples for an independence hypothesis. Let us assume that we randomly draw two subsets U and V in T with respectively n_a and n_b transactions. Then, $N_{a\bar{b}} = \left|U \bigcap \overline{V}\right|$ is the random variable associated with the number of negative examples in this random model.

Definition 6. The implicative intensity II of the rule $a \rightarrow b$ is defined by

$$II\,(a \rightarrow b) = 1 - p\left(N_{a\bar{b}} \leq n_{a\bar{b}}\right)$$

if $n_a \neq n$; otherwise

$$II(a \rightarrow b) = 0.$$

In practice, the distribution of $N_{a\overline{b}}$ depends on the random drawing pattern. We here consider a hyper-geometric law: $p\left(N_{a\overline{b}} = k\right) = \frac{C_{n_{\overline{a}}}^{n_{\overline{b}}-k} C_{n_a}^{k}}{C_n^{n_b}}$. The effective value of II can be easily computed with this recursive formula. Other models based on the binomial law and the Poisson distribution have been proposed.

A.2 Entropic Implication Intensity

Definition 6 essentially measures the surprisingness of the rule $a \rightarrow b$. However, taking the contrapositive $\overline{b} \rightarrow \overline{a}$ into account could reinforce the assertion of the implication between a and b. Moreover, it could improve the quality of discrimination of II when the transaction set T increases: if A and B are small compared to T, their complementary sets are large and vice-versa.

For these reasons, we have introduced a weighted version of the implication intensity $(E(a,b).II(a \rightarrow b))^{1/2}$ where $E(a,b)$ measures the disequilibrium between n_{ab} and $n_{a\overline{b}}$ – associated with $a \rightarrow b$ –, and the disequilibrium between $n_{a\overline{b}}$ and $n_{\overline{a}\overline{b}}$ – associated with its contrapositive – [9]. Intuitively, the surprise must be softened (respectively confirmed) when the number of negative examples $n_{a\overline{b}}$ is high (respectively small) for the rule and its contrapositive considering the observed cardinalities n_a and $n_{\overline{b}}$.

A well-known index for taking the cardinalities into account non-linearly is the Shannon conditional entropy. The conditional entropy $H_{b/a}$ of cases (a and b) and (a and $\overline{b}$) given a is defined by

$$H_{b/a} = -\frac{n_{ab}}{n_a} log_2 \frac{n_{ab}}{n_a} - \frac{n_{ab}}{n_a} log_2 \frac{n_{a\overline{b}}}{n_a}$$

and, similarly, we obtain the conditional entropy $H_{\overline{a}/\overline{b}}$ of cases ($\overline{a}$ and $\overline{b}$) and (a and $\overline{b}$) given $\overline{b}$. The complements of 1 for these uncertainties $1 - H$ can be interpreted as the average information collected by the realization of these experiments; the higher this information, the stronger the quality of the implication and its contrapositive.

The expected behavior of the weighted version of II is determined in three stages: *(i)* a slow reaction to the first negative examples (robustness to noise), *(ii)* an acceleration of the rejection in the neighborhood of the equilibrium, *(iii)* an increasing rejection beyond the equilibrium. The adjustment of $1 - H$ proposed in definition 6 satisfies these requirements.

Definition 7. Let $\alpha > 1$ be a fixed number. The disequilibriums are measured by $E(a,b)$, is defined by

$$E(a,b) = \left(\left(1 - H_{b/a}\right)^{\alpha} . \left(1 - H_{\overline{a}/\overline{b}}\right)^{\alpha}\right)^{1/2\alpha}$$

if $\frac{n_{a\overline{b}}}{n} \in \left[0, \frac{n_a}{2n}\right[\bigcap \left[0, \frac{n_b}{2n}\right[$;

$$E(a,b) = 0$$

otherwise.

And, the weighted version of the implication intensity – called the entropic implication intensity – is given by

$$EII\,(a \to b) = (E\,(a,b)\,.\,II\,(a \to b))^{1/2}$$

Raising the conditional entropies to the power α reinforces the contrast between the different stages presented above.

A.3 TIC

In [6], we introduced DIR (Directed Information Ratio), a new rule IM which is based on information theory. DIR is the entropy decrease rate of the consequent due to the truth of the antecedent, but it is not calculated with a classical entropy function. We use an asymmetric entropy function which considers that the uncertainty is maximal (entropy = 1) when the studied modality is not the more likely. This allows DIR to differentiate two opposite rules $a \to b$ and $a \to \overline{b}$, which is not possible with the other information-theoretic measures of rule interestingness. Moreover, to our knowledge, DIR is the only rule IM which rejects both independence and equilibrium, i.e. it discards both the rules whose antecedent and consequent are negatively correlated, and the rules which have more negative examples than examples.

In [8], we proposed another IM, derived from DIR, which assesses the rules by taking their contrapositives into account. This new IM called TIC (*Taux Informationnel modulé par la Contraposée, in French*) is the geometric mean of the values of DIR for a rule and its contrapositive (if one of the two values of DIR is negative, then TIC is worth zero). Considering both the rule and its contrapositive allows to discover rules that are closer to logical implication.

A.4 IPEE

As there was no statistical IMs evaluating the deviation from equilibrium, we proposed the new measure IPEE in [7]. Following II, IPEE is based on a probabilistic model. However, while II evaluates the statistical significance of the deviation from independence, IPEE evaluates the statistical significance of the deviation from equilibrium.

B Formulas of IMs

N	Interestingness measure	$f(n, n_a, n_b, n_{a\bar{b}})$	Reference
0	Causal Confidence	$1 - \frac{1}{2}(\frac{1}{n_a} + \frac{1}{n_{\bar{b}}})n_{a\bar{b}}$	[18]
1	Causal Confirm	$\frac{n_a + n_{\bar{b}} - 4n_{a\bar{b}}}{n}$	[18]
2	Causal Confirmed-Confidence	$1 - \frac{1}{2}(\frac{3}{n_a} + \frac{1}{n_{\bar{b}}})n_{a\bar{b}}$	[18]
3	Causal Support	$\frac{n_a + n_{\bar{b}} - 2n_{a\bar{b}}}{n}$	[18]
4	Collective Strength	$\frac{(n_a - n_{a\bar{b}})(n_{\bar{b}} - n_{a\bar{b}})(n_a n_{\bar{b}} + n_b n_{\bar{a}})}{(n_a n_b + n_{\bar{a}} n_{\bar{b}})(n_b - n_a + 2n_{a\bar{b}})}$	[30]
5	Confidence	$1 - \frac{n_{a\bar{b}}}{n_a}$	[2]
6	Conviction	$\frac{n_a n_{\bar{b}}}{n n_{a\bar{b}}}$	[30]
7	Cosine	$\frac{n_a - n_{a\bar{b}}}{\sqrt{n_a n_b}}$	[30]
8	Dependency	$\lvert \frac{n_{\bar{b}}}{n} - \frac{n_{a\bar{b}}}{n_a} \rvert$	[18]
9	Descriptive Confirm	$\frac{n_a - 2n_{a\bar{b}}}{n}$	[18]
10	Descriptive Confirmed-Confidence	$1 - 2\frac{n_{a\bar{b}}}{n_a}$	[18]
11	EII ($\alpha = 1$)	$\sqrt{\varphi \times I^{\frac{1}{2\alpha}}}$	[9]
12	EII ($\alpha = 2$)	$\sqrt{\varphi \times I^{\frac{1}{2\alpha}}}$	[9]
13	Example & Contra-Example	$1 - \frac{n_{a\bar{b}}}{n_a - n_{a\bar{b}}}$	[13]
14	Gini-index	$\frac{(n_a - n_{a\bar{b}})^2 + n_{a\bar{b}}^2}{n n_a} + \frac{(n_b - n_a + n_{a\bar{b}})^2 + (n_{\bar{b}} - n_{a\bar{b}})^2}{n n_{\bar{a}}} - \frac{n_b^2}{n^2} - \frac{n_{\bar{b}}^2}{n^2}$	[30]
15	II	$1 - \sum_{k=max(0, n_a - n_b)}^{n_{a\bar{b}}} \frac{C_{n_b}^{n_a - k} C_{n_{\bar{b}}}^{k}}{C_n^{n_a}}$	[13]
16	IPEE	$1 - \frac{1}{2^{n_a}} \sum_{k=0}^{n_{a\bar{b}}} C_{n_a}^{k}$	[7]
17	Jaccard	$\frac{n_a - n_{a\bar{b}}}{n_b + n_{a\bar{b}}}$	[30]
18	J-measure	$\frac{n_a - n_{a\bar{b}}}{n} log_2 \frac{n(n_a - n_{a\bar{b}})}{n_a n_b} + \frac{n_{a\bar{b}}}{n} log_2 \frac{n n_{a\bar{b}}}{n_a n_{\bar{b}}}$	[30]
19	Kappa	$\frac{2(n_a n_{\bar{b}} - n n_{a\bar{b}})}{n_a n_{\bar{b}} + n_{\bar{a}} n_b}$	[30]
20	Klosgen	$\sqrt{\frac{n_a - n_{a\bar{b}}}{n}}(\frac{n_{\bar{b}}}{n} - \frac{n_{a\bar{b}}}{n_a})$	[30]
21	Laplace	$\frac{n_a + 1 - n_{a\bar{b}}}{n_a + 2}$	[30]
22	Least Contradiction	$\frac{n_a - 2n_{a\bar{b}}}{n_b}$	[4]
23	Lift	$\frac{n(n_a - n_{a\bar{b}})}{n_a n_b}$	[26]
24	Lerman	$\frac{n_a - n_{a\bar{b}} - \frac{n_a n_b}{n}}{\sqrt{\frac{n_a n_b}{n}}}$	[13]
25	Loevinger	$1 - \frac{n n_{a\bar{b}}}{n_a n_{\bar{b}}}$	[22]
26	Odds Ratio	$\frac{(n_a - n_{a\bar{b}})(n_{\bar{b}} - n_{a\bar{b}})}{n_{a\bar{b}}(n_b - n_a + n_{a\bar{b}})}$	[30]
27	Pavillon/Added Value	$\frac{n_{\bar{b}}}{n} - \frac{n_{a\bar{b}}}{n_a}$	[30]
28	Phi-Coefficient	$\frac{n_a n_{\bar{b}} - n n_{a\bar{b}}}{\sqrt{n_a n_b n_{\bar{a}} n_{\bar{b}}}}$	[30]
29	Putative Causal Dependency	$\frac{3}{2} + \frac{4n_a - 3n_b}{2n} - (\frac{3}{2n_a} + \frac{2}{n_{\bar{b}}})n_{a\bar{b}}$	[18]
30	Rule Interest	$\frac{n_a n_{\bar{b}}}{n} - n_{a\bar{b}}$	[25]
31	Sebag & Schoenauer	$\frac{n_a}{n_{a\bar{b}}} - 1$	[28]
32	Support	$\frac{n_a - n_{a\bar{b}}}{n}$	[1]
33	TIC	$\sqrt{DIR(a \rightarrow b) \times DIR(\bar{b} \rightarrow a)}$	[8] [6]
34	Yule's Q	$\frac{n_a n_{\bar{b}} - n n_{a\bar{b}}}{n_a n_{\bar{b}} + (n_b - n_{\bar{b}} - 2n_a)n_{a\bar{b}} + 2n_{a\bar{b}}^2}$	[30]
35	Yule's Y	$\frac{\sqrt{(n_a - n_{a\bar{b}})(n_{\bar{b}} - n_{a\bar{b}})} - \sqrt{n_{a\bar{b}}(n_b - n_a + n_{a\bar{b}})}}{\sqrt{(n_a - n_{a\bar{b}})(n_{\bar{b}} - n_{a\bar{b}})} + \sqrt{n_{a\bar{b}}(n_b - n_a + n_{a\bar{b}})}}$	[30]

Association Rule Interestingness Measures: Experimental and Theoretical Studies

Philippe Lenca[1], Benoît Vaillant[1], Patrick Meyer[2], and Stéphane Lallich[3]

[1] GET/ENST Bretagne, LUSSI Department, TAMCIC, UMR CNRS 2872, France
philippe.lenca@enst-bretagne.fr, benoit.vaillant@enst-bretagne.fr
[2] University of Luxembourg, Applied Mathematics Unit, Luxembourg
patrick.meyer@uni.lu
[3] University of Lyon 2, ERIC Laboratory, France
stephane.lallich@univ-lyon2.fr

Summary. It is a common problem that KDD processes may generate a large number of patterns depending on the algorithm used, and its parameters. It is hence impossible for an expert to assess these patterns. This is the case with the well-known APRIORI algorithm. One of the methods used to cope with such an amount of output depends on using association rule interestingness measures. Stating that selecting interesting rules also means using an adapted measure, we present a formal and an experimental study of 20 measures. The experimental studies carried out on 10 data sets lead to an experimental classification of the measures. This study is compared to an analysis of the formal and meaningful properties of the measures. Finally, the properties are used in a multi-criteria decision analysis in order to select amongst the available measures the one or those that best take into account the user's needs. These approaches seem to be complementary and could be useful in solving the problem of a user's choice of measure.

Key words: association rule, interestingness measure, interestingness criteria, measure classification, measure selection.

Introduction

One of the main objectives of Knowledge Discovery in Databases (KDD) is to produce interesting patterns. This notion of interest highly depends on the user's goals. This user is not assumed to be a data mining expert, but rather an expert in the field being mined. Moreover, it is well known that the interestingness of a pattern is difficult to evaluate objectively. Indeed, this estimation greatly depends on the expert user's interests [48], [37]. Ideally, a pattern should be *valid, new and comprehensive* [24], but these generic terms cover a large number of situations when examined in a precise context. It is

Philippe Lenca et al.: *Association Rule Interestingness Measures: Experimental and Theoretical Studies*, Studies in Computational Intelligence (SCI) **43**, 51–76 (2007)
www.springerlink.com

a common problem that data mining algorithms produce a huge amount of output, and that the end user is then unable to analyse it individually. What is more, a large part of this output is uninteresting [72]. Thus, when dealing with pattern selection one has to face two problems: the quantity and the quality of rules. This is particularly true when mining association rules with the well-known algorithms of the APRIORI family, within a support-confidence framework [2], and this is the issue that we will assess.

In this context, different solutions, more or less involving the user [83], can been considered. Visual data mining uses human visual capabilities to explore the data and patterns discovered (e.g. [78], [84], [47], [79], [43]). Human centred approaches emphasize the cooperation between the user and learning algorithms (e.g. [67], [54], [56], [9]).

Finally, interestingness measures can be used in order to filter and/or sort discovered rules (e.g. [37], [88], [38], [39]). Generally, one distinguishes between objective and subjective interestingness measures. Objective measures are said to be *data-driven* and only take into account the data cardinalities. Subjective measures are *user-driven* in the sense that they take into account the user's *a priori* knowledge and goals. For a discussion about subjective aspects of rule interestingness measures, the reader can refer to [82], [65] and [66].

It should be noted that, in practice, both objective and subjective approaches should be used to select interesting rules [26], the objective ones serving as a kind of first filter to select potentially interesting rules, while the subjective ones can be used as a final filter to retain only the truly interesting rules, depending on the applicative context.

We will focus on objective interestingness measures and take into account both user preferences or goals for association rule discovery and the nature of the data being mined. Such rules were defined in [2]: given a typical market-basket (transactional) database E, the association rule $A \rightarrow B$ means *if someone buys the set of items* A, *then he/she probably also buys item* B. It is of importance to make the distinction between the association rule $A \rightarrow B$, which focuses on cooccurrence and gives asymmetric meaning to A and B, and logical implication $A \Rightarrow B$ or equivalence $A \Leftrightarrow B$ [51].

Interestingness measures play an essential role, reducing the number of discovered rules and retaining only the *best ones*, in a post-processing step.

In order to improve the selection of rules, many classical measures have been used, like the Chi-square test for independency, or the correlation coefficient. Due to specific needs, additional measures have been proposed, such as the lift [17], the M_{GK} measure [33], relative interestingness [41], general measure [44], the entropic intensity of implication [31], the probabilistic discriminant index [63], the maximal participation index [40], or the h-confidence [94], information theoretic based measures [12], parametrised measures [52]. As a consequence, a large number of measures are available (see for example [34] for an extensive list of classical measures).

Depending on the user's goals, data mining experts may propose the use of an appropriate interestingness measure, but this selection task cannot be done by the expert user, if left on his own.

This choice is hard, since rule interestingness measures have many different qualities or flaws, since there is no *optimal* measure. One way to solve this problem is to try to find good compromises [59]. A well-known example of such a controversial measure is the support. On the one hand, it is heavily used for filtering purposes in APRIORI algorithms [2], [73], as its anti-monotonicity property simplifies the large lattice that has to be explored. On the other hand, it has almost all the flaws a user would like to avoid, such as variability of the value under the independence hypothesis or the value for a logical rule [75]. Finally, one should be very careful when using the support-confidence framework in defining the interestingness of a rule [76], [16]. To bypass this difficulty different works look for highly correlated items, like as in the CORCLASS algorithm [96] and in the algorithms presented in [21].

It is then relevant to study interestingness measures, so that rules are selected according to the user's needs and context [59]. Interestingness measures have to support KDD process through system-human interaction [71], [1]. Many works (for instance [6], [53], [36], [37], [87], [85], [51], [17], [60], [89], [86], [70]) have formally extracted and studied several specificities of various measures, and the importance of objective evaluation criteria of interestingness measures has already been focused on by [75] and [26].

In this chapter, we will assess the issue of selecting an adapted interestingness measure faced with an applicative context and user's aims.

First, we introduce a set of 20 classical measures which seem applicable in an association rule mining context [57]. In the second section, these measures are analyzed through eight formal properties that make sense from an end user's point of view. In order to highlight the wide variety of measures and have a case based overview of their behaviour, the third section focuses on a tool we have developed, HERBS [90], and an empirical classification of the measures is built out of experimental campaigns [92]. This classification is then compared to another clustering of the measures, based on their theoretical studies. Out of theoretical properties, we finally propose a multi-criteria decision aid (MCDA) approach assessing the issue of selecting an measure adapted to the user's context (aims, goals, nature of the data, etc.) [59]. Finally, we conclude and outline some perspectives that are to be studied.

1 Interestingness measures

In this section, we present the 20 objective association rules interestingness measures that we studied. These measures are usually defined using the 2×2 contingency table presented in figure 1, and is a classical way of measuring association in the case of paired attributes [23], such as in the GUHA method [21], in the 4FT-MINER tool [80] and in the APRIORI algorithm [2].

Given a rule A → B, we note:

- $n = |E|$ the total number of records in the database E
- $n_a =$ the number of records satisfying A
- $n_b =$ the number of records satisfying B
- $n_{ab} =$ the number of records satisfying both A and B (the examples of the rule)
- $n_{a\bar{b}} = n_a - n_{ab}$ the number of records satisfying A but not B (the counter-examples of the rule)

For any X, we note p_x instead of n_x/n when we consider relative frequencies rather than absolute frequencies on the data set E. It is clear that, given n, n_a and n_b, or p_a and p_b, knowing one cell of the contingency table in figure 1 is enough to deduce the other ones.

A\B	0	1	total
0	$p_{\bar{a}\bar{b}}$	$p_{\bar{a}b}$	$p_{\bar{a}}$
1	$p_{a\bar{b}}$	p_{ab}	p_a
total	$p_{\bar{b}}$	p_b	1

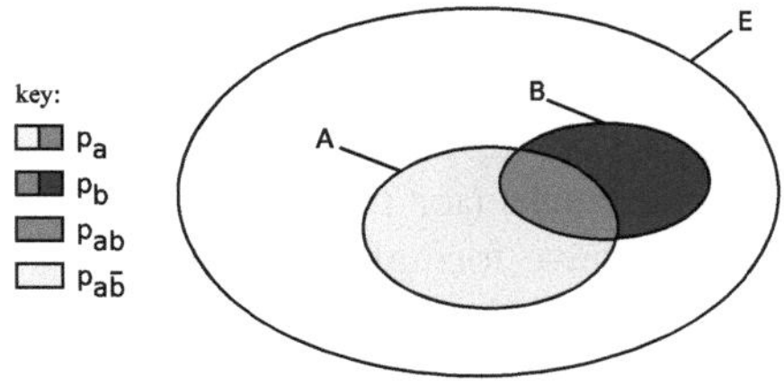

Fig. 1. Notations

We restricted the list of measures to decreasing ones, with respect to $n_{a\bar{b}}$, all marginal frequencies being fixed. This choice reflects the common assertion that the fewer counter-examples (A true and B false) to the rule there are, the higher the interestingness of the rule. Thus some measures like χ^2, Pearson's r^2, Goodman and Smyth's J-measure or Pearl's measure are not considered in this study. The selected measures are listed in table 1, which also includes bibliographical references. Their definition and co-domain, using absolute frequencies, is given in table 2. At first glance, table 2 shows important variations between the formulae. This is due to the fact that measures do not tell the same story. These variations are also noticeable since co-domains are quite different ([0, 1], [0, +∞[,]-∞, 1] and others with bounds depending on n_a, n_b and/or n_{ab}). For taking into account such variations one may use aggregation operators of valued relations [5] or normalized measures [25].

For a given decreasing monotonic measure μ (with respect to $n_{a\bar{b}}$ margins n_a and n_b being fixed), the selection of interesting rules is done by positioning a threshold α and keeping only the rules satisfying $\mu(A \rightarrow B) \geq \alpha$. The value of this threshold α has to be fixed by the expert, and the same threshold is considered for all the rules extracted during the data mining process. Thus, fixing α is an important issue [16].

Table 1. List of selected measures

	Name	References
BF	Bayes factor	[45]
CENCONF	centred confidence	
CONF	confidence	[2]
CONV	conviction	[18]
ECR	examples and counter-examples rate	
EII	entropic intensity of implication	[31]
IG	information gain	[20]
- IMPIND	implication index	[64]
INTIMP	intensity of implication	[29]
KAPPA	Kappa coefficient	[22]
LAP	Laplace	[28]
LC	least contradiction	[3]
LIFT	Lift	[17]
LOE	Loevinger	[36]
PDI	probabilistic discriminant index	[63]
PS	Piatetsky-Shapiro	[75]
R	Pearson's correlation coefficient	[74]
SEB	Sebag and Schoenauer	[81]
SUP	support	[2]
ZHANG	Zhang	[95]

In our set of measures, we kept the well-known support and confidence: these are the two most frequently used measures in algorithms based on the selection of frequent itemsets for association rule extraction [2], [73].

Many other measures are linear transformations of the confidence, enhancing it, by enabling comparisons with p_b. This transformation is generally achieved by centering the confidence on p_b, using different scale coefficients (centered confidence, Piatetsky-Shapiro's measure, Loevinger's measure, Zhang's measure, correlation, implication index, least contradiction). It is also possible to divide the confidence by p_b (lift).

Other measures, like Sebag and Schoenauer's or the rate of examples and counter-examples, are monotonically increasing transformations of confidence, while the information gain is a monotonically increasing transformation of the lift. Thus, these measures will rank rules in the same order and differ, for example, from their semantic meaning [28].

Some measures focus on counter-examples, like the conviction or the above-cited implication index. This latter measure is the basis of several different probabilistic measures like the probabilistic discriminant index, the intensity of implication, or its entropic version, which takes into account an entropic coefficient, enhancing the discriminant power of the intensity of implication. These last two measures were adapted in order to let them have the desired property of being constant under a null hypothesis (this property is discussed in section 2). For the intensity of implication, the statistical law was

Table 2. Association rule quality measures

	Definition	Co-domain
BF	$\frac{n_{ab}n_{\bar b}}{n_b n_{a\bar b}}$	$[0, +\infty[$
CenConf	$\frac{nn_{ab}-n_a n_b}{nn_a}$	$[-\frac{n_b}{n}, \frac{n_{\bar b}}{n}]$
Conf	$\frac{n_{ab}}{n_a}$	$[0, 1]$
Conv	$\frac{n_a n_{\bar b}}{nn_{a\bar b}}$	$[\frac{n_{\bar b}}{n}, +\infty[$
ECR	$\frac{n_{ab}-n_{a\bar b}}{n_{ab}} = 1 - \frac{1}{\frac{n_a}{n_{a\bar b}}-1}$	$]-\infty, 1]$
EII	$\left\{[(1-h_1(\frac{n_{a\bar b}}{n})^2)(1-h_2(\frac{n_{a\bar b}}{n})^2)]^{1/4}\text{IntImp}\right\}^{1/2}$	$[0, 1]$
IG	$\log(\frac{nn_{ab}}{n_a n_b})$	$]-\infty, \log\frac{n}{n_b}]$
-ImpInd	$\frac{n_a n_b - nn_{ab}}{\sqrt{nn_a n_{\bar b}}}$	$[-\frac{\sqrt{n_a}n_b}{\sqrt{nn_{\bar b}}}, \sqrt{\frac{n_a n_{\bar b}}{n}}]$
IntImp	$P\left[N(0,1) \geq \text{ImpInd}\right]$	$[0, 1]$
Kappa	$2\frac{nn_{ab}-n_a n_b}{nn_a+nn_b-2n_a n_b}$	$[-2\frac{n_a n_b}{n_a n_{\bar b}+n_{\bar a}n_b}, 2\frac{n_a n_{\bar b}}{n_a n_{\bar b}+n_{\bar a}n_b}]$
Lap	$\frac{n_{ab}+1}{n_a+2}$	$[\frac{1}{n_a+2}, \frac{n_a+1}{n_a+2}]$
LC	$\frac{n_{ab}-n_{a\bar b}}{n_b}$	$[-\frac{n_a}{n_b}, \frac{n_a}{n_b}]$
Lift	$\frac{nn_{ab}}{n_a n_b}$	$[0, \frac{n}{n_b}]$
Loe	$\frac{nn_{ab}-n_a n_b}{n_a n_{\bar b}}$	$[-\frac{n_b}{n_{\bar b}}, 1]$
PDI	$P\left[\mathcal{N}(0,1) > \text{ImpInd}^{CR/\mathcal{B}}\right]$	$]0, 1[$
PS	$n_{ab} - \frac{n_a n_b}{n}$	$[-\frac{n_a n_b}{n}, \frac{n_a n_{\bar b}}{n}]$
R	$\frac{nn_{ab}-n_a n_b}{\sqrt{nn_a n_b n_{\bar a}\cdot n_{\bar b}}}$	$[-\sqrt{\frac{n_a n_b}{nn_{\bar a}n_{\bar b}}}, \sqrt{\frac{n_a n_{\bar b}}{nn_{\bar a}n_b}}]$
Seb	$\frac{n_{ab}}{n_{a\bar b}}$	$[0, +\infty[$
Sup	$\frac{n_{ab}}{n}$	$[0, \frac{n_a}{n}]$
Zhang	$\frac{nn_{ab}-n_a n_b}{\max\{n_{ab}n_{\bar b}, n_b n_{a\bar b}\}}$	$[-1, 1]$

$\text{ImpInd}^{CR/\mathcal{B}}$ corresponds to ImpInd, centred reduced (CR) for a rule set $\mathcal{B}$.

$h_1(t) = -(1-\frac{n\cdot t}{n_a})\log_2(1-\frac{n\cdot t}{n_a}) - \frac{n\cdot t}{n_a}\log_2(\frac{n\cdot t}{n_a})$ if $t \in [0, n_a/(2n)[$; else $h_1(t) = 1$

$h_2(t) = -(1-\frac{n\cdot t}{n_{\bar b}})\log_2(1-\frac{n\cdot t}{n_{\bar b}}) - \frac{n\cdot t}{n_{\bar b}}\log_2(\frac{n\cdot t}{n_{\bar b}})$ if $t \in [0, n_{\bar b}/(2n)[$; else $h_2(t) = 1$

$\mathcal{N}(0,1)$ stands for the centered and reduced normal repartition function approximated using the centred and reduced normal distribution function. The entropic intensity of implication was modified, according to the definition of the truncated entropic intensity of implication, TEII, as presented in [52].

The bayesian factor, also called sufficiency in [26] or odd-multiplier by [28], is a kind of odd-ratio, based on the comparison of the odd of A and B on B rather than the odd of A and $\overline{A}$ on B. It has been thoroughly studied in [32].

Finally, Laplace's measure is a variant of the confidence, taking the total number of records n into account.

2 Evaluation properties

In this section, we propose a list of eight meaningful properties to evaluate the previous list of measures. We present each property, explaining its interest and the modalities it can take.

Two actors take part in this analysis: the user who is an expert of the data mined, whose problem is to select the *best rules*, and the analyst, a specialist of MCDA and KDD, who tries to help the expert. We call the former E_r and the latter E_a.

For some properties, a preference order on the modalities they can take is straightforward. These properties can be considered as criteria by E_a without the intervention of E_r, namely g_1, g_2, g_3, g_4 and g_7, and will be called normative. In addition to these, the properties g_5, g_6 and g_8 need E_r to express his preferences on the values they can take, and will be called subjective [60].

For normative properties, we note yes if the measure has the desired property and no otherwise.

Table 3 recalls the semantics and the number of modalities of the 8 properties. The results of the evaluations are summarized in table 4.

Property g_1: **asymmetric processing of A and B [26].** Since the head and the body of a rule may have a very different signification, it is desirable to distinguish measures that give different evaluations of rules A $\rightarrow$ B and B $\rightarrow$ A from those that do not. We note no if the measure is symmetric, yes otherwise.

Property g_2: **decrease with n_b [75].** Given n_{ab}, $n_{a\overline{b}}$ and $n_{\overline{a}\overline{b}}$, it is of interest to relate the interestingness of a rule to the size of B. In this situation, if the number of records verifying B (i.e. verifying B but not A) increases, the interestingness of the rule should decrease. We note yes if the measure is a decreasing function with n_b, no otherwise.

Property g_3: **reference situations, independence [75].** To avoid keeping rules that contain no information, it is necessary to eliminate the A $\rightarrow$ B rule when A and B are independent, which means that the probability of obtaining B is independent of the fact that A is true or not. A comfortable way of dealing with this is to require that a measure's value at independence should be constant. We note yes if the measure's value at independence is constant and no otherwise.

Property g_4: **reference situations, logical rule [57].** Similarly, the second reference situation we consider is related to the value of the measure when there is no counter-example. Depending on the co-domain (see table 2), three cases arise. First, the measure takes a value independent of the marginal frequencies (see table in figure 1) and thus takes a constant and maximal value[4]. A second case is considered when the measure takes an infinite value when $n_{a\bar{b}}$ is null. Finally, a third and more uncomfortable case arises when the value taken by the measure depends on the marginal frequencies when $n_{a\bar{b}} = 0$. It is desirable that the value should be constant or possibly infinite. We note `yes` in the cases of a constant or infinite value, `no` otherwise.

We do not take into account the value for the incompatibility situation. The latter reference situation is obtained when $\mathtt{A} \cap \mathtt{B} = \emptyset$, and expresses the fact that `B` cannot be realized if `A` already is. Our choice is based on the fact that incompatibility is related to the rule $\mathtt{A} \rightarrow \overline{\mathtt{B}}$ and not $\mathtt{A} \rightarrow \mathtt{B}$.

Property g_5: **linearity with $p_{a\bar{b}}$ around 0^+ [17].** Some users express the desire to have a weak decrease in the neighborhood of a logic rule rather than a fast or even linear decrease (as with confidence or its linear transformations). This reflects the fact that the user may tolerate a few counter-examples without significant loss of interest, but will definitely not tolerate too many of them. However, the opposite choice may be preferred as a convex decrease with $n_{a\bar{b}}$ around the logic rule increases the sensitivity to a false positive. We hence note `convex` if the measure is convex with $n_{a\bar{b}}$ near 0, `linear` if it is linear and `concave` if it is concave.

Property g_6: **sensitivity to n (total number of records) [51], [17].** Intuitively, if the rates of presence of `A`, $\mathtt{A} \rightarrow \mathtt{B}$, `B` are constant, it may be interesting to see how the measure reacts to a global extension of the database (with no evolution of rates).

If the measure increases with n and has a maximum value, then there is a risk that all the evaluations might come close to this maximum. The measure would then lose its discrimination power. The preference of the user might be indifferent to having a measure which is invariant or not with the dilatation of data. We note `desc` (for descriptive measures) if the measure is invariant and `stat` (for statistical ones) if it increases with n.

Property g_7: **easiness to fix a threshold [57].** Even if properties g_3 and g_4 are valid, it is still difficult to decide the best threshold value that separates interesting from uninteresting rules. This property allows us to identify measures whose threshold is more or less difficult to locate. To establish this property, we propose to proceed in the following (and very conventional) way by providing a sense of the strength of the evidence against the null hypothesis, that is, the p-value. Due to the high number of tests, this probability should not be interpreted as a statistical risk, but rather as a control parameter [51]. In some cases, the measure is defined as such a probability. More

[4] Recall that due to our eligibility criterion, we restrict our study to decreasing measures with respect to $n_{a\bar{b}}$, all marginal frequencies being fixed.

generally, we can define such a threshold from one of the three types of models proposed by [62] to establish the law followed by $n_{a\bar{b}}$ under the hypothesis of link absence. We note `yes` if the measure easily supports such an evaluation, and `no` otherwise.

Property g_8: **intelligibility [57].** Intelligibility denotes the ability of the measure to express a comprehensive idea of the interestingness of a rule. We will consider that a measure is intelligible if its semantics is easily understandable by the expert of the data E_r[5]. We assign the value `yes` to this property if the measure can be expressed in that way, `avg` if the measure can be estimated with common quantities, and `no` if it seems impossible to give any simple concrete explanation of the measure.

Table 3. Properties of the measures

Property	Semantics	Modalities
g_1	asymmetric processing of `A` and `B`	2
g_2	decrease with n_b	2
g_3	reference situations: independence	2
g_4	reference situations: logical rule	2
g_5	linearity with $n_{a\bar{b}}$ around 0^+	3
g_6	sensitivity to n	2
g_7	easiness to fix a threshold	2
g_8	intelligibility	3

The extension of this list is currently being studied, and in particular discrimination, antimonotonicity, and robustness to noise. Discrimination is quite interesting since it might be related to criteria g_6 (sensitivity to the cardinality of the total space), which generally occurs simultaneously with a loss of discrimination. Antimonotonicity also is an interesting property from the computing point of view, both for APRIORI algorithms and Galois lattice based methods [73]. Robustness to noise has been focused on in [4] and [61].

Finally, different alternatives could be proposed for property g_3 (independence). It could be interesting to replace the independence condition ($p_{b/a} = p_b$) by the equilibrium condition ($p_{b/a} = 0.5$) that corresponds to predictive purposes [10]. More generally, a confidence threshold θ ($p_{b/a} = \theta, p_b < \theta < 1$) could be taken into account, especially for targeting purposes [52].

[5] It is obvious that this property is subjective. The evaluations of the measures on this property given hereafter can be commonly accepted. Nevertheless, depending on E_r, our evaluations could be revised.

Table 4. Evaluation matrix

	g_1	g_2	g_3	g_4	g_5	g_6	g_7	g_8
BF	yes	yes	yes	yes	convex	desc	yes	yes
CenConf	yes	yes	yes	no	linear	desc	yes	yes
Conf	yes	no	no	yes	linear	desc	yes	yes
Conv	yes	yes	yes	yes	convex	desc	yes	avg
ECR	yes	no	no	yes	concave	desc	yes	avg
TEII	yes	yes	yes	no	concave	stat	no	no
IG	no	yes	yes	no	concave	desc	yes	no
- ImpInd	yes	yes	yes	no	linear	stat	yes	no
IntImp	yes	yes	yes	no	concave	stat	yes	no
Kappa	no	yes	yes	no	linear	desc	yes	no
Lap	yes	no	no	no	linear	desc	yes	no
LC	yes	yes	no	no	linear	desc	yes	avg
Lift	no	yes	yes	no	linear	desc	yes	yes
Loe	yes	yes	yes	yes	linear	desc	yes	avg
PDI	yes	yes	yes	no	concave	stat	yes	no
PS	no	yes	yes	no	linear	stat	yes	avg
R	no	yes	yes	no	linear	desc	yes	avg
Seb	yes	no	no	yes	convex	desc	yes	avg
Sup	no	no	no	no	linear	desc	yes	yes
Zhang	yes	yes	yes	yes	concave	desc	no	no

3 Interestingness measure classifications

Beyond a formal analysis, based on meaningful properties, it is interesting to observe the behavior of the measures on data. We present an experimental classification based on preorder comparisons, these preorders being induced by interestingness measures on rule sets. This classification is carried out using our experimentation tool, Herbs. A formal classification based on the formal properties is proposed using a hierarchical ascendent clustering. Finally, we compare the two classifications.

3.1 An overview of Herbs, an experimentation tool

The aim of Herbs [90], [46] is to analyse rule sets and compare or investigate interestingness measures through concrete experiments. It has been designed as an interactive *post-analysis* tool, and hence data sets, rule sets and interestingness measures are considered as inputs. Various useful experimentation schemes are implemented in Herbs, from simple descriptive statistics about rule sets, to comparative overviews of the evaluation of a rule set by several measures.

We here propose an experimental analysis and comparison of measures, based on their application to 10 pairs of data sets and rule sets. A synthetic comparison of the rankings of a rule set by the measures is given by computing a preorder agreement coefficient, τ_1 which is derived from Kendall's τ (see [27]). This agreement compares a pair of preorders induced by two measures, and its value is in the range $[-1; 1]$. The maximum value is obtained when the two pre-orders are equal, whereas the minimum value is obtained in various cases, and especially for reversed preorders.

From a computational point of view, using such a coefficient can be seen as complex since its evaluation is done in $\mathcal{O}(\eta^2)$, where η is the number of rules in the rule set considered, when a correlation analysis can be done in $\mathcal{O}(\eta)$ (the correlation index between interestingness measures is used in the ARQAT tool [42] for example). Still, from the numerous coefficients presented in [27], the τ_1 coefficient best suits our needs. What is more, HERBS uses a relational database in order to store the experimental results. Building an index on these values greatly optimizes the computation of this coefficient. Finally, only a slight modification of the formula is required in order to return to more classical agreement coefficients, such as Kendall's τ or Spearman's ρ.

3.2 Experimental classification

Experiments were carried out on databases retrieved from the UCI Repository (`ftp://ftp.ics.uci.edu/` [8]). When there is no ambiguity, we will refer indifferently to the pair formed by a data set and a rule set, or to the single data set or rule set, using their names in the Repository. We denote by *BCW* the *breast-cancer-wisconsin* database. The parameters of the APRIORI algorithm [9] were fixed experimentally in order to obtain rule sets of an acceptable size in terms of computational cost (see table 5). The great differences in size of the rule sets is related to the number of modalities of the different attributes of the case databases. A particular option was used in order to compute *Cmc*: APRIORI, which usually explores a restricted number of nodes of the lattice formed by the different modalities of the attributes, was forced to explore the entire lattice. *Cmc2* was obtained by filtering *Cmc*, with a minimum lift of 1.2. The *Solarflare* database is divided into two case sets, $\mathcal{SF}_1$ and $\mathcal{SF}_2$, described by the same attributes. $\mathcal{R}_1$ (resp. $\mathcal{R}_2$) is the rule set coming from $\mathcal{SF}_1$ (resp. $\mathcal{SF}_2$). We filtered $\mathcal{R}_1$, with the method exposed in [91] following the results of [50] in order to keep only rules that are significant from a statistical point of view. Using $\mathcal{SF}_1$ (resp. $\mathcal{SF}_2$), we obtained the rule set $\mathcal{R}_1^1$ (resp. $\mathcal{R}_1^2$). The characteristics of the sets are summarized in table 5.

We generated 10 preorder comparison matrices, which are presented in table 6 (the value of τ_1 is proportional to the radius of the corresponding portion of disc, a radius null corresponding to an agreement of -1, and a radius of 1 corresponding to an agreement value of 1). The AMADO method [19] was applied to the average matrix of the results in order to reorganize the rows and the columns of this matrix, and highlight the block structures. The results are quite in agreement, and we can make out 3 main groups of measures, and in two of these groups we can distinguish two subgroups (see tables 6 and 7).

Table 5. Summary of the different sets used, and APRIORI parameters

name	n	sup_{min}	$conf_{min}$	η
Autompg	392	5	50	49
BCW	683	10	70	3095
Car	1728	5	60	145
Cmc	1473	5	60	2878
Cmc2	n/a	n/a	n/a	766

name	n	sup_{min}	$conf_{min}$	η
$(\mathcal{SF}_1, \mathcal{R}_1)$	323	20	85	5402
$(\mathcal{SF}_2, \mathcal{R}_2)$	1066	20	85	6312
$\mathcal{R}_1^1$	n/a	n/a	n/a	4130
$\mathcal{R}_1^2$	n/a	n/a	n/a	2994

Table 6. Preorder comparisons of 20 measures on 10 experiments.

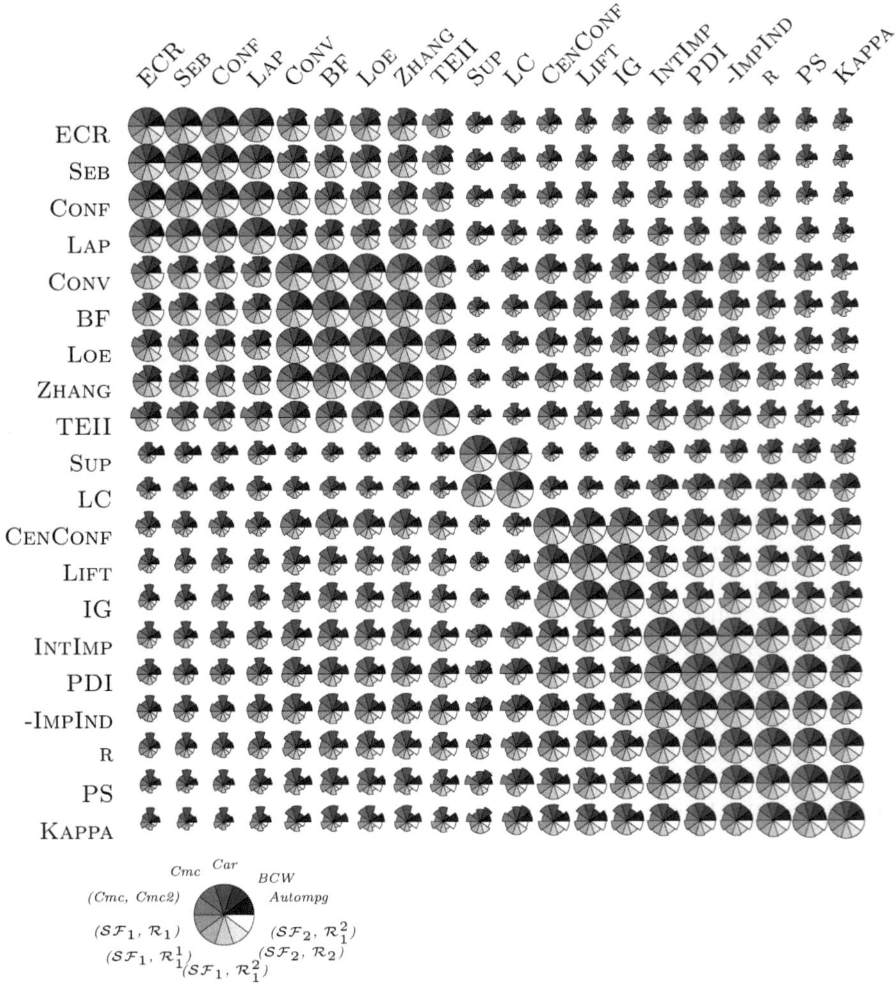

The first group consists of {ECR, SEB,CONF, LAP, CONV, BF, LOE, ZHANG, TEII} and can be sub-categorized into two subgroups: E_1 ={ECR, SEB,CONF, LAP} and E_2 ={CONV, BF, LOE, ZHANG, TEII}. The second main group consists of E_3 ={SUP, LC}, behaving very differently from

the previous measures. The third group, {CenConf, Lift, IG, IntImp, PDI, -ImpInd, r, PS, Kappa}, can be split into two, as was the first one, and leads to the two following subgroups: E_4 ={CenConf, Lift, IG} and E_5 ={IntImp, PDI, -ImpInd, r, PS, Kappa}.

3.3 Formal classification

The formal approach can be synthetized with a 20×8 matrix, containing the evaluation of the 20 measures on the 8 properties. We kept only 6 of the properties for the comparison between experimental and formal approaches, as two of them – namely g_7 (easiness to fix a threshold) and g_8 (intelligibility) – do not influence the experimental results at all.

All these properties are bivaluate except g_5 which is trivaluate. The 20×6 matrix formally obtained was re-encoded in a 20×6 matrix composed of real values, 0 or 1 in the binary cases, and 0, 0.5 or 1 for g_5. These values do not represent any judgement on the measures, but only list the properties shared by the different measures.

The typology in 5 classes, F_i, $i = 1 \ldots 5$ (see table 7) coming from this matrix is obtained with a hierarchical ascendant clustering, using the average linkage, applied to the Manhattan distance.

3.4 Comparison of the two classifications

Table 7 shows that both approaches globally lead to similar clusterings, but some shifts are interesting. The main differences concern {Sup, LC} and TEII.

The experimental classification leads to two main classes, $E_1 \cup E_2$ and $E_4 \cup E_5$. The coherence between the two classifications is underlined by the fact that apart from the three above-mentioned measures, $E_1 = F_1 \cup F_2$, $F_3 \subset E_2$ and $E_4 \cup E_5 \subset F_4 \cup F_5$.

From a formal point of view, Sup and LC are quite close, forming class F_2 together with Lap. There also is a strong link between the classes F_1 and F_2. Apart from Sup and Lap, the measures belonging to these classes are those sharing the property of making reference to indetermination when evaluating the quality of a rule (i.e. measures having a constant value when $n_{ab} = n_{a\bar{b}} = n_a/2$, [11], [10]), although this property was not taken into account in our formal classification.

The formal class F_5 is made out of the measures built on the implication index, namely -ImpInd itself, IntImp which is derived from the former through the use of the normal distribution, and the two discriminent measures, TEII and PDI. In our formal approach no dinstiction can be made between IntImp, TEII and PDI, since none of the criteria g_1 to g_6 take into account the discriminating power of the measures. We are currently working on such a criterion. Apart from TEII, these measures make up the same experimental class, which also includes r, Kappa and PS. The altered behavior of TEII is

Table 7. Cross-classification of the measures

Formal \ Experimental	Class E_1	Class E_2	Class E_3	Class E_4	Class E_5
Class F_1	CONF, SEB, ECR				
Class F_2	LAP		SUP, LC		
Class F_3		CONV, BF, LOE, ZHANG			
Class F_4				LIFT, IG, CENCONF	R, KAPPA, PS
Class F_5		TEII			INTIMP, -IMPIND, PDI

due to the fact that it is derived from INTIMP through the use of an inclusion index. This inclusion index plays a major role in the evaluation of the quality of a rule and thus accounts for the experimental differences. Experimentally, TEII thus shifts to LOE, ZHANG, BF and CONV (class E_2).

Formally, LAP shifts to LC and SUP (class F_2). A reason for this shift is that although it is really close to SUP in our formal study, LAP can differ from CONF experimentally only for values of n_a close to 0 (nuggets). The minimum thresholds of the APRIORI algorithms make this impossible, and this can be seen as an algorithmic bias [92].

Property g_4 has an important impact on experimental results. When it is verified, all the logical rules are evaluated with a maximal value, no matter what the conclusion is. BF, CONV, LOE, ZHANG, and ECR, SEB, CONF, *i.e.* the measures for which $g_4 =$ yes, make the experimental group $E_1 \cup E_2$. Only TEII and LAP, also belonging to these classes, do not share this property.

4 A multi-criteria decision approach towards measure selection

In this section, we will analyze and evaluate the measures described earlier and summarized in table 2. This analysis was done by a few MCDA procedures, in particular the TOMASO method for sorting [69], a ranking procedure based on kernels of digraphs [7] and the PROMETHEE method [15]. These three methods have produced very similar results. In this chapter, we focus on the analysis by the PROMETHEE method to obtain a ranking. A formalization of the decision problem is discussed in [58]. This approach has been used in a real context by [77].

4.1 A few words on the PROMETHEE method

Its objectives are to build partial and complete rankings on alternatives (in this case, the measures) and to visualize the structure of the problem in a plane called the GAIA plane, similarly to a principal component analysis. The PROMETHEE method requires information about the importance of the criteria (a criteria is a property on which a preference modeling is known) to be given by a set of weights. Several tools allow these weights to be fixed in order to represent the decision maker's preferences (E_r in our context). The first step of the method is to make pairwise comparisons on the measures within each criterion. This means that for small (large) deviations, E_r will allocate a small (large) preference to the best measure. This is done through the concept of preference functions. Then, each measure is confronted with the other ones in order to define outranking flows. The positive (negative) outranking flow expresses to what degree a measure a is outranking (outranked by) the others. Finally, partial and complete rankings are generated from these outranking flows. The GAIA plane provides information on the conflicting character of the criteria and on the impact of the weights on the final decision. It is a projection, based on a net flow ϕ derived from the outranking flows, of the measures and the criteria in a common plane. For a more detailed description of this method, the reader can refer to [14], for example.

4.2 Analysis of the quality measures

We consider the following two realistic scenarios for the analysis:

Sc1: The expert E_r tolerates *the appearance of a certain number of counter-examples* to a decision rule. In this case, the rejection of a rule is postponed until enough counter-examples are found. The shape of the curve representing the value of the measure versus the number of counter examples should ideally be concave (at least in the neighbourhood of the maximum); the order on the values of criterion g_5 (non-linearity with respect to the number of counter-examples) is therefore `concave` $\succ$ `linear` $\succ$ `convex`, where $\succ$ means "is preferred to".

Sc2: The expert E_r refuses *the appearance of too many counter-examples* to a decision rule. The rejection of the rule must be done rapidly with respect to the number of counter-examples. The shape of the curve is therefore ideally convex (in the neighbourhood of the maximum at least) and the order on the values of criterion g_5 is `convex` $\succ$ `linear` $\succ$ `concave`.

For both scenarios, for criterion g_6 we assume that the expert prefers a measure which increases with n, the size of the data. Thus, the order on the values of criterion g_6 is `stat` $\succ$ `desc`. For the other criteria which are assumed to be normative, the expert has no influence on the order of the values.

We start by analysing the problem with equal weights for the criteria to get a first view of the structure of the problem. The total rankings for the two scenarios are given in table 8.

First, we notice that both scenarios reflect the preferences of E_r on the shape of the curve. We can see that for **Sc1** the two leading measures are IntImp and PDI which are both concave. Similarly, for **Sc2**, the two leading measures are BF and Conv which are both convex. This first analysis also shows that the linear measure Loe is a very interesting measure as it is well placed in both scenarios. It stands for a good compromise.

Sensitivity analyses on the weights systems show that small changes in the weights affect the rankings. Nevertheless a closer look shows that these

Table 8. Total rankings for scenarios **Sc1** and **Sc2**.

Rank:	1	2	3	4	5	6	7
Sc1:	IntImp, PDI		Loe	BF	CenConf	Conv	-ImpInd
Sc2:	BF	Conv	Loe	CenConf	-ImpInd	PS	Seb
Rank:	8	9	10	11	12	13	14
Sc1:	Zhang, TEII		PS	ECR	Lift	Conf	IG
Sc2:	Lift	Conf	IntImp, PDI		r, LC		Zhang
Rank:	15	16	17	18	19	20	
Sc1:	r, LC		Seb	Kappa	Sup	Lap	
Sc2:	TEII	Kappa	ECR	Sup	IG	Lap	

modifications only occur locally and that the first positions of the rankings remain stable.

Therefore one can say that for an expert E_r who has no particular opinion on the importance of the different criteria, or who considers that the criteria are equally important, the rankings of table 8 are quite acceptable.

An analysis of the Gaia planes gives us further indications about the measures. Figure 2 shows the Gaia planes for **Sc1** and **Sc2**.

Let us first note that the percentage of cumulated variance for the first two factors represented by the Gaia plane is 60.20%. The information taken from the Gaia plane should therefore be considered as approximative and conclusions be drawn with great care. First we observe that the measures (triangles in the figure) are distributed homogeneously in the plane. Then we can see that the Gaia plane is well covered by the set of criteria (axes with squares in the figure). We conclude that the description of the measures selected by the criteria is discriminant and only slightly redundant.

The Gaia plane furthermore helps to detect independent and conflicting criteria. The decision axis π (axis with a circle) indicates in what direction the best alternatives are situated for a given weights system.

For **Sc1** we can see that several couples of criteria are independent: (g_4, g_5), (g_4, g_8), (g_5, g_3), (g_5, g_2), (g_8, g_3), (g_1, g_6) and (g_8, g_2)[6]. We can also observe conflicting criteria. For example g_4 conflicts with g_3 and g_2; and criteria g_5 and g_6 conflict with g_7 and g_8. This type of information gives hints on the behaviour and the structure of the problem. For example, measures which are good for criterion g_5 (concave) will tend to be bad for criterion g_8 (unintelligible).

For **Sc2** similar observations can be made. The major difference lies in criterion g_5 which represents similar preferences to criteria g_7 and g_8 but is conflicting with g_6.

For **Sc1**, the decision axis π is moderately long and heads in the opposite direction of g_7 and g_8. This means that measures which allow us to fix the threshold easily and which are easily understandable (and which are quite bad on the remaining criteria) can appear in the leading positions of the ranking only if the relative weights of g_7 and g_8 are very high. However we think that the importance of criterion g_3 (independence hypothesis) should not be

[6] If g_i and g_j are independent, we write that the couple (g_i, g_j) is independent.

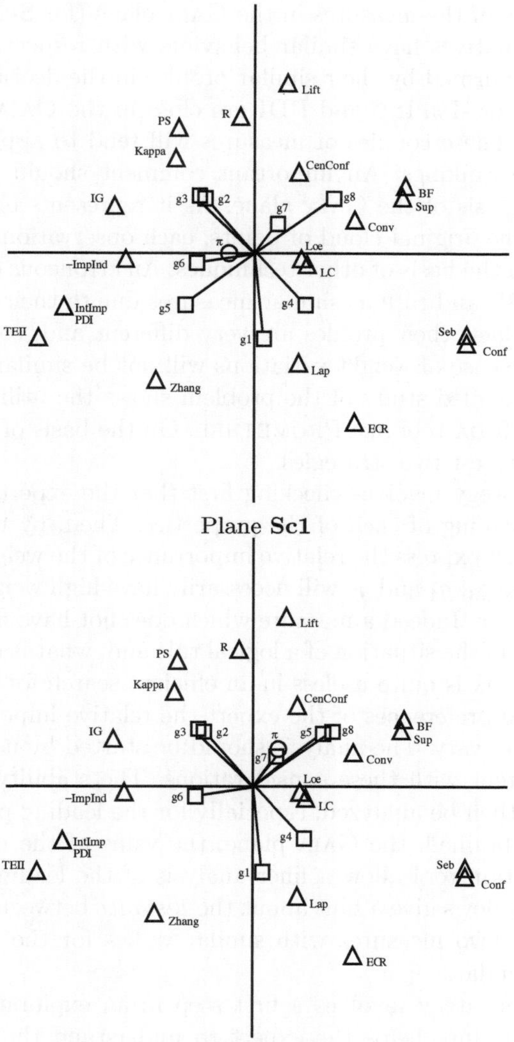

Plane **Sc2**

Fig. 2. GAIA planes for **Sc1** and **Sc2**

neglected compared to a criterion like g_8 (intelligibility). Thus, if the expert is aware of the impact of his weights system on the result, we can suppose that a measure like SUP, exclusively good on g_7 and g_8, will never appear in the leading positions of the ranking. For **Sc2** the decision axis π is also moderately long. It points in direction of g_7, g_5 and g_8. This partly explains the ranking of table 8.

The positions of the measures in the GAIA plane (for **Sc1** and **Sc2**) show that many alternatives have similar behaviors with respect to weight variations. This is confirmed by their similar profiles in the decision matrix. Thus SEB and CONF, or -IMPIND and PDI are close in the GAIA plane and have similar profiles. These couples of measures will tend to appear in neighbour positions in the rankings. An important comment should be made at this point of the analysis of the GAIA plane. As it represents only a part of the information of the original cloud of points, each observation must be verified in the data or on the basis of other techniques. An erroneous conclusion would be to consider BF and SUP as similar measures due to their proximity in the GAIA plane. In fact, their profiles are very different and, consequently, their behaviour in the case of weight variations will not be similar.

This quite detailed study of the problem shows the utility of an analysis by means of a MCDA tool like PROMETHEE. On the basis of the observations above we can suggest two strategies.

The first strategy involves checking first that the expert E_r has well understood the meaning of each of the properties. Then, by means of a set of questions, he must express the relative importance of the weights of each criterion. Criteria like g_3, g_4 and g_7 will necessarily have high weights to guarantee a certain coherence. Indeed a measure which does not have fixed values at independence and in the situation of a logical rule and, what is more, a threshold which is hard to fix is quite useless in an efficient search for interesting rules. According to the preferences of the expert the relative importance of criteria like g_1 and g_8 can vary. The analysis should be started by using an initial set of weights coherent with these considerations. The stability of the resulting ranking should then be analyzed, especially for the leading positions. If a stable ranking is obtained, the GAIA plane, the value of the net flows and the profile visualization tool allow a finer analysis of the leading measures. The values of the net flows give a hint about the *distance* between two alternatives in the ranking. Two measures with similar values for the net flows can be considered as similar.

The second strategy involves a first step in an exploration of the GAIA plane. This procedure helps the expert to understand the structure of the problem and to detect similar and different measures. Furthermore, the visualization of the criteria in the same plane as the alternatives make it possible to detect the influence of the modification of the weights on the final ranking. This exploratory strategy should be applied with an expert E_r who has a priori knowledge about certain measures. He will be able to determine a preorder on the importance of the criteria by detecting some well known measures in the GAIA plane. By using this first approximate weights system, the first strategy can be applied. An a posteriori validation can be done by determining the positions of the well known measures in the final ranking.

5 Conclusion and perspectives

Association rule quality measures play a major role within a KDD process, but they have a large diversity of properties, which have to be studied both on formal aspects and on real data in order to use a measure adapted to the user's context. In this chapter, we have studied 20 association rule interestingness measures evaluated on 8 properties, and 10 data sets.

The experimental results we present come from a tool we developed, HERBS briefly presented. We were then able to identify 3 main groups of measures in the two approaches, which may be refined in 5 smaller classes. The resulting clusterings are globally in agreement, and the discordancies discussed. The experimental approach seems to be an important addition to the formal approach. Indeed, it first confirmed the validity of the list of formal properties we thought were worth studying. What is more, it has also led to a new reflection on the importance of these properties. For example, requiring that a rule quality measure should have a fixed value for a logical rule has the bias of favouring logical rules with a large conclusion. From the formal study, we proposed a multicriteria decision aid approach illustrating how to help expert users choose an adapted interestingness measure in the context of association rule mining. We present the use of the PROMETHEE decision aid method.

Our approach is a first step to improving the quality of a set of rules that will effectively be presented to the user. Other factors, beyond interestingness measures, can be used. Among them, attribute costs and misclassification costs [26], and cognitive constraints [55].

In addition to the interest of having such a list of properties for a large number of measures, the use of the PROMETHEE method has confirmed the fact that the expert's preferences have some influence on the ordering of the interestingness measures, and that there are similarities between different measures. Moreover, the PROMETHEE method allows us to make a better analysis of the user's preferences (the GAIA plane makes it easy to identify different clusters of criteria and measures).

Our set of criteria covers a large range of the user's preferences, but it is clearly not exhaustive. New criteria could also lead to a better distinction between measures which are similar at the present time. We are confident that some important criteria may also arise from experimental evaluation (such as the discrimination strength and the robustness).

Finally, we would like to point out that even if SUP is poorly rated in both scenarios it is a mandatory measure in algorithms like APRIORI since its antimonotonicity property drives and simplifies the exploration of the lattice of itemsets. In our set of 20 measures, SUP is the only one to have this property.

Acknowledgment

Benot Vaillant would like to thanks Brest Métropole Océane for financial support of his Ph.D. thesis. The authors would like to thank the members of the CNRS working group GAFOQUALITÉ for interesting discussions about evaluation properties.

References

1. H. Abe, S. Tsumoto, M. Ohsaki, and T. Yamaguchi. Evaluating model construction methods with objective rule evaluation indices to support human experts. In V. Torra, Y. Narukawa, A. Valls, and J. Domingo-Ferrer, editors, *Modeling Decisions for Artificial Intelligence*, volume 3885 of *Lecture Notes in Computer Science*, pages 93–104, Tarragona, Spain, 2006. Springer-Verlag.
2. R. Agrawal, T. Imielinski, and A.N. Swami. Mining association rules between sets of items in large databases. In P. Buneman and S. Jajodia, editors, *ACM SIGMOD International Conference on Management of Data*, pages 207–216, 1993.
3. J. Azé and Y. Kodratoff. Evaluation de la résistance au bruit de quelques mesures d'extraction de règles d'assocation. In D. Hérin and D.A. Zighed, editors, *Extraction des connaissances et apprentissage*, volume 1, pages 143–154. Hermes, 2002.
4. J. Azé and Y. Kodratoff. A study of the effect of noisy data in rule extraction systems. In *The Sixteenth European Meeting on Cybernetics and Systems Research*, volume 2, pages 781–786, 2002.
5. J. P. Barthélemy, A. Legrain, P. Lenca, and B. Vaillant. Aggregation of valued relations applied to association rule interestingness measures. In V. Torra, Y. Narukawa, A. Valls, and J. Domingo-Ferrer, editors, *Modeling Decisions for Artificial Intelligence*, volume 3885 of *Lecture Notes in Computer Science*, pages 203–214, Tarrogona, Spain, 2006. Springer-Verlag.
6. R. J. Bayardo and R. Agrawal. Mining the most interesting rules. In *ACM SIGKDD International Conference on Knowledge Discovery and Data Mining*, pages 145–154, 1999.
7. R. Bisdorff. Bipolar ranking from pairwise fuzzy outrankings. *Belgian Journal of Operations Research, Statistics and Computer Science*, 37 (4) 97:379–387, 1999.
8. C.L. Blake and C.J. Merz. UCI repository of machine learning databases. http://www.ics.uci.edu/~mlearn/MLRepository.html, 1998.
9. J. Blanchard, F. Guillet, and H. Briand. A virtual reality environment for knowledge mining. In R. Bisdorff, editor, *Human Centered Processes*, pages 175–179, Luxembourg, 2003.
10. J. Blanchard, F. Guillet, H. Briand, and R. Gras. Assessing the interestingness of rules with a probabilistic measure of deviation from equilibrium. In J. Janssen and P. Lenca, editors, *The XIth International Symposium on Applied Stochastic Models and Data Analysis*, pages 191–200, Brest, France, 2005.
11. J. Blanchard, F. Guillet, H. Briand, and R. Gras. IPEE : Indice probabiliste d'écart à l'équilibre pour l'évaluation de la qualité des règles. In *Atelier Qualité des Données et des Connaissances (EGC 2005)*, pages 26–34, 2005.

12. J. Blanchard, F. Guillet, R. Gras, and H. Briand. Using information-theoretic measures to assess association rule interestingness. In *The 5th IEEE International Conference on Data Mining*, pages 66–73, Houston, Texas, USA, 2005. IEEE Computer Society Press.
13. C. Borgelt and R. Kruse. Induction of association rules: APRIORI implementation. In *Compstat'02*, pages 395–400, Berlin, Germany, 2002. Physica Verlag.
14. J.P. Brans and B. Mareschal. PROMETHEE-GAIA – *Une méthode d'aide à la décision en présence de critères multiples.* Ellipses, 2002.
15. J.P. Brans and P. Vincke. A preference ranking organization method. *Management Science*, 31(6):647–656, 1985.
16. T. Brijs, K. Vanhoof, and G. Wets. Defining interestingness for association rules. *International journal of information theories and applications*, 10(4):370–376, 2003.
17. S. Brin, R. Motwani, and C. Silverstein. Beyond market baskets: generalizing association rules to correlations. In *ACM SIGMOD/PODS'97 Joint Conference*, pages 265–276, 1997.
18. S. Brin, R. Motwani, J.D. Ullman, and S. Tsur. Dynamic itemset counting and implication rules for market basket data. In J. Peckham, editor, *ACM SIGMOD International Conference on Management of Data*, pages 255–264, Tucson, Arizona, USA, 1997. ACM Press.
19. J.-H. Chauchat and A. Risson. *Visualization of Categorical Data*, chapter 3, pages 37–45. Blasius J. & Greenacre M. ed., 1998. New York: Academic Press.
20. K.W. Church and P. Hanks. Word association norms, mutual information an lexicography. *Computational Linguistics*, 16(1):22–29, 1990.
21. E. Cohen, M. Datar, S. Fujiwara, A. Gionis, P. Indyk, R. Motwani, J. Ullman, and C. Yang. Finding interesting associations without support pruning. In *The 16th International conference on Data engineering*, 2000.
22. J. Cohen. A coefficient of agreement for nominal scale. *Educational and Psychological Measurement*, 20:37–46, 1960.
23. A.W.F. Edwards. The measure of association in a 2 x 2 table. *Journal of the Royal Statistical Society, Series A*, 126(1):109–114, 1963.
24. U.M. Fayyad, G. Piatetsky-Shapiro, P. Smyth, and R. Uthurusamy, editors. *Advances in Knowledge Discovery and Data Mining.* AAAI/MIT Press, 1996.
25. D. Feno, J. Diatta, and A. Totohasina. Normalisée d'une mesure probabiliste de la qualité des règles d'association : étude de cas. In *Atelier Qualité des Données et des Connaissances (EGC 2006)*, pages 25–30, 2006.
26. A. Freitas. On rule interestingness measures. *Knowledge-Based Systems journal*, pages 309–315, 1999.
27. V. Giakoumakis and B. Monjardet. Coefficients d'accord entre deux préordres totaux. *Statistique et Analyse des Données*, 12(1 et 2):46–99, 1987.
28. I.J. Good. The estimation of probabilities: An essay on modern bayesian methods. The MIT Press, Cambridge, MA, 1965.
29. R. Gras, S. Ag. Almouloud, M. Bailleuil, A. Larher, M. Polo, H. Ratsimba-Rajohn, and A. Totohasina. *L'implication Statistique, Nouvelle Méthode Exploratoire de Données. Application à la Didactique, Travaux et Thèses.* La Pensée Sauvage, 1996.
30. R. Gras, R. Couturier, J. Blanchard, H. Briand, P. Kuntz, and P. Peter. Quelques critères pour une mesure de qualité de règles d'association - un exemple: l'intensité d'implication. *Revue des Nouvelles Technologies de l'Information (Mesures de Qualité pour la Fouille de Données)*, (RNTI-E-1):3–31, 2004.

31. R. Gras, P. Kuntz, R. Couturier, and F. Guillet. Une version entropique de l'intensité d'implication pour les corpus volumineux. In H. Briand and F. Guillet, editors, *Extraction des connaissances et apprentissage*, volume 1, pages 69–80. Hermes, 2001.
32. S. Greco, Z. Pawlak, and R. Slowinski. Can bayesian confirmation measures be useful for rough set decision rules? *Engineering Applications of Artificial Intelligence*, 17(4):345–361, 2004.
33. S. Guillaume. *Traitement des données volumineuses, Mesures et algorithmes d'extraction de règles d'association et règles ordinales.* PhD thesis, Université de Nantes, 2000.
34. F. Guillet. Mesures de la qualité des connaissances en ECD. Atelier, Extraction et gestion des connaissances, 2004.
35. P. Hajek, I. Havel, and M. Chytil. The GUHA method of automatic hypotheses determination. *Computing*, (1):293–308, 1966.
36. R.J. Hilderman and H.J. Hamilton. Applying objective interestingness measures in data mining systems. In *Fourth European Symposium on Principles of Data Mining and Knowledge Discovery*, pages 432–439. Springer Verlag, 2000.
37. R.J. Hilderman and H.J. Hamilton. Evaluation of interestingness measures for ranking discovered knowledge. *Lecture Notes in Computer Science*, 2035:247–259, 2001.
38. R.J. Hilderman and H.J. Hamilton. *Knowledge Discovery and Measures of Interest.* Kluwer Academic Publishers, 2001.
39. R.J. Hilderman and H.J. Hamilton. Measuring the interestingness of discovered knowledge: A principled approach. *Intelligent Data Analysis*, 7(4):347–382, 2003.
40. Y. Huang, H. Xiong, S. Shekhar, and J. Pei. Mining confident co-location rules without a support threshold. In *The 18th Annual ACM Symposium on Applied Computing.* ACM, 2003.
41. F. Hussain, H. Liu, E. Suzuki, and H. Lu. Exception rule mining with a relative interestingness measure. In T. Terano, H. Liu, and A.L.P. Chen, editors, *The Fourth Pacific-Asia Conference on Knowledge Discovery and Data Mining*, volume 1805 of *Lecture Notes in Artificial Intelligence*, pages 86–97. Springer-Verlag, 2000.
42. X-H. Huynh, F. Guillet, and H. Briand. ARQAT: An exploratory analysis tool for interestingness measures. In J. Janssen and P. Lenca, editors, *The XIth International Symposium on Applied Stochastic Models and Data Analysis*, pages 334–344, Brest, France, 2005.
43. A. Iodice D'Enza, F. Palumbo, and M. Greenacre. Exploratory data analysis leading towards the most interesting binary association rules. In J. Janssen and P. Lenca, editors, *The XIth International Symposium on Applied Stochastic Models and Data Analysis*, pages 256–265, Brest, France, 2005.
44. S. Jaroszewicz and D.A. Simovici. A general measure of rule interestingness. In *The 5th European Conference on Principles of Data Mining and Knowledge Discovery*, pages 253–265, London, UK, 2001. Springer-Verlag.
45. H.J. Jeffreys. Some tests of significance treated by the theory of probability. In *Proceedings of the Cambridge Philosophical Society*, number 31, pages 203–222, 1935.
46. M. Kamber and R. Shingal. Evaluating the interestingness of characteristic rules. In *The Second International Conference on Knowledge Discovery and Data Mining*, pages 263–266, Portland, Oregon, August 1996.

47. D. A. Keim. Information visualization and visual data mining. *IEEE Transactions On Visualization And Computer Graphics*, 7(1):100–107, 2002.
48. M. Klemettinen, H. Mannila, P. Ronkainen, H. Toivonen, and A.I. Verkamo. Finding interesting rules from large sets of discovered association rules. In N.R. Adam, B.K. Bhargava, and Y. Yesha, editors, *Third International Conference on Information and Knowledge Management*, pages 401–407. ACM Press, 1994.
49. S. Lallich. Mesure et validation en extraction des connaissances à partir des données. Habilitation à Diriger des Recherches – Université Lyon 2, 2002.
50. S. Lallich, E. Prudhomme, and O. Teytaud. Contrôle du risque multiple en sélection de règles d'association significatives. In G. Hébrail, L. Lebart, and J.-M. Petit, editors, *Extraction et gestion des connaissances*, volume 1-2, pages 305–316. Cépaduès Editions, 2004.
51. S. Lallich and O. Teytaud. Évaluation et validation de l'intérêt des règles d'association. *Revue des Nouvelles Technologies de l'Information (Mesures de Qualité pour la Fouille de Données)*, (RNTI-E-1):193–217, 2004.
52. S. Lallich, B. Vaillant, and P. Lenca. Parametrised measures for the evaluation of association rule interestingness. In J. Janssen and P. Lenca, editors, *The XIth International Symposium on Applied Stochastic Models and Data Analysis*, pages 220–229, Brest, France, 2005.
53. N. Lavrac, P. Flach, and B. Zupan. Rule evaluation measures: A unifying view. In S. Dzeroski and P. Flach, editors, *Ninth International Workshop on Inductive Logic Programming*, volume 1634 of *Lecture Notes in Computer Science*, pages 174–185. Springer-Verlag, 1999.
54. E. Le Saux, P. Lenca, J-P. Barthélemy, and P. Picouet. Updating a rule basis under cognitive constraints: the COMAPS tool. In *The Seventeenth European Annual Conference on Human Decision Making and Manual Control*, pages 3–9, December 1998.
55. E. Le Saux, P. Lenca, and P. Picouet. Dynamic adaptation of rules bases under cognitive constraints. *European Journal of Operational Research*, 136(2):299–309, 2002.
56. R. Lehn, F. Guillet, P. Kuntz, H. Briand, and J. Philippé. Felix: An interactive rule mining interface in a KDD process. In P. Lenca, editor, *Human Centered Processes*, pages 169–174, Brest, France, 1999.
57. P. Lenca, P. Meyer, P. Picouet, B. Vaillant, and S. Lallich. Critères d'évaluation des mesures de qualité en ECD. *Revue des Nouvelles Technologies de l'Information (Entreposage et Fouille de Données)*, (1):123–134, 2003.
58. P. Lenca, P. Meyer, B. Vaillant, and S. Lallich. A multicriteria decision aid for interestingness measure selection. Technical Report LUSSI-TR-2004-01-EN, Département LUSSI, ENST Bretagne, 2004.
59. P. Lenca, P. Meyer, B. Vaillant, and P. Picouet. Aide multicritère à la décision pour évaluer les indices de qualité des connaissances – modélisation des préférences de l'utilisateur. In M.-S. Hacid, Y. Kodratoff, and D. Boulanger, editors, *Extraction et gestion des connaissances*, volume 17 of *RSTI-RIA*, pages 271–282. Lavoisier, 2003.
60. P. Lenca, P. Meyer, B. Vaillant, P. Picouet, and S. Lallich. Évaluation et analyse multicritère des mesures de qualité des règles d'association. *Revue des Nouvelles Technologies de l'Information (Mesures de Qualité pour la Fouille de Données)*, (RNTI-E-1):219–246, 2004.

61. P. Lenca, B. Vaillant, and S. Lallich. On the robustness of association rules. In *IEEE International Conference on Cybernetics and Intelligent Systems*, Bangkok, Thailand, 2006.
62. I.C. Lerman. *Classification et analyse ordinale des données*. Dunod, 1970.
63. I.C. Lerman and J. Azé. Une mesure probabiliste contextuelle discriminante de qualité des règles d'association. In M.-S. Hacid, Y. Kodratoff, and D. Boulanger, editors, *Extraction et gestion des connaissances*, volume 17 of *RSTI-RIA*, pages 247–262. Lavoisier, 2003.
64. I.C. Lerman, R. Gras, and H. Rostam. Elaboration d'un indice d'implication pour les données binaires, i et ii. *Mathématiques et Sciences Humaines*, (74, 75):5–35, 5–47, 1981.
65. B. Liu, W. Hsu, and S. Chen. Using general impressions to analyze discovered classification rules. In *Third International Conference on Knowledge Discovery and Data Mining*, pages 31–36, 1997.
66. B. Liu, W. Hsu, S. Chen, and Y. Ma. Analyzing the subjective interestingness of association rules. *IEEE Intelligent Systems*, 15(5):47–55, 2000.
67. B. Liu, W. Hsu, K. Wang, and S. Chen. Visually aided exploration of interesting association rules. In *Third Pacific-Asia Conference on Methodologies for Knowledge Discovery and Data Mining*, pages 380–389. Springer Verlag, 1999.
68. J. Loevinger. A systemic approach to the construction and evaluation of tests of ability. *Psychological monographs*, 61(4), 1947.
69. J.-L. Marichal, P. Meyer, and M. Roubens. Sorting multi-attribute alternatives: The TOMASO method. *Computers & Operations Research*, (32):861–877, 2005.
70. K. McGarry. A survey of interestingness measures for knowledge discovery. *Knowledge Engineering Review Journal*, 20(1):39–61, 2005.
71. M. Ohsaki, Y. Sato, S. Kitaguchi, H. Yokoi, and T. Yamaguchi. Comparison between objective interestingness measures and real human interest in medical data mining. In R. Orchard, C. Yang, and M. Ali, editors, *The 17th international conference on Innovations in Applied Artificial Intelligence*, volume 3029 of *Lecture Notes in Artificial Intelligence*, pages 1072–1081. Springer-Verlag, 2004.
72. B. Padmanabhan. The interestingness paradox in pattern discovery. *Journal of Applied Statistics*, 31(8):1019–1035, 2004.
73. N. Pasquier, Y. Bastide, R. Taouil, and L. Lakhal. Discovering frequent closed itemsets for association rules. In C. Beeri and P. Buneman, editors, *The 7th International Conference on Database Theory*, volume 1540 of *Lecture Notes in Computer Science*, pages 398–416, Jerusalem, Israel, 1999. Springer.
74. K. Pearson. Mathematical contributions to the theory of evolution. iii. regression, heredity and panmixia. *Philosophical Transactions of the Royal Society*, A, 1896.
75. G. Piatetsky-Shapiro. Discovery, analysis and presentation of strong rules. In G. Piatetsky-Shapiro and W.J. Frawley, editors, *Knowledge Discovery in Databases*, pages 229–248. AAAI/MIT Press, 1991.
76. P. Picouet and P. Lenca. *Bases de données et internet*, chapter Extraction de connaissances à partir des données, pages 395–420. Hermes Science, 2001.
77. M. Plasse, N. Niang, G. Saporta, and L. Leblond. Une comparaison de certains indices de pertinence des règles d'association. In G. Ritschard and C. Djeraba, editors, *Extraction et gestion des connaissances*, volume 1-2, pages 561–568. Cépaduès-Éditions, 2006.

78. F. Poulet. Visualization in data-mining and knowledge discovery. In P. Lenca, editor, *Human Centered Processes*, pages 183–191, Brest, France, 1999.
79. F. Poulet. Towards visual data mining. In *6th International Conference on Enterprise Information Systems*, pages 349–356, 2004.
80. J. Rauch and M. Simunek. Mining for 4ft association rules by 4ft-miner. In *Proceeding of the International Conference On Applications of Prolog*, pages 285–294, Tokyo, Japan, 2001.
81. M. Sebag and M. Schoenauer. Generation of rules with certainty and confidence factors from incomplete and incoherent learning bases. In J. Boose, B. Gaines, and M. Linster, editors, *The European Knowledge Acquisition Workshop*, pages 28–1–28–20. Gesellschaft für Mathematik und Datenverarbeitung mbH, 1988.
82. A. Silberschatz and A. Tuzhilin. On subjective measures of interestingness in knowledge discovery. In *Knowledge Discovery and Data Mining*, pages 275–281, 1995.
83. A. Silberschatz and A. Tuzhilin. User-assisted knowledge discovery: How much should the user be involved. In *ACM-SIGMOD Workshop on Research Issues on Data Mining and Knowledge Discovery*, 1996.
84. S.J. Simoff. Towards the development of environments for designing visualisation support for visual data mining. In S.J. Simoff, M. Noirhomme-Fraiture, and M.H. Böhlen, editors, *International Workshop on Visual Data Mining in cunjunction with ECML/PKDD'01*, pages 93–106, 2001.
85. E. Suzuki. In pursuit of interesting patterns with undirected discovery of exception rules. In S. Arikawa and A. Shinohara, editors, *Progresses in Discovery Science*, volume 2281 of *Lecture Notes in Computer Science*, pages 504–517. Springer-Verlag, 2002.
86. E. Suzuki. Discovering interesting exception rules with rule pair. In *ECML/PKDD Workshop on Advances in Inductive Rule Learning*, pages 163–178, 2004.
87. P.-N. Tan, V. Kumar, and J. Srivastava. Selecting the right interestingness measure for association patterns. In *The Eighth ACM SIGKDD International Conference on KDD*, pages 32–41, 2002.
88. P-N. Tan, V. Kumar, and J. Srivastava. Selecting the right objective measure for association analysis. *Information Systems*, 4(29):293–313, 2004.
89. A. Totohasina, H. Ralambondrainy, and J. Diatta. Notes sur les mesures probabilistes de la qualité des règles d'association: un algorithme efficace d'extraction des règles d'association implicative. In *7ème Colloque Africain sur la Recherche en Informatique*, pages 511–518, 2004.
90. B. Vaillant. Evaluation de connaissances: le problème du choix d'une mesure de qualité en extraction de connaissances à partir des données. Master's thesis, Ecole Nationale Supérieure des Télécommunications de Bretagne, 2002.
91. B. Vaillant, P. Lenca, and S. Lallich. Association rule interestingness measures: an experimental study. Technical Report LUSSI-TR-2004-02-EN, Département LUSSI, ENST Bretagne, 2004.
92. B. Vaillant, P. Lenca, and S. Lallich. A clustering of interestingness measures. In E. Suzuki and S. Arikawa, editors, *Discovery Science*, volume 3245 of *Lecture Notes in Artificial Intelligence*, pages 290–297, Padova, Italy, 2004. Springer-Verlag.
93. B. Vaillant, P. Picouet, and P. Lenca. An extensible platform for rule quality measure benchmarking. In R. Bisdorff, editor, *Human Centered Processes*, pages 187–191, 2003.

94. H. Xiong, P. Tan, and V. Kumar. Mining strong affinity association patterns in data sets with skewed support distribution. In *Third IEEE International Conference on Data Mining*, pages 387–394, Melbourne, Florida, 2003.
95. T. Zhang. Association rules. In T. Terano, H. Liu, and A.L.P. Chen, editors, *4th Pacific-Asia Conference Knowledge Discovery and Data Mining, Current Issues and New Applications*, volume 1805 of *Lecture Notes in Computer Science*, Kyoto, Japan, 2000. Springer.
96. A. Zimmermann and L. De Raedt. CorClass: Correlated association rule mining for classification. In E. Suzuki and S. Arikawa, editors, *Discovery Science*, volume 3245 of *Lecture Notes in Artificial Intelligence*, pages 60–72, Padova, Italy, 2004. Springer-Verlag.

On the Discovery of Exception Rules: A Survey

Béatrice Duval[1], Ansaf Salleb[2], and Christel Vrain[3]

[1] LERIA, UFR Sciences, 2 Bd Lavoisier, 49045 Angers Cedex 01, France
Beatrice.Duval@univ-angers.fr

[2] CCLS, Columbia University, 475 Riverside Dr., New York, NY 10115, U.S.A
Ansaf@ccls.columbia.edu

[3] LIFO, Université d'Orléans, B.P. 6759, F-45067 Orléans, France
Christel.Vrain@univ-orleans.fr

Summary. In this chapter, we present a survey of different approaches developed for mining exception rules. Exception rules are interesting in the context of quality measures since such rules are intrinsically satisfied by few individuals in the database and many criteria relying on the number of occurrences, such as for instance the support measure, are no longer relevant. Therefore traditional measures must be coupled with other criteria. In that context, some works have proposed to use the expert's knowledge: she/he can provide the system either with constraints on the syntactic form of the rules, thus reducing the search space, or with commonsense rules that have to be refined by the data mining process. Works that rely on either of these approaches, with their particular quality evaluation are presented in this survey. Moreover, this presentation also gives ideas on how numeric criteria can be intertwined with user-centered approaches.

Key words: association rules, exception rules, quality measure, interestingness, unexpectedness, belief system.

1 Introduction

Knowledge Discovery in Databases (KDD) is a very active and diversified research field, which aims at finding unknown, implicit, interesting and useful knowledge. Several domains, as for instance Statistics or Machine Learning, have already addressed the problem of analyzing or exploring data in order to build models allowing either data explanation (as for instance in Descriptive Statistics) or inferences (for instance in Inductive Machine Learning). The key problems are: efficiency of the algorithms developed so far, and evaluation of the quality of the models. The first point is related to the nature and to the size of the databases: the size of the search space (the set of all possible models) increases with the number of attributes and their types, the test for evaluating hypotheses depends on the number of tuples stored in the database. This problem is all the more important in KDD that the databases

Béatrice Duval et al.: *On the Discovery of Exception Rules: A Survey*, Studies in Computational Intelligence (SCI) **43**, 77–98 (2007)
www.springerlink.com

are huge. To deal with these problems, a typical method in KDD employs heuristics to prune the search space. One of the most commonly used heuristics is the support heuristics: only models that are satisfied by a sufficiently high percentage of data - expressed by a minimum support threshold [1] - are explored. Concerning the second point, the classical accuracy criterion is no longer appropriate: it is required for the models to be interesting and useful, i.e. they should help an expert in his/her work, mainly in decision making. Such notions are very difficult to capture, and they are either expressed by *a priori* constraints on the models, or quantified through a quality measure, as for instance the confidence measure.

In this paper, we consider models expressed by rules, which are considered as quite easily understandable by an expert. We mainly focus on mining association rules, as introduced by [1], and expressing correlations on data. In the following, we consider data sets that contain a great number of tuples; each tuple represents an example and is described by qualitative (also called nominal) attributes. We call an *item* the assignment of a value to an attribute and an *itemset* is a set of items. Extracting association rules aims at finding rules of the form $\mathcal{X} \rightarrow \mathcal{Y}$, where $\mathcal{X}$ and $\mathcal{Y}$ are disjoint itemsets. Many works have already addressed this task. Because of the size of the search space, they often rely on the support heuristics, allowing to find only rules satisfied by a certain number of tuples of the database. Nevertheless, even when coupled with quality measures, many rules are generated, and it is difficult for the expert to analyze all these rules. Moreover, many of these rules model commonsense knowledge that is already known by the expert and therefore not very useful, whereas she/he is looking for unexpected and surprising rules. So unexpectedness is a main characteristic of interesting rules. For those reasons, some works have addressed the problem of learning exception rules, as for instance [9, 11, 14, 15, 16, 19, 20, 21, 24, 25]. A general rule, also called *a commonsense rule*, describes regularities in data, observed on a large number of tuples. *An exception rule* expresses some regularity observed on few tuples and that contradicts a commonsense rule; therefore an exception rule is generally unexpected and reveals interesting knowledge.

Let us for instance consider an example inspired by the bank credit domain. A commonsense rule could express that an unemployed person cannot obtain a credit. Nevertheless, an unemployed person that aims at creating a new enterprise can under some conditions obtain a credit for realizing his/her project. This example illustrates the framework of that study: we have two rules with contradictory conclusions that can be used in the context of an unemployed person aiming at building a new company. The second rule represents the exception rule, since it can be applied in a more specific situation and it is that rule that has to be applied when all its conditions are fulfilled.

We conduct here a survey of different approaches for finding exception rules. On one hand, the fact that few tuples satisfy such a rule makes it surprising and allows a finer study of data. On the other hand, from the point of view of algorithms the search space can no longer be pruned only by the support heuristics. Therefore, two points must be developed: the criteria that model the notions of utility but that also allow to prune the search space and the algorithms for exploring the search space.

Discovery methods for exception rules can be classified into two approaches, called objective and subjective. They differ on the way commonsense knowledge is provided to the system (either automatically learned or provided by the user) and on the criteria used to measure the degree of interestingness of exception rules. More precisely, in the subjective approach, also called directed approach, the domain expert gives background knowledge, often called *beliefs*, which forms a system of commonsense rules; the interestingness of exception rules to these beliefs is evaluated according to subjective criteria depending on the user as for instance the *unexpectedness* and the *actionability* [17] that measures the actions that can be triggered when an exception is detected. On the other hand, the objective approach, also called undirected approach, does not use *a priori* knowledge and searches both commonsense rules and exception rules. Rules are evaluated by statistical measures, such as support, confidence, entropy, ... that depend only on data. Such criteria have to evaluate both a commonsense rule and its exceptions. Since an exhaustive search is not possible, most systems reduce the search space by putting syntactic constraints on the forms of the expected rules. For example, search can be restricted to rules that have a limited number of premises. This can be justified by the fact that to be useful for decision making, a rule must be understandable and simple.

This paper classifies the different approaches for mining exception rules according to the objective (Section 2)/subjective criterion (Section 3). The first one has attracted more attention. We present in Section 2 four methods: the first one described in 2.1 searches for rule-exception pairs, the second one (2.2) involves the reference rule in the evaluation of interestingness. The two last methods (2.3 and 2.4) are both based on deviation analysis. In Section 3, a theoretical belief-driven framework is presented in 3.1 and a belief-driven system is described in 3.2. We conclude in Section 4 with a comparative summary of the works presented in this survey.

2 Objective Approaches

2.1 Undirected search of rule-exception pairs

This section presents the works realized by Suzuki [19, 20, 21, 24, 25]. Undirected search of exception rules aims at discovering, at the same time, a rule, called *a commonsense rule* CSR and a rule, called *an exception rule* EXC, that describes exceptions to the commonsense rule. In the following, such a piece of knowledge will be called a *rule-exception pair*.

More precisely, this undirected search considers pairs of rules of the following form:

$$Y_\mu \rightarrow x \qquad \text{(CSR)}$$
$$Y_\mu \wedge Z_\nu \rightarrow x' \qquad \text{(EXC)}$$

where x and x' are items with the same attribute but with different values and $Y_\mu = y_1 \wedge y_2 ... \wedge y_\mu$, $Z_\nu = z_1 \wedge z_2 ... \wedge z_\nu$ are conjunctions of items.
For example, x and x' can be items that instantiate the same class attribute with two different class values: a rule-exception pair then gives a common sense rule and an exception rule that can be applied in the same situations but that conclude on different class decisions.

One central aspect of such works is to propose some criterion to evaluate the interestingness of the different rule-exception pairs that can be obtained. The criterion can be given by a unique measure that evaluates the interestingness of the rule-exception pair. In this case, the obvious total pre-order on the different pairs induced by this measure enables to define the maximum number of pairs that must be returned by the search process.
The criterion can also be composed of several measures, as for example the support and the confidence, with a specific threshold for each measure. In this case, there is no obvious total pre-order on the pairs and it is more difficult to obtain a suitable number of interesting rule-exception pairs. In fact, the number of computed pairs crucially depends on the thresholds given by the user; so a way to correctly adjust these thresholds must be proposed.

In either case, undirected discovery of exceptions is seen as an exploration of a search space where each node describes a rule-exception pair and this exploration is necessarily limited.

In the following of this section, we briefly describe the algorithm used to discover rule-exception pair, then we present the different quality measures that have been proposed in Suzuki's works.

Algorithm

A rule-exception pair

$$Y_\mu \rightarrow x \qquad \text{(CSR)}$$
$$Y_\mu \wedge Z_\nu \rightarrow x' \qquad \text{(EXC)}$$

is viewed as a node $r(x, x', \mu, \nu)$ in a search tree. The nodes of depth 1 are the pairs of conclusions (x, x') that must be considered. These nodes can be seen as rules without premises, which are characterized by $\mu = 0$ and $\nu = 0$. A node of depth 2 corresponds to $\mu = 1$ and $\nu = 0$ and represents a pair where the commonsense rule has only one premise and a node of depth 3 corresponds to $\mu = 1$ and $\nu = 1$. When depth increases by 1, an item is added either to the premises of the commonsense rules or to the premises of the exception rules. More generally a node of depth l ($l \geq 4$) satisfies $\mu + \nu = l - 1$ ($\mu, \nu \geq 1$). Consequently, a node $r(x, x', \mu', \nu')$ that is a descendant of a node $r(x, x', \mu, \nu)$ corresponds to a rule for which $\mu' \geq \mu$ and $\nu' \geq \nu$. This tree is explored

by a depth first search, limited by a maximum depth M; so the algorithm only examines the rule pairs that satisfy $\mu + \nu \leq M - 1$. For example, the experiments reported in the papers use a maximum depth of 8. As previously mentioned, such a constraint is justified since usually the experts do not easily understand complex rules that involve many premises.
For each evaluation criterion that he uses, Suzuki proposes properties that enables tree pruning during the exploration.

Evaluation measures of the rule-exception pairs

The first systems MEPRO and MEPROUX proposed by Suzuki [25, 19] use an information-based measure to determine the interestingness of a pair. In order to characterize the interestingness of a rule induced from a data set, Smyth and Goodman [18] have defined the *J-measure*, also called Average Compressed Entropy (denoted by ACE in the following). Given Y_μ and x such that $p(x|Y_\mu) \geq 0.5$, the ACE of the rule $Y_\mu \rightarrow x$ is defined by:

$$ACE(x, Y_\mu) = p(x, Y_\mu) \log_2 \left(\frac{p(x|Y_\mu)}{p(x)} \right) + p(\neg x, Y_\mu) \log_2 \left(\frac{p(\neg x|Y_\mu)}{p(\neg x)} \right)$$

which can be also written:

$$ACE(x, Y_\mu) = p(Y_\mu) \times \left[p(x|Y_\mu) \log_2 \left(\frac{p(x|Y_\mu)}{p(x)} \right) + p(\neg x|Y_\mu) \log_2 \left(\frac{p(\neg x|Y_\mu)}{p(\neg x)} \right) \right]$$

So the ACE is composed of a factor $p(Y_\mu)$ that evaluates the generality of the rule and of a term $p(x|Y_\mu) \log_2 \left(\frac{p(x|Y_\mu)}{p(x)} \right) + p(\neg x|Y_\mu) \log_2 \left(\frac{p(\neg x|Y_\mu)}{p(\neg x)} \right)$ that we shall note here $j(x, Y_\mu)$. The term $j(x, Y_\mu)$, also called cross entropy or Kullback-Leibler measure in the literature, evaluates the dissimilarity between two probability distributions. Here $j(x, Y_\mu)$ measures the difference between our *a priori* knowledge on x and our *a posteriori* knowledge when Y_μ is known. An exception rule $Y_\mu \wedge Z_\nu \rightarrow x'$ with a large ACE may be not interesting if the ACE of the associated commonsense rule is low. That is why the evaluation of the rule-exception pair must take into account the ACE of each rule. Suzuki [19] shows that the arithmetic mean is inappropriate to evaluate the combined interestingness of a pair; in fact, for $p(x)$ and $p(x')$ given, the arithmetic mean of the ACE reaches a maximum when $ACE(x, Y_\mu) = 0$ or when $ACE(x', Y_\mu \wedge Z_\nu) \approx 0$, two situations that do not correspond to interesting cases. The geometric mean of the ACE, noted GACE, does not have such shortcomings and it is retained as an appropriate measure of interestingness for a rule-exception pair:

$$GACE(x, Y_\mu, x', Z_\nu) = \sqrt{ACE(x, Y_\mu) ACE(x', Y_\mu \wedge Z_\nu)}$$

Besides, a rule $Y_\mu \wedge Z_\nu \rightarrow x'$ cannot be considered as an unexpected exception to the commonsense rule $Y_\mu \rightarrow x$ when the rule $Z_\nu \rightarrow x'$, which will be called the *reference rule* (REF), has a high confidence (i.e. when $p(x'|Z_\nu)$ is high). To take into account such situations, Suzuki points out that he has explored several possible constraints and finally in the MEPROUX system, the reference rule must satisfy the following property:

$$p(x'|Z_\nu) \leq p(x') + \frac{1 - p(x')}{2}$$

Following Section 2.2 proposes another way to involve the reference rule in the evaluation.
Moreover, to evaluate the relevance of the combination of premises Y_μ and Z_ν, the system also requires that

$$p(x'|Y_\mu, Z_\nu) > p(x'|Z_\nu)$$

This allows to find rule-exception pairs that are interesting and for which the exception rule is unexpected.
Table 1 reports an example of rules extracted from the data set "mushroom" [10], where the class attribute, edible(e) or poisonous(p), is the only item that can appear in the conclusion (other experiments with no restrictions on the conclusion have also been done). This table gives first the commonsense rule, then the exception rule and finally the reference rule. For each rule, the probability of the conclusion (*co*), and the probability of the premise (*pre*) are given, together with the conditional probability of the conclusion given the premise and the ACE and GACE of the rule.

Table 1. An example of a rule-exception pair and the associated reference rule

		$p(co\|pre)$	$p(co)$	$p(pre)$	ACE	GACE
bruises=f, ring-number=o	→ class= p	0.74	0.48	0.54	0.107	
{bruises=f, ring-number=o, ss-aring=f}	→ class=e	1.00	0.52	0.05	0.048	0.0713
ss-aring=f	→ class=e	0.74	0.52	0.07		

These rules express the following knowledge: 74% of the mushrooms whose "bruises" are "f" and whose "ring-number" is "o" are poisonous, but among them, those that have "ss-aring=f" are all edible. This exception is really unexpected since it cannot be predicted by the reference rule that has a conditional probability of 74%.

In his other works [20, 21], Suzuki uses several criteria to evaluate the degree of interestingness of a rule-exception pair. The problem is again to discover a rule-exception pair of the following form:

$$Y_\mu \rightarrow x \qquad \text{(CSR)}$$

$$Y_\mu \wedge Z_\nu \rightarrow x' \qquad \text{(EXC)}$$

In order to evaluate the quality of these rules, the user must indicate 5 thresholds θ_1^S, θ_1^F, θ_2^S, θ_2^F and θ_2^I and the computed rules must satisfy the following conditions:

$$p(Y_\mu) \geq \theta_1^S \tag{1}$$
$$p(x|Y_\mu) \geq \theta_1^F \tag{2}$$
$$p(Y_\mu, Z_\nu) \geq \theta_2^S \tag{3}$$
$$p(x'|Y_\mu, Z_\nu) \geq \theta_2^F \tag{4}$$
$$p(x'|Z_\nu) \leq \theta_2^I \tag{5}$$

The thresholds θ_1^S and θ_1^F express the generality and the confidence of the commonsense rule. In the same way, θ_2^S and θ_2^F assure the generality and the confidence of the exception rule. The exception rule is observed in few cases but has a high confidence, so we expect the user to give thresholds that verify $\theta_1^S > \theta_2^S$ and $\theta_1^F < \theta_2^F$. The equation (5) means that the reference rule $Z_\nu \rightarrow x'$ is not a strong regularity, which is necessary for the exceptions to be interesting.

In [21], the probabilities $p(Y_\mu)$, $p(x|Y_\mu)$, ... are estimated from the frequencies of the items in the data set, while in [20], true probabilities are used and estimated under normality assumptions with a certain significance level.

Of course, using several thresholds leads to the difficult problem of choosing their appropriate values; it requires a good knowledge of the data but also a fine understanding of the search process, and so it heavily relies on the user decision. Too restrictive thresholds prevent the discovery of interesting exception rules, even though tolerant values can produce too many pairs. To deal with this difficulty, [21] proposes a method to dynamically update thresholds in order to find a suitable number of rule-exception pairs.

In [24], another measure is used to focus on unexpected exception rules. Intensity of implication [5] is a criterion that measures the unexpectedness of a rule: this index measures the degree of surprise that a rule has so few counter examples. It has good properties to evaluate the quality of a rule [6]. In the problem of discovering rule-exception pairs, Suzuki uses intensity of implication to measure the fact that the exception rule has so few counter examples in a universe that contains only the objects that satisfy the premise of the commonsense rule. Among the pairs that satisfy equations (1-5), the interesting pairs are those for which intensity of implication of the exception rule is the highest.

In [22], Suzuki reports a series of experiments that have been realized with the different systems. They prove the effectiveness of the pruning strategies proposed in the algorithms. Moreover these experiments, especially those on medical data sets, prove that the rule-exception framework really enables the discovery of interesting patterns.

2.2 Involving the reference rule in the evaluation of interest

The work presented in [8] follows a very similar approach but relies on a very different search algorithm.
The authors consider boolean attributes in rule conclusions and search exception rules that correspond to the following schema:

$Y_\mu \rightarrow x$	high support, high confidence	commonsense rule	CSR
$Y_\mu \wedge Z_\nu \rightarrow \neg x$	low support, high confidence	exception rule	EXC
$Z_\nu \rightarrow \neg x$	low support and/or low confidence	reference rule	REF

As explained in section 2.1, the notion of reference rule has been considered in [19, 24, 20]. In [8], the reference rule is computed from an associated commonsense rule. Indeed, if $Z_\nu \rightarrow x$ is a commonsense rule with a support and confidence higher than the required minimum thresholds, then the rule $Z_\nu \rightarrow \neg x$ is considered as a reference rule, because it has a low support or a low confidence.

So [8] proposes an algorithm that relies on commonsense rules to search the exceptions. The user specifies the minimum thresholds for the support and confidence of the commonsense rules. An Apriori algorithm computes all the frequent itemsets and all the commonsense rules. For each pair of commonsense rules ($Y_\mu \rightarrow x$, $Z_\nu \rightarrow x$), if $Y_\mu \cup Z_\nu$ is not a frequent itemset, then $Y_\mu, Z_\nu \rightarrow \neg x$ is a candidate exception rule (see Figure 1). When the set of

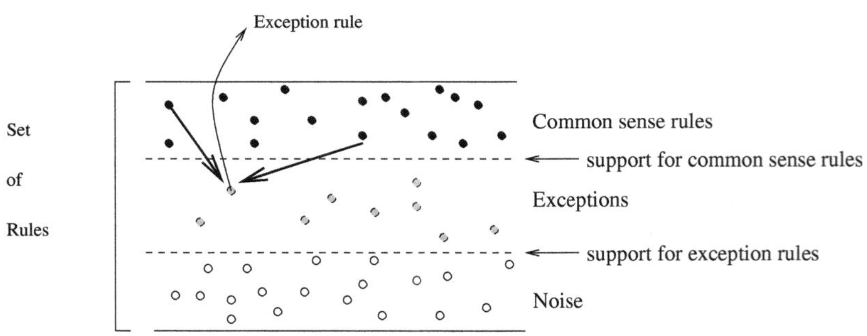

Fig. 1. Digging out the exceptions [8]

possible exception rules is built, the algorithm scans the database once to obtain the support and confidence of each candidate rule and to retain only those that satisfy the exception thresholds given by the user. The final step is to evaluate the interestingness of these possible exceptions. The authors propose a relative interestingness measure that evaluates each exception rule $Y_\mu, Z_\nu \rightarrow \neg x$ with respect to its commonsense rule $Y_\mu \rightarrow x$ and its reference rule $Z_\nu \rightarrow \neg x$. This measure relies on the cross entropy or Kullback Leibler distance (see 2.1). Let us recall that the relative entropy between two probability functions is defined as

$$D(p(x)||q(x)) = \sum_{x \in X} p(x) \log \frac{p(x)}{q(x)}$$

This non symmetric measure evaluates the inefficiency of assuming that the distribution is $q(x)$ when it is in fact $p(x)$ [18]. To compare the distribution related to the rule $Y_\mu, Z_\nu \rightarrow \neg x$ with the distributions related to the rules $Y_\mu \rightarrow x$, $Z_\nu \rightarrow x$, the authors propose to add the cross entropies $D(Y_\mu, Z_\nu \rightarrow \neg x || Y_\mu \rightarrow x)$ and $D(Y_\mu, Z_\nu \rightarrow \neg x || Z_\nu \rightarrow x)$. Moreover they want to compare the information associated to these rules, both for the support point of view and for the confidence point of view. The relative interestingness related to confidence, RI_c, is defined by considering the conditional probabilities in the cross entropy formula, which gives:

$$RI_c = p(x|Y_\mu, Z_\nu) \log_2 \left(\frac{p(x|Y_\mu, Z_\nu)^2}{p(x|Y_\mu)p(x|Z_\nu)} \right) + p(\neg x|Y_\mu, Z_\nu) \log_2 \left(\frac{p(\neg x|Y_\mu, Z_\nu)^2}{p(\neg x|Y_\mu)p(\neg x|Z_\nu)} \right)$$

In the same way, the relative interestingness related to the support, RI_s, is the sum of two cross entropies calculated with the joint probabilities, which gives:

$$RI_s = p(x, Y_\mu, Z_\nu) \log_2 \left(\frac{p(x, Y_\mu, Z_\nu)^2}{p(x, Y_\mu)p(x, Z_\nu)} \right) + p(\neg x, Y_\mu, Z_\nu) \log_2 \left(\frac{p(\neg x, Y_\mu, Z_\nu)^2}{p(\neg x, Y_\mu)p(\neg x, Z_\nu)} \right)$$

The measure of interestingness is finally $RI = RI_c + RI_s$.

This work can be compared to [19]. We have seen, in section 2.1, that [19] uses a unique measure called $GACE(x, Y_\mu, x', Z_\nu)$ to evaluate a rule-exception pair; this value combines the ACE of the rules $Y_\mu \rightarrow x$ and $Y_\mu \wedge Z_\nu \rightarrow x'$, but we have seen that an extra constraint must be imposed on $p(x', Z_\nu)$. To enhance this approach, the interestingness measure of [8] involves the reference rule $Z_\nu \rightarrow x'$ in the evaluation, and so the extra constraint on $p(x', Z_\nu)$ is no longer necessary. Besides experiments on various data sets show that the GACE measure and the RI measure behave similarly if the interestingness RI is computed without considering the reference rule. When reference rules are considered, the interestingness measure enables a finer evaluation of the rules by giving in a unique value the quality of the triplet (commonsense rule, exception rule, reference rule). In their paper, the authors report examples of rules discovered from Irvine data sets [10]. On these examples, we can see that two rule-exception pairs that have the same GACE value can be discriminated by the RI measure that involves the reference rule. This measure RI tries to evaluate both the interestingness and the unexpectedness of an exception rule.

2.3 Deviation analysis

In the work proposed by Liu et al. [9], deviation analysis is used in order to address the problem of the discovery of exception rules where commonsense knowledge is provided by the expert or learned by a system. Their approach is based on three observations. First, any exception would have a low support, otherwise it is considered as a strong pattern. Second, a reasonable induction algorithm can summarize data and learn rules and finally, attributes in the rules are salient features. The proposed method is four-step:

Identifying relevant attributes This step aims at identifying few salient attributes to focus on. Such attributes can be identified by using an induction method (such as association rule discovery), or domain knowledge. They may be also chosen by the expert of the domain or a user. For instance, if we consider a credit database storing information about people that have or have not been granted a credit, one can focus on the two attributes *Jobless* and *Credit*.

Building contingency tables In this step, a contingency table is built for each couple of relevant attributes. If we consider the couple of attributes (A, B), supposing that the attribute A takes its values in the domain $A_1, A_2, \ldots, A_r$ and the attribute B in $B_1, B_2, \ldots, B_s$, then the associated contingency table is given in Table 2. In this table, x_{ij} denotes the frequency of a couple (A_i, B_j) in the database. We denote by:

- n the total sum of the frequencies, computed by $n = \sum_{i=1}^{r} \sum_{j=1}^{s} x_{ij}$
- $n_{i.}$ the marginal horizontal frequency $n_{i.} = \sum_{j=1}^{s} x_{ij}$
- $n_{.j}$ the marginal vertical frequency $n_{.j} = \sum_{i=1}^{r} x_{ij}$
- n_{ij} the expected frequency for the occurrence (A_i, B_j) computed by $n_{ij} = \frac{n_{i.} \times n_{.j}}{n}$

The deviation according to the expected frequency as defined in [9] is given by:

$$\delta_{ij} = \frac{x_{ij} - n_{ij}}{n_{ij}}$$

Table 2. Contingency table for two attributes A and B

	B				
A	B_1	B_2	$\cdots$	B_s	total
A_1	$(n_{11})x_{11}$	$(n_{12})x_{12}$	$\cdots$	$(n_{1s})x_{1s}$	$n_{1.}$
A_2	$(n_{21})x_{21}$	$(n_{22})x_{22}$	$\cdots$	$(n_{2s})x_{2n}$	$n_{2.}$
$\cdots$	$\cdots$	$\cdots$	$\cdots$	$\cdots$	$\cdots$
A_r	$(n_{r1})x_{r1}$	$(n_{r2})x_{r2}$	$\cdots$	$(n_{rs})x_{rs}$	$n_{r.}$
total	$n_{.1}$	$n_{.2}$	$\cdots$	$n_{.s}$	n

For example, if we consider the couple of attributes (Credit, Jobless) of the previous example, its corresponding contingency table is given by Table 3 below.

Table 3. Contingency table for the attributes Jobless and Credit

	Jobless		
Credit	No	Yes	total
No	(35.5) 28	(4.5) 12	40
Yes	(75.5) 83	(9.5) 2	85
total	111	14	125

Identifying remarkable negative deviations In this step, deviations as defined above are computed and negative deviations pointed out. The expected frequencies are those that would be observed in case of independence of the two considered attributes. A user-specified deviation threshold is given in order to keep only significant deviations. Among these deviations, the positive ones correspond to frequent itemsets, which confirm a strong association between the two attributes whereas the negative ones may correspond to infrequent itemsets leading probably to exceptions. The deviations concerning the credit example are given in Table 4. Considering the deviation threshold of $\delta_t = -0.5$, the only negative remarkable deviation concerns the couple (Jobless=Yes, Credit=Yes) which says that Jobless persons are given a credit which is intuitively quite unexpected. For this deviation, we need to verify if an exception is associated to this occurrence of attribute value pair. This verification is made in the next step.

Table 4. Deviation analysis for the attributes Jobless and Credit

Jobless	Credit	x_{ij}	n_{ij}	δ
Yes	No	12	4.5	+1.68
Yes	Yes	2	9.5	**-0.79**
No	No	28	35.5	-0.21
No	Yes	83	75.5	+0.10

Finding reliable exceptions After the identification of relevant negative deviations, this step focuses on the part of the database that verifies the attribute value pair concerned by the deviation. Further data mining is performed on this database-window as for instance mining frequent itemsets. More precisely, in the window concerning the pair $(A = A_i, B = B_j)$, all frequent itemsets will contain $A = A_i$ and $B = B_j$ and will lead to rules of the form $(A = A_i) \wedge X \rightarrow (B = B_j)$ of confidence 1. Since a strong association found in that window may be also a strong rule in the whole database, a further step is to discard any itemset found in such rules which is frequent globally. Following the example of the credit database, and considering the negative deviation Jobless=Yes and Credit=Yes, applying the algorithm Apriori on the tuples of the database concerning jobless people who have been granted a credit, one gets the exception rule "A female jobless with little experience is given credit" which has a high confidence and a very low support.

The method proposed by Liu et al. differs from the previous ones developed in sections 2.1 and 2.2 since it does not use a support measure: the couples $(A = A_i, B = B_j)$ that are studied are those that deviate from the expected frequencies assuming the independence between attributes. Let us notice that it is possible to examine couples with very low support and probably to consider noise as exceptions.

This work relies on the study of negative correlations measured by the notion of deviation. Other works are also based on the notion of Independence between items, as for instance, the use of the chi-squared test in [4]. Deviation is also used in the context of actionability issue. Indeed, Shapiro and Matheus [14] developed a health-care system named KEFIR that suggests corrective actions in response to some relevant deviations discovered in the data.

2.4 Negative association rules

In [15], the authors consider the problem of mining negative associations in a large database of customer transactions. It is well known that a "naive" approach cannot be applied to the search of negative association rules, because negative associations concern itemsets with low frequency and there are too many candidate itemsets. So the idea proposed in [15] is to rely on previously discovered positive associations and on domain knowledge to constrain the search of negative associations. In fact, an exception in this context means finding that two items are not associated while two very similar items are associated, the similarity between items being given by a taxonomy of the domain. In the context of transactional database, this approach enables to extract negative rules that specify what items a customer is not likely to buy given that he or she buys a certain set of items. For example, such a rule can express that "60% of the customers who buy potato chips do not buy bottled water".

Finding such negative rules means finding itemsets that are not frequent in the database; as there is a great number of such itemsets, one can obtain a huge number of negative rules, and a lot of them are really uninteresting. So one key problem is to propose interestingness measure for negative rules. We find in this context the same idea as previously: the unexpectedness of a situation is a crucial notion to evaluate its interestingness. A rule is interesting if it contradicts or significantly deviates from our earlier expectations.

In order to discover such interesting deviations, the authors propose to use domain knowledge in the form of taxonomies that enable to group similar items. Given some knowledge about positive associations in the data, the idea is that similar items must be linked by similar associations. For example, suppose that two brands A and B of the same soft drink appear in the database, and we find that customers that buy chips also usually buy the drink A. In this case, we expect that chips are also associated to the drink B. If this is not the case, we can speak of an interesting negative association between chips and drink B.

Let us consider for example the domain knowledge (from [15]) described in Figure 2.

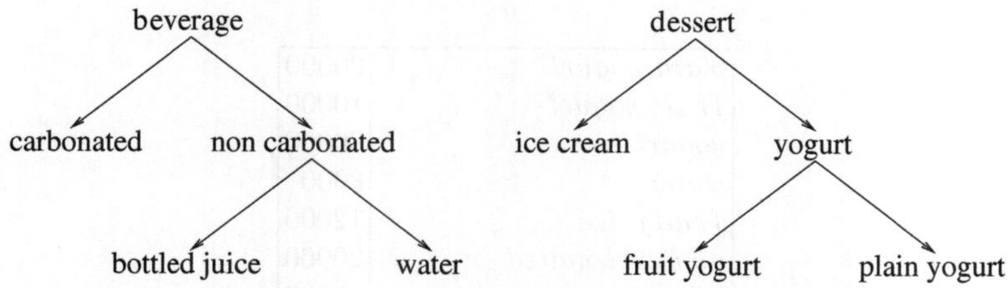

Fig. 2. A taxonomy of items

If a positive association states that people who buy non carbonated beverages also buy yogurts, then, according to the closeness of the items in the taxonomy, we expect the same association between bottled juices and yogurts. An interesting negative association rule contradicts this expectation and so its support is much smaller that the expected support.

More formally, a negative association rule is an implication of the form $X \not\rightarrow Y$ where X and Y are itemsets with $X \cap Y = \emptyset$.

The interestingness measure of a negative association rule $X \not\rightarrow Y$ is defined by:

$$RI = \frac{\mathcal{E}[support(X \cup Y)] - support(X \cup Y)}{support(X)}$$

where $\mathcal{E}[support(X \cup Y)]$ is the expected support of an itemset $X \cup Y$. RI is negatively related to the actual support of $X \cup Y$: it is highest when the actual support is 0 and it equals 0 if the actual support is the same as the expected support.

The algorithm relies on the following points:

- The basic assumption is that items that have a common parent in the taxonomy must have similar associations with other items.
- The aim is to discover negative association rules such that support of X and support of Y are greater than $MinSup$ and such that the degree of interestingness of the rule is greater than $MinRI$.
- Only negative rules for which the expected support can be calculated according to the taxonomy are considered. For example, the support of $\{p_1, \ldots, p_k\}$, where each p_i is an immediate child of $\hat{p_i}$ is estimated by the formula:

$$\mathcal{E}[sup(p_1 \cup \ldots \cup p_k)] = \frac{sup(\hat{p_1} \cup \ldots \cup \hat{p_k}) \times sup(p_1) \times \ldots \times sup(p_k)}{sup(\hat{p_1}) \times \ldots \times sup(\hat{p_k})}$$

To illustrate this, let us consider the taxonomy given in Figure 2 and the supports given in Table 5.

Table 5. Support of the different itemsets

plain yogurt	20000
fruit yogurt	10000
yogurt	30000
water	8000
fruitjuice	12000
non carbonated	20000
yogurt $\wedge$ *non carbonated*	15000

From the known support of the itemset *yogurt* $\wedge$ *non carbonated*, a rule of proportionality is applied to estimate the support of an itemset with a child of *yogurt* and a child of *non carbonated*.
So the expected support of *plain yogurt* $\wedge$ *water* equals:

$$15000 \times \frac{20000}{30000} \times \frac{8000}{20000} = 4000$$

If the actual support of *plain yogurt* $\wedge$ *water* is much lower, for example 800, then the negative association rule *water* $\not\rightarrow$ *plain yogurt* has an interestingness measure of $\frac{4000-800}{8000} = 0.4$ and it is considered as interesting if the minimum threshold forRI is set to 0.4

In this work, the notion of negative association rule is original but the main criticism is that the semantics of such rules is not clearly defined. In fact, the semantics depends of the interestingness measure but it is not clear how a rule like $X \not\rightarrow Y$ could be interpreted and used. Moreover, this search method can be applied only when a taxonomy on the domain exists and this kind of information is not always available.

3 Subjective Approaches

3.1 A theoretical belief-based framework

This section is a survey of the subjective-based methods developed in the data mining literature to address the unexpectedness issue. In [17], Silberschatz and Tuzhilin used the term *unexpectedness* in the context of interestingness measures for patterns evaluation. They classify such measures into objective (data-driven) and subjective (user-driven) measures. According to them, from the subjective point of view, a pattern is interesting if it is:

- Actionable: the end-user can act on it to her/his advantage.
- Unexpected: the end-user is surprised by such findings.

As pointed out by the authors, the actionability is subtle and difficult to capture; they propose rather to capture it through unexpectedness, arguing that unexpected patterns are those that lead the expert of the domain to make some actions. Readers interested by the notion of actionability can refer to the state of the art proposed in [7].

In the framework presented in [17], evaluating the unexpectedness of a discovered pattern is done according to a *Belief System* that the user has: the more the pattern disagrees with a belief system, the more unexpected it is. There are two kinds of beliefs. On one hand, *hard beliefs* are those beliefs that are always true and that cannot be changed. In this case, detecting a contradicting pattern means that something is wrong with the data used to find this pattern. On the other hand, *soft beliefs* are those that the user is willing to change with a new evidence. Each soft belief is assigned with a *degree* specifying how the user is confident in it. In their work, the authors proposed five approaches to affect such degrees: Bayesian, Dempster-Shafer, Frequency, Cyc's and Statistical approaches. The authors claim that the Bayesian one is the most appropriate for defining the degree of beliefs even if any other approach they have defined can be used. In the following, we detail the Bayesian-based approach where degrees are measured with the conditional probability.

Let $\alpha \in \mathcal{B}$ be a soft belief and ξ the previous evidence supporting that belief. The degree of the belief α given the evidence ξ, denoted by $d(\alpha|\xi)$, is the conditional probability $P(\alpha|\xi)$ that α holds given ξ. Given a new evidence E, $d(\alpha|\xi)$ is updated using the Bayes rule:

$$d(\alpha|E,\xi) = P(\alpha|E,\xi) = \frac{P(E|\alpha,\xi)P(\alpha|\xi)}{P(E|\alpha,\xi)P(\alpha|\xi) + P(E|\neg\alpha,\xi)P(\neg\alpha|\xi)}$$

Given a new pattern p, we wonder how strongly it "affects" the belief system $\mathcal{B}$. To quantify this notion, a weight w_i is assigned to each soft belief $\alpha_i \in \mathcal{B}$ so that $\sum_{\alpha_i \in \mathcal{B}} w_i = 1$. The following formula measures how much a new pattern p changes the degrees of soft beliefs in $\mathcal{B}$ knowing the evidence ξ:

$$I(p,\mathcal{B},\xi) = \sum_{\alpha_i \in \mathcal{B}} w_i \left| d(\alpha_i|p,\xi) - d(\alpha_i|\xi) \right|$$

As pointed out by the authors, the proposed framework is especially useful in applications where data change rapidly with time, as in some of the On-Line Transaction Processing systems (airline reservation, banking, ...). Indeed, if it exists in $\mathcal{B}$ a belief α such that $d(\alpha|E,\xi) \neq d(\alpha|\xi)$, where E is new data, then there must exist in E a pattern p such that $I(p,\mathcal{B},\xi) \neq 0$. When new data is available, belief degrees are updated in order to take into consideration this new data. If many changes in these degrees occur (according to a given threshold), then this means that there are interesting patterns hidden in the data and it is pertinent in this case to trigger a knowledge discovery process. Let us notice that in this framework, the notion of surprise is not only used to discover exception rules, but also to follow the evolution of a belief system.

In our point of view, such a belief-driven discovery method is quite interesting; in applications where data evolve quickly, it must be useful to point out to the user when it is appropriate to restart a knowledge discovery process. However, to the best of our knowledge, no work has reported an application of this subjective theoretical approach. We guess that it is not so easy to compute belief degrees with the Bayes formula. Indeed, computing *a posteriori* probabilities from *a priori* probabilities is expensive and could make the applicability of such a method difficult. Furthermore, stating beliefs and weights, on which this approach relies completely, may be a hard task for the expert of the domain.

3.2 A belief-based system

Using a belief system is also the approach adopted by Padmanabhan and Tuzhilin [11, 12, 13] for discovering exception rules that contradict belief rules. Consider a belief $X \to Y$ and a rule $A \to B$, where both X and A are conjunctions of atomic conditions and both Y and B are single atomic conditions on boolean attributes. In their work presented in [11, 12], a rule $A \to B$ is unexpected with respect to the belief $X \to Y$ on the dataset $\mathcal{D}$ if the following conditions hold:

1. B and Y logically contradict each other.
2. $X \wedge A$ holds on a statistically large subset of tuples in $\mathcal{D}$.
3. $A, X \to B$ holds and since B and Y logically contradict each other, it follows that $A, X \to \neg Y$ also holds.

The authors have presented an algorithm called ZoomUR (stating for Zoom to Unexpected Association Rules) which uses the support and the confidence as interestingness measures in order to detect exception rules that contradict user beliefs. ZoomUR is a two-step algorithm:

- ZoomInUR: explores the space of exception rules with an Apriori-like method [1]. Given a dataset $\mathcal{D}$ and a belief $X \to Y$, ZoomInUR discovers all exception rules of the form $X, A \to C$, where C contradicts Y. More precisely, for a belief $X \to Y$, ZoomInUR first generates all frequent itemsets, according to a given minimum support threshold, that are constrained to contain X. This is achieved by a level-wise algorithm as Apriori. Then, ZoomInUR uses the discovered frequent itemsets to generate exception rules that contain X in their left-hand sides and such that their confidences exceed a given minimum confidence threshold.
- ZoomOutUR: determines all the exception rules more general than those already discovered by ZoomInUR. For each exception $X \wedge A \to C$, ZoomOutUR looks for the more general exceptions of the form $X' \wedge A \to C$ with $X' \subset X$.

The authors have reported also in [11, 12] two applications of their approach to real-life data concerning a web log-file data and a commercial purchase data. For instance, considering the second application and given the belief *"Professionals tend to shop more on weekends than on weekdays"*, ZoominUR detected the exception *"In December, professionals tend to shop more on weekdays than on weekends"*.
Such an exception may be explained by the fact that December is a holiday period which makes professionals shop more often in weekdays. Also, ZoomoutUR discovered that *"In December, shoppers tend to shop more on weekdays than on weekends"*, which means that shoppers in general are shopping in weekdays during December. This rule is a generalization of the exception rule given below. It turns out that the first exception is not really interesting since this is true not only for professionals. This reminds us the notion of *reference rule* evoked in the objective methods (Sections 2.1, 2.2) and used to discard uninteresting exception rules. So, starting from a set of belief rules, exception rules as discovered by ZoominUR are just intermediate rules to find more general ones. However, as pointed out by the authors, such more general rules do not actually contradict the belief rule.

The marketing database used to conduct the experiments in [11, 12] contains about 90 000 tuples defined on 36 qualitative attributes, each one having between 2 and 15 modalities. Considering 15 belief rules, the algorithm generates 600 exception rules considered as interesting by the expert of the domain. This number of rules is very low in comparison to the 40 000 association rules generated by Apriori algorithm for the same support and confidence thresholds. However, considering belief rules separately must lead to a set of redundant rules. This idea was addressed later by Padmanabhan and Tuzhilin in [13] where they extended ZoomUR to MinZoomUR to extract a minimum set of exception rules.

By focusing on user beliefs, the method proposed so far reduces drastically the number of association rules thus facilitating their post analysis. But in our opinion, this method should be improved by using other interestingness measures than the usual support and confidence.

4 Conclusion

The different works that have been presented in this survey look for rules that express a very strong regularity between the items, but which are satisfied by a small number of individuals. These rules, called exception rules, are defined relatively to a commonsense rule, which they contradict. In Sections 2.1, 2.2 and 3.2, the matter is to study rule-exception pairs of the following form:

$X \rightarrow Y$	called	*commonsense rule*	noted	CSR
$X \wedge A \rightarrow Z$	called	*exception rule*	noted	EXC

where X and A are itemsets and Y and Z are contradictory items. When dealing with boolean attributes, contradictory items consist of an item and its negation, and in the case of discrete attributes, contradictory items are atoms with the same attribute but different values. In order to assess the degree of interestingness of exception rules, some works introduce a third rule $A \rightarrow Z$, called a reference rule, denoted by REF. Taking into account this reference rule allows to discard expected exception rules; in fact, the exception $X \wedge A \rightarrow Z$ is not surprising when the reference rule $A \rightarrow Z$ has a high confidence. Table 6 is a summary of the main characteristics of the different approaches. Suzuki in [23] gives a complementary view to this survey by proposing an evaluation scheme of the methods enlarged to group discovery methods. In our chapter, we emphasize on the quality measures and works are compared according to the three following points:

- Quality measures: support, confidence, entropy-based measures, ...
- Information provided by the user: measures parameters, commonsense rules, reference rules, background knowledge
- Search methods

In order to apply the methods developed in that framework, it is important to distinguish the objective approaches from the subjective ones.

The objective approaches have the advantage that no prior knowledge is needed, since commonsense rules and exception rules are discovered during the process. In that context, two search methods can be applied: either a commonsense rule and its exception are searched simultaneously, or in a two-step process searching first for the commonsense rules, for instance using an Apriori-like algorithm and then searching for exceptions to these rules. On the other hand, since the objective approaches rely only on data, it is very important to define relevant measures to assess the quality of the rules that are extracted.

Except for the works described in [20, 21] that rely on several support and confidence thresholds, the different measures that are used are quite similar since they rely on the entropy notion and they aim at capturing in the same measure both the interestingness and the surprising nature of a pair of rules. These measures have been applied on classical datasets from Irvine site, but the papers do not relate experiments on artificial datasets that would allow for instance to evaluate their behaviors on noisy datasets. This would be very interesting, since exception rules have by definition low supports and are thus very sensitive to noise.

In the subjective approaches, commonsense rules model knowledge, called beliefs, given by an expert of the domain. The expert has also to quantify his/her degree of certainty in each of his/her beliefs. When certain knowledge is contradicted by data, this allows to express doubts on the reliability of data. On the other hand, when "soft" knowledge is contradicted by data, this allows to infer exception rules. Some might be very interesting when they are new rules that the expert did not know in a domain he/she already knows very well.

Table 6. Characteristics of some methods for mining exception rules. CSR denotes a commonsense rule, EXC an exception rule, and REF a reference rule

	Informations provided by the user	**Search method**	**Measures**
Suzuki et al. [19] [20] [21] [24] [25]	Support and confidence thresholds for CSR and EXC *or* no required information parameter to control rule complexity	Exhaustive computation of pairs (CSR,EXC)	Support and confidence significance level *or* A unique entropy-based measure for the pair (CSR, EXC): GACE(CSR, EXC)
Hussain et al. [8]	Support and confidence thresholds for CSR and REF Support threshold for EXC	Apriori to find CSR and REF Computation of the candidate rules EXC Evaluation of the possible triplets	A unique entropy-based measure for the triplet (CSR, EXC, REF): RI(CSR, EXC, REF)
Liu et al. [9]	Pairs of interesting attributes are given or computed Deviation threshold	Deviation analysis to find negative deviations Windowing on the deviations to exhibit strong rules No requirement for rules CSR, REF	Deviation measures on contingency tables
Savasere et al. [15]	Support threshold Interestingness threshold Taxonomy	Computation of CSR $A \rightarrow B$ Search for negative rules $X \not\rightarrow Y$, with X "close" to A and Y "close" B according to the taxonomy	Deviation analysis between expected support and real support
Silberschatz and Tuzhilin [17]	A belief system Belief degrees	Evaluation of the unexpectedness of discovered patterns according to a belief system	A measure assessing how much a new pattern changes the degrees of a belief system
Padmanabhan et al. [11]	A set of beliefs CSR Support threshold Confidence threshold	Search of frequent patterns that contradict a belief CSR Confidence filtering Generalization of EXC to find rules REF	Support Confidence

In fact, the underlying idea of all the works presented in this paper is that frequent association rules are often already known by the user. As a consequence, it seems that searching association rules with low supports seems more promising for getting interesting knowledge. Since such a search cannot be performed exhaustively by methods like Apriori, all the works rely on knowledge either already known, or previously computed (the commonsense rules) in order to limit the search space. Nevertheless, let us mention the work of [2] that differ from that framework since it tries to extract rules with a low support, without references to rules already given; it relies on a new quality measure that does not take into account a minimum support threshold.

Let us notice that learning exception rules to commonsense rules is only a special case of the more general problem of global consistency of a set of rules. Truly, the problem could be viewed as studying whether a set of rules obtained by an Apriori-like algorithm is consistent, *i.e.*, contains schemes of rules as those studied in this paper, and it would be interesting to show such sets of rules to the user. Of course, such schemes are all the more likely to appear that the minimum support threshold is low.

Another research topic which is interesting in that context is discovering association rules with negations, like $L_1, \ldots, L_p \to L_{p+1}, \ldots, L_k$, where each L_i is a literal, *i.e.*, an item or the negation of an item. Mining such rules requires to extend the notion of itemset to literalset, where a literal is either an item or its negation. This leads to a search space with a size of 3^n literalsets, where n is the number of items. Tractable approaches [3, 26] must propose constraints to reduce the search space. This problem is partially addressed in the works presented in that survey, since exception rules allow the introduction of negation in the conclusion of a rule. Works on negative association rules $X \not\to Y$ are also an interesting topic, but the important point on the underlying semantics of the rules has to be studied and particularly, the links between the rule $X \not\to Y$ and the rule $X \to \neg Y$.

Acknowledgments

The authors would like to thank the reviewers for their comments and suggestions.

References

1. R. Agrawal, T. Imielinski, and A. N. Swami. Mining association rules between sets of items in large databases. In Peter Buneman and Sushil Jajodia, editors, *Proceedings of the 1993 ACM SIGMOD International Conference on Management of Data*, pages 207–216, Washington, D.C., 26–28 1993.
2. J. Azé. Une nouvelle mesure de qualité pour l'extraction de pépites de connaissances. *Extraction des connaissances et apprentissage*, 17(1):171–182, 2003.

3. J.-F. Boulicaut, A. Bykowski, and B. Jeudy. Towards the tractable discovery of association rules with negations. In *Proceedings of the Fourth International Conference on Flexible Query Answering Systems, FQAS 2000*, pages 425–434, 2000.
4. S. Brin, R. Motwani, and C. Silverstein. Beyond market baskets: Generalizing association rules to correlations. In *Proceedings of the ACM SIGMOD International Conference on Management of Data*, volume 26,2 of *SIGMOD Record*, pages 265–276, New York, May13–15 1997. ACM Press.
5. R. Gras and A. Lahrer. L'implication statistique: une nouvelle méthode d'analyse des données. *Mathématiques, Informatique et Sciences Humaines*, 120:5–31, 1993.
6. S. Guillaume, F. Guillet, and J. Philippe. Improving the discovery of association rules with intensity of implication. In *PKDD '98: Proceedings of the Second European Symposium on Principles of Data Mining and Knowledge Discovery*, pages 318–327, London, UK, 1998. Springer-Verlag.
7. Z. He, X. Xu, and S. Deng. Data mining for actionable knowledge: A survey. Technical report, Harbin Institute of Technology China, 2005.
8. F. Hussain, H. Liu, E. Suzuki, and H. Lu. Exception rule mining with a relative interestingness measure. In *PAKDD: Pacific-Asia Conference on Methodologies for Knowledge Discovery and Data Mining*, pages 86–97. LNCS, 2000.
9. H. Liu, H. Lu, L. Feng, and F. Hussain. Efficient search of reliable exceptions. In *Pacific-Asia Conference on Knowledge Discovery and Data Mining*, pages 194–203, 1999.
10. P. M. Murphy and D. W. Aha. *UCI Repository of Machine Learning Databases*. Machine-readable collection, Dept of Information and Computer Science, University of California, Irvine, 1995. [Available by anonymous ftp from `ics.uci.edu` in directory `pub/machine-learning-databases`].
11. B. Padmanabhan and A. Tuzhilin. A belief-driven method for discovering unexpected patterns. In *Proceedings of the Fourth International Conference on Knowledge Discovery and Data Mining*, pages 94–100, 1998.
12. B. Padmanabhan and A. Tuzhilin. Unexpectedness as a measure of interestingness in knowledge discovery. *Decis. Support Syst.*, 27(3):303–318, 1999.
13. B. Padmanabhan and A. Tuzhilin. Small is beautiful: discovering the minimal set of unexpected patterns. In *Proceedings of the Sixth International Conference on Knowledge Discovery and Data Mining*, pages 54–63, 2000.
14. G. Piatetsky-Shapiro and C. J. Matheus. The interestingness of deviations. In *Proceedings of the Knowledge Discovery in Databases Workshop (KDD-94)*, pages 25 – 36, 1994.
15. A. Savasere, E. Omiecinski, and S. B. Navathe. Mining for strong negative associations in a large database of customer transactions. In *International Conference on Data Engineering (ICDE 1998)*, pages 494–502, 1998.
16. A. Silberschatz and A. Tuzhilin. On subjective measures of interestingness in knowledge discovery. In *Proceedings of the First International Conference on Knowledge Discovery and Data Mining*, pages 275–281, 1995.
17. A. Silberschatz and A. Tuzhilin. What makes patterns interesting in knowledge discovery systems. *IEEE Trans. On Knowledge And Data Engineering*, 8:970–974, 1996.
18. P. Smyth and R. M. Goodman. An information theoretic approach to rule induction from databases. *IEEE Trans. Knowledge And Data Engineering*, 4:301–316, 1992.

19. E. Suzuki. Discovering unexpected exceptions: A stochastic approach. In *Proceedings of the fourth international workshop on RSFD*, pages 225–232, 1996.
20. E. Suzuki. Autonomous discovery of reliable exception rules. In David Heckerman, Heikki Mannila, Daryl Pregibon, and Ramasamy Uthurusamy, editors, *Proceedings of the Third International Conference on Knowledge Discovery and Data Mining (KDD-97)*, page 259. AAAI Press, 1997.
21. E. Suzuki. Scheduled discovery of exception rules. In Setsuo Arikawa and Koichi Furukawa, editors, *Proceedings of the 2nd International Conference on Discovery Science (DS-99)*, volume 1721 of *LNAI*, pages 184–195, Berlin, December 6–8 1999. Springer.
22. E. Suzuki. Discovering interesting exception rules with rule pair. In *PKDD Workshop on Advances in Inductive Rule Learning*, pages 163–177, 2004.
23. E. Suzuki. Evaluation Scheme for Exception Rule/Group Discovery. In Ning Zhong and Jiming Liu, editors, *Intelligent Technologies for Information Analysis*, pages 89–108, Berlin, 2004. Springer-Verlag.
24. E. Suzuki and Y. Kodratoff. Discovery of surprising exception rules based on intensity of implication. In Jan M. Żytkow and Mohamed Quafafou, editors, *Proceedings of the 2nd European Symposium on Principles of Data Mining and Knowledge Discovery (PKDD-98)*, volume 1510 of *LNAI*, pages 10–18, Berlin, September 23–26 1998. Springer.
25. E. Suzuki and M. Shimura. Exceptional knowledge discovery in databases based on information theory. In Evangelos Simoudis, Jia Wei Han, and Usama Fayyad, editors, *Proceedings of the Second International Conference on Knowledge Discovery and Data Mining (KDD-96)*, pages 275–278. AAAI Press, 1996.
26. X. Wu, C. Zhang, and S. Zhang. Efficient mining of both positive and negative association rules. *ACM Trans. Inf. Syst.*, 22(3):381–405, 2004.

Part II

From data to rule quality

Measuring and Modelling Data Quality for Quality-Awareness in Data Mining

Laure Berti-Équille

IRISA, Campus Universitaire de Beaulieu, Rennes, France
Laure.Berti-Equille@irisa.fr

Summary. This chapter presents an overview of data quality management, data linkage and data cleaning techniques that can be advantageously employed for improving the quality awareness of the knowledge discovery process. Based on this database-oriented overview of data quality management, this chapter also presents a pragmatic step-by-step framework for data quality awareness and enhancement before warehousing and during the knowledge discovery process. Each step may use, combine and exploit the data quality characterization, measurement and management methods, and the related techniques proposed in the literature.

Key words: Data Quality Management, Record Linkage, Data Cleaning, Data Quality Metadata

1 Introduction

The quality of data mining results and the validity of results interpretation essentially rely on the data preparation process and on the quality of the analyzed datasets. Indeed, data mining processes and applications require various forms of data preparation, correction and consolidation, combining complex data transformation operations and cleaning techniques. This is because the data input to the mining algorithms is assumed to conform to "nice" data distributions, containing no missing, inconsistent or incorrect values. This leaves a large gap between the available "dirty" data and the available machinery to process and analyze the data for discovering added-value knowledge and decision making. Data quality is a multidimensional, complex and morphing concept [19]. In the last decade, there has been a significant amount of work in the area of information and data quality management initiated by several research communities (database, statistics, workflow management, knowledge engineering), ranging from techniques that assess information quality to build large-scale data integration systems over heterogeneous data sources with different degrees of quality and trust. Many data quality definitions, metrics, models,

Laure Berti-Équille: *Measuring and Modelling Data Quality for Quality-Awareness in Data Mining*, Studies in Computational Intelligence (SCI) **43**, 101–126 (2007)
www.springerlink.com

and methodologies have been proposed by academics and practitioners with the aim to tackle the main classes of data quality problems:

- Duplicate detection and record matching known under various names: data or record linkage [38], merge/purge problem [58], object matching [13, 63], duplicate elimination [8, 100, 4], citation matching [12, 3], identity uncertainty [36], entity identification [50], entity resolution [9], string matching [30], or approximate string join [49],
- instance conflict resolution [44] using data source selection [16, 26] or data cleaning techniques [23],
- missing values [84] and incomplete data [42],
- staleness of data [14, 17].

In error-free data warehouses or database-backed information systems with perfectly clean data, knowledge discovery techniques (such as clustering or mining association rules) can be relevantly used as decision making processes to automatically derive new knowledge patterns and new concepts from data. Unfortunately, most of the time, this data is neither rigorously chosen from the various heterogeneous information sources with different degrees of quality and trust, nor carefully controlled for quality [5]. Deficiencies in data quality still are a burning issue in many application areas, and become acute for practical applications of knowledge discovery and data mining techniques [85]. An example using association rules discovery is used to illustrate this basic idea. This will be completed by other application examples in Section 2. Among traditional descriptive data mining techniques, association rules discovery [78] identifies intra-transaction patterns in a database and describes how much the presence of a set of attributes in a database's record (*i.e.*, a transaction) implicates the presence of other distinct sets of attributes in the same record (respectively in the same transaction). The quality of association rules is commonly evaluated by the support and confidence measures [78, 67, 77]. The support of a rule measures the occurrence frequency of the pattern in the rule while the confidence is the measure of the strength of implication. The problem of mining association rules is to generate all association rules that have support and confidence greater than the user-specified minimum support and confidence thresholds. Besides support and confidence, other measures for knowledge quality evaluation (called interestingness measures) have been proposed in the literature with the purpose of supplying alternative indicators to the user in the understanding and use of the new discovered knowledge [67, 77]. But, to illustrate the impact of low-quality data over discovered association rule quality, one might legitimately wonder whether a so-called "interesting" rule noted $LHS \rightarrow RHS$ (with the following semantics: Left-Hand Side implies Right-Hand Side) iis meaningful when 30% of the LHS data are not up-to-date anymore, 20% of the RHS data are not accurate, and 15% of the LHS data come from a data source that is well-known for its bad reputation and lack of credibility.

The main contribution of this chapter is twofold: first, an exhaustive overview of data quality management is given which can be advantageously employed for improving the data quality awareness of knowledge discovery and data mining techniques; secondly, a step-by-step framework for data quality enhancement and awareness for the KDD process is proposed.

The rest of the chapter is organized as follows. Section 2 discusses motivations for data quality awareness and management in three mining examples. Section 3 gives an exhaustive overview on data quality characterization, modeling, management and cleaning techniques. In Section 4, a quality-aware KDD framework is presented. Section 5 provides concluding remarks.

2 Motivation

Three mining application areas are presented to show the importance of data quality awareness with considering some of its various dimensions [11, 11]: *i.e.*, data accuracy, precision, completeness, currency, trustworthiness, non-duplication, and data source reputation.

Example 1: In life sciences, researchers extensively collaborate with each other, sharing biomedical and genomic data and their experimental results. This necessitates dynamically integrating different databases or warehousing them into a single repository [21]. Overlapping data sources may be maintained in a controlled way, such as replication of data on different sites for load balancing or for security reasons. But uncontrolled overlaps are very frequent cases. Moreover, scientists need to know how reliable the data is if they are to base their research on it, because pursuing incorrect theories and experiments costs time and money. The current solution to ensure data quality is verification by human experts. The two main drawbacks are: *i)* data sources are autonomous and as a result, sources may provide excellent reliability for one specific area, or for some data subsets, and *ii)* verification is a manual process of data accreditation by specialists that slows the incorporation of data and that is not free from conflicts of interest. Biological databank providers will not directly support data quality evaluations to the same degree since there is no motivation for them to do so, and there is currently no standard for evaluating and comparing biomedical data quality. Mining for patterns in contradictory biomedical data has been proposed in [35], but other automatic, impartial, and independent data quality evaluation techniques are needed for all types of data before any biomedical mining applications. Moreover, chips and micro-array data are now becoming standard tools for the high-throughput analysis of gene expression. The power of micro-array analysis and array-based gene expression image mining lies in the ability to compare large sets of genes from different tissues or in various experimental conditions to identify pathways and regulatory networks. As micro-array laboratories are scaling up their facilities, manual assessment of chip images becomes cumbersome and prone to subjective criteria. Several systems exist to standardize data quality control.

But others need to be developed for the highly automated assessment of experimental micro-array data quality and, therefore, for monitoring a variety of quality parameters and offering means to minimize noise, remove systematic measurement errors before the bio-mining processes.

Example 2: In business or technological intelligence gathering efforts, data is often collected from many heterogeneous information sources, such as technical reports, human assets, transcripts, commercial documents, competitive studies, knowledge-sharing Web sites, newsgroups, etc. It is obvious that each of these data sources have different degrees of trust and quality that can be advantageously (or maliciously) used (e.g., in viral marketing). With the variety of data sources to consider and incorporate, it is currently time-consuming to sift through each of these sources to determine which are the most accurate. To make the correct decisions based on the intelligence available in a timely manner, automatic means are needed to determine accurate data sources and to be able to detect malicious or compromised data sources to prevent them from influencing the decision making processes. Thus, mining techniques should be aware of completeness, accuracy, trustworthiness and inter-dependency of data sources for ensuring critical data and decisions.

Example 3: Multimedia content (combining text, image, audio, video) is now prevailing for all data exchanges, transferred through various heterogeneous networks to several kinds of terminal devices (TV set, set top box, mobile phone, smart phone, PDA, PC). Ensuring end-to-end quality of service along the whole audio-visual delivery chain and integrity of multimedia content through adaptation to network and terminal characteristics is the commonly shared goal of all the content, service and network providers today in the market [89]. These characteristics are mainly captured by sensors and may be contradictory or erroneous. Human involvement in the raw data capturing process can also introduce noise. On the other side of the multimedia delivery chain, multimedia mining techniques try to decipher the multimedia data deluge trying to bridge the "semantic" gap. High-dimension reduction techniques, such as pre-clustering multimedia databases rely on media description accuracy and freshness because these low-level descriptive features (usually sensitive to the dataset size) are pre-computed only once over very large multimedia databases. In most cases, they are not synchronously updated and re-computed each time the database is modified (problem of data freshness). Clustering and association rules discovery on multimedia databases [10, 12, 72, 69] generally require the manual creation of a thesaurus of labels (e.g., "animal", "plant", etc.) for semantically describing groups of images or videos, and then inferring the mapping between the keywords labels (at the semantic level) and the clusters of low-level descriptive features (at the pixel level). Human subjectivity may lead to misinterpretation or bad precision and recall for the content-based information retrieval scenarios. Low-quality of multimedia mining results is therefore due to the lack of completeness and precision of the chosen keywords for characterizing the richness of multimedia contents.

Table 1. Problems and Current Solutions for Data Quality Management

Processing Step	Data Quality Problems	Potential Solutions and References
Data Creation, Capture, Gathering or Import	Manual entry, OCR, speech recognition Complex data type (multimedia) No standardized format or data schema Duplicates Approximations, surrogates Measurement/Sensor Errors Hardware or software constraints Automatic massive data import	**Preemptive Approaches:** - Workflow Management Methodologies [47, 93, 81, 82] - Architectures for Data Quality Management [3, 28] [45, 14, 18, 33] - Data audits, data stewardship [2] **Retrospective and Corrective Approaches:** - Data Diagnosis : error and outliers detection [15, 22] - Data Cleaning: record linkage [38], merge/purge problem [58], object matching [13, 63], duplicate elimination [8, 100, 4], citation matching [12, 3], identity uncertainty [36], entity identification [50], entity resolution [9], or approximate string join [49], address and string matching [30]
Data Delivery	Information destruction or mutilation by inappropriate pre-processing Data Loss: buffer overflows, transmission problems No Checks	Data quality control [29] Data editing Data publishing Data aggregation and data squashing [95] Use checksum Monitor data transmission, data integrity data format Data mining techniques to check correctness of data transmissions
Data Storage	Metadata paucity and staleness Inappropriate data models and schemas Ad hoc Modifications Hardware constraints Software limitations	Metadata Management [19, 32, 40] Plan ahead and customize for domain: Data Profiling, data browsing and monitoring [10, 102]
Data Integration	Multiple heterogeneous data sources Time synchronization Atypical Data Legacy systems Sociological factors Ad hoc Joints Random Matching Heuristics	Mandate accurate timestamps, data lineage [6] Commercial tools for data migration Data scrubbing, profiling [4] Academic tools and language extensions for data cleaning (ETL): Potter's Wheel [94], Ajax [34], Bellman [10], Arktos [76], Febrl [70], and ClueMaker [17] and for quality-driven query processing: HIQIQ [26, 25] Academic tools for approximate join matching
Data Retrieval	Human errors Computational constraints Software constraints, limitations, incompatibility	Recall / Precision significance Feedback loop
Statistical Analysis and Data Mining	Issues of scale Performance and confidence guarantees Belief in black boxes and dart boards Attachment to a family of models Insufficient domain expertise Lack of familiarity with the data	Data Preparation for mining [85, 15] Exploratory Data Mining - (EDM) [19] Greater accountability from analysts Continuous, ongoing analysis rather than one-shot solutions Sampling vs. full analysis Feedback loops

As briefly shown above, these three examples are conditioned to various data quality problems occurring along the entire data processing continuum [19]. Data preparation [15] is crucial and consists of several necessary operations such as cleaning data, normalizing, handling noisy, uncertain or untrustworthy information, handling missing values, transforming and coding data in such a way that it becomes suitable for the data mining process. As shown in Table 1, synthesizing Dasu and Johnson's vision [19], several academic and commercial solutions have been proposed to tackle more or less pragmatically the continuous problems of data quality at each step of data processing.

3 An Overview of Data Quality Management

Maintaining a certain level of quality of data is challenging and cannot be limited to one-shot approaches addressing simpler, more abstract versions of the problems of dirty or low-quality data [5, 19, 29]. Solving these problems requires highly domain- and context-dependent information and human expertise. Classically, the database literature refers to data quality management as ensuring: *i)* syntactic correctness (e.g., constraints enforcement that prevent "garbage data" from being entered into the database) and *ii)* semantic correctness (*i.e.*, data in the database that truthfully reflects the real world entities and situations). This traditional approach of data quality management has lead to techniques such as integrity constraints, concurrency control and schema integration for distributed and heterogeneous systems. A broader vision of data quality management is presented in this chapter (but still with a database orientation). In the last decade, literature on data and information quality across different research communities (including databases, statistics, workflow management and knowledge engineering) have proposed a plethora of:

- Data quality dimensions with various definitions depending on authors and application contexts [11, 28, 11, 83]
- Data quality dimension classifications that are depending on the audience type: practical and technical [93], more general [8, 82] or depending on the system architecture type: see [25] for integrated information systems, [59, 64] for data warehouse systems, or [52] for cooperative information systems (CIS)
- Data quality metrics [19, 97, 25, 98]
- Conceptual models and methodologies [71, 61, 60, 62, 73, 96, 81]
- Frameworks to improve or assess data quality in databases [7, 45, 93, 83, 81], in (cooperative) information systems [18, 3, 14, 82] or in data warehouse systems [47, 59, 75, 73].

To give a detailed overview, the rest of this section will present the different paradigms for data quality characterization, modeling, measurement and management methodologies.

3.1 Data Quality Characterization

Since 1980 with Brodie's proposition in [1], more than 200 dimensions have been collected to characterize data quality in the literature [11, 45, 93, 96]. The most frequently mentioned data quality dimensions in the literature are accuracy, completeness, timeliness and consistency, with various definitions:

- Accuracy is the extent to which collected data is free of errors [51] or it can be measured by the quotient of the number of correct values in a source and the number of the overall number of values [25].
- Completeness is the percentage of the real-world information entered in the sources and/or the data warehouse [59] or it can be measured by the quotient of the number of non-null values in a source and the size of the universal relation [25].

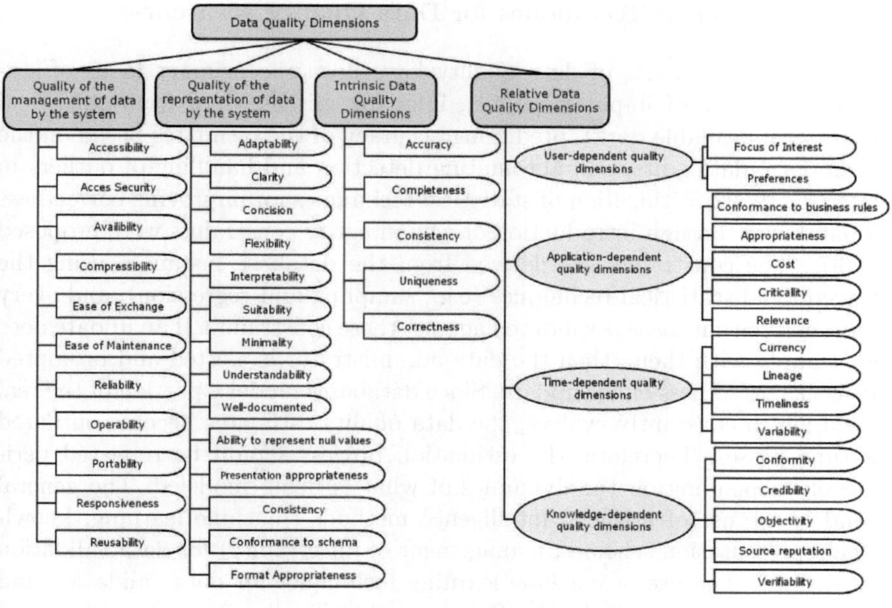

Fig. 1. Map of Data Quality Dimensions

- Timeliness is the extent to which data are sufficiently up-to-date for a task [51]; the definitions of freshness, currency, volatility are reported in [14].
- Consistency is the coherence of the same data represented in multiple copies or different data with respect of the pre-defined integrity constraints and rules [11].

Figure 1 presents a classification of data quality dimensions that are divided into the following four categories:

1. Quality dimensions describing the quality of the management of data by the system based on the satisfaction of technical and physical constraints (e.g., accessibility, ease of maintenance, reliability, etc.),
2. Quality dimensions describing the quality of the representation of data in the system based on the satisfaction of conceptual constraints on modeling and information presentation (e.g., conformance to schema, appropriate presentation, clarity, etc.),
3. Intrinsic data quality dimensions (e.g., accuracy, uniqueness, consistency, etc.),
4. Relative data quality dimensions with dependence on the user (e.g., user preferences), or on the application (e.g., criticality, conformance to business rules, etc.), or time-dependent (e.g., variability, volatility, freshness, etc.) or with dependence on a given knowledge-state (e.g., data source reputation, verifiability).

3.2 Measurement Techniques for Data Quality Awareness

The statistical aspects of data quality have been the primary focus of statistical methods of imputation (*i.e.*, inferring missing data from statistical patterns of available data), predicting accuracy of the estimates based on the given data, data edits, and automating detection and handling of outliers in data [15, 19, 22]. Utilization of statistical techniques for improving correctness of databases through introduction of new integrity constraints were proposed in [97]. The constraints are derived from the database instances using the conventional statistical techniques (e.g., sampling and regression), and every update of the database is validated against these constraints. If an update does not comply with them, then the data administrator is alerted and prompted to check correctness of the update. Since databases model a portion of the real world which constantly evolves, the data quality estimates become outdated as time passes. Therefore, the estimation process should be repeated periodically depending on the dynamics of what is being modeled. The general trend is the use of artificial intelligence methods (machine learning, knowledge representation schemes, management of uncertainty) for data validation [19, 10, 98]. The use of machine learning techniques for data validation and correction was first presented by Parsaye and Chignell: rules inferred from the database instances by machine learning methods were used to identify outliers in data and facilitate the data validation process. Another similar approach was proposed by [41].

EDM (Exploratory Data Mining) [19] is a set of statistical techniques providing summaries that characterize data with typical values (e.g., medians and averages), variance, range, quantiles and correlations. Used as a first pass, EDM methods can be advantageously employed for data pre-processing before carrying out more expensive analyses. EDM aims to be widely applicable while dealing with unfamiliar datasets. These techniques have a quick response time, and have results which are easy to interpret, to store, and to update. EDM can either be driven by the models to facilitate the use of parametric methods (model log-linear, for instance), or be driven by the data without any prior assumptions about inter-relationships between data. Well-known non-parametric techniques for exploring multivariate distributions (such as clustering, hierarchical or neural networks) can be used. The EDM summaries (e.g., averages, standard deviations, medians or other quantiles) can be used to characterize the data distribution, the correlations between attributes, or the center of the value distribution of a representative attribute. They can also be used to quantify and describe the dispersion of the attribute values around the center (form, density, symmetry, etc.). Other techniques are used to detect and cope with other problems on data, such as missing values, improbable outliers and incomplete values. Concerning the techniques of analysis on missing data, the method of imputation through regression described by Little and Rubin [84] is generally used. Other methods such as Markov Chain Monte Carlo (MCMC) [42] are used to simulate data under the multivariate normal distribution assumption. Other references related to the problem of missing values are described in the tutorial by Pearson [85]. Concerning the outliers: the techniques of detection are mainly control charts and various techniques based: *i)* on a mathematical model, *ii)* on geometrical methods for distance measurement in the dataset (called geometric outliers), or *iii)* on the distribution (or the density) of data population [22] extended by the concept of local exception (called local distributional outliers) [15]. Other tests of "goodness-of-fit" such as Chi^2 check the independence of the attributes. The Kolmogorov-Smirnov test provides a measure of the maximum distance between the supposed distribution of the data and the empirical distribution computed from the data. These univariate tests are very useful to validate the analysis techniques and the assumptions on the used models. Moreover, complex and multivariate tests can be used such as pyramids, hyper-pyramids and Mahalanobis test for distances between multivariate averages [92]. The interested reader is invited to read the survey of Pyle [15], in particular for the use of entropy as a preliminary data characterization measure ([15], Section 11.3), and [19] for the description of these techniques.

3.3 Data Quality Models

In practice, assessing data quality in database systems has mainly been conducted by professional assessors with more and more cost-competitive auditing practices. Well-known approaches from industrial quality management and software quality assessment have been adapted for data quality and have proposed an extension of metadata management for data quality [93, 82]. The use of metadata for data quality evaluation and improvement has been advocated by many authors [19, 32, 40]. Rothenberg argued that information producers should perform *Verification, Validation, and Certification (VV&C)* of their data and that they should provide data quality metadata along with the datasets [40]. Several propositions fully integrate the modeling and the management of quality metadata into the database design. Among these process-oriented approaches, the TDQM program (*Total Data Quality Management*) proposed by Wang *et al.* [96, 81] provides a methodology including the modeling of data quality in the Entity-Relationship conceptual database model. It also proposes guidelines for adding step-by-step data quality metadata on each element of the model (entity, attribute, association). Other works have taken related (still similar) approaches in modeling and capturing the quality of relational data [71, 52] and of semi-structured data (e.g., D^2Q [62]). Figure Fig. 2 presents a generic model of data quality with UML formalism that synthesizes the common aspects of the approaches. One or more data quality instances may be associated to one data item (*i.e.*, attribute value, set of values, record, table, domain, etc.). Data quality is composed of one or more dimensions with public attributes representing the type and the category of the quality dimension which is composed of one or more measures

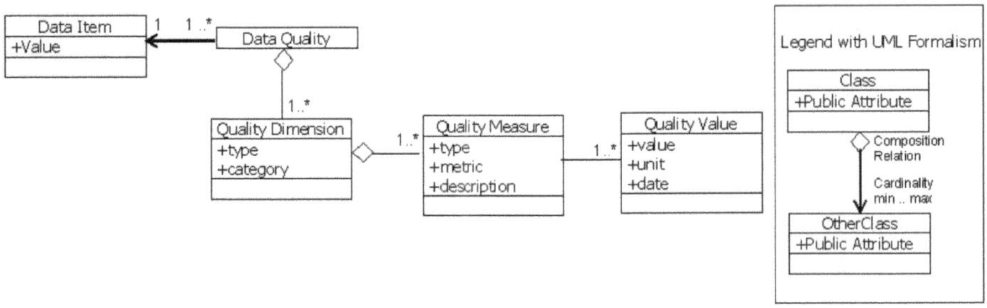

Fig. 2. Data Quality Model

characterized by its type, metric and description. Each measure has one or more values with the corresponding unit and measurement date.

Most of the proposed data quality models rely on data quality metadata being available. Unfortunately, these approaches rely on precise and accurate metadata. However, such metadata are not always available and no commonly accepted standard describing data quality dimensions currently exists. Although considerable efforts have been invested in the development of metadata standard vocabularies for the exchange of information across different application domains (mainly for geographic information systems [57]), including substantial work on data quality characterization in theses domains, the obvious fact is that in practice the quality metadata in many application domains remains a luxury. More specifically for data warehouse systems, many propositions concern the definition of quality models [59, 64, 75, 73] with particular attention paid to data transformation logs or history (also called lineage) [6]. This metadata is very useful for the analysis and the interpretation of the probability distributions on the truncated data, for instance, and also for debugging, implementing quality feedback loops and analyzing the causes of data errors.

3.4 Data Quality Methodologies

This overview is focused on the state of the art in data quality management with a wide description range of related works from data quality characterization and modeling (with quality dimension definitions, metrics and models) to higher level methodological approaches. In this last trend of research, there has been considerable work on methodologies for measuring, improving or assessing the quality of data in database and information systems: e.g., *TIQM - Quality Management Methodology Implementation* [47], Redman's methodology [93], and *IP-MAP - Information Product Map* framework [62, 31]. These three methodologies were proposed for traditional management information systems (MIS), and not especially for data warehouses or heterogeneous cooperative information systems [47]. All methodologies deal with both process-centric and data-centric activities. Process-centric activities propose to

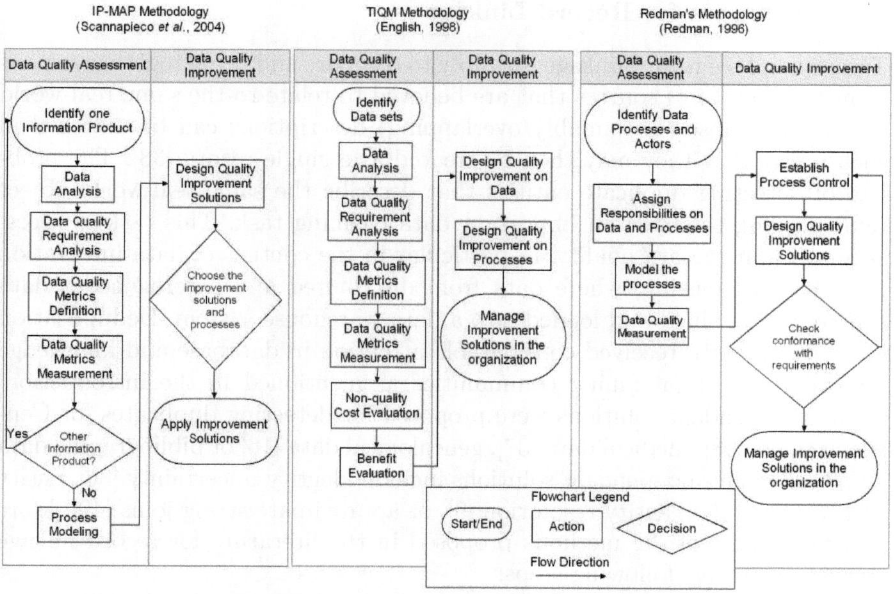

Fig. 3. IP-MAP, TIQM and Redman's Methodology for Data Quality Management

improve data quality, modifying some aspects of the processes that manage data and intra-organizational data flows. Data-centric activities improve data quality independently from the processes. Only the TIQM methodology provides guidelines for economic aspects related to data quality, specifically for evaluation of the cost of quality loss, the costs of the data improvement process and the benefits and savings resulting from data quality improvement. Figure Fig. 3 presents the three methodologies as flowcharts covering both aspects of data quality assessment and improvement with several common steps.

Redman's methodology [93] provides a huge number of case studies that provide evidence of the high cost related with low quality in information systems. A methodology specifically focused on costs and savings is described in [14]. Avenali *et al.* in [7] describe an algorithm for optimal choice of sources in terms of improvement of quality/cost balancing for a given demand of data and corresponding quality in a cooperative information system. However, such methodologies rely on human assessment of data, which is often time-consuming and possibly error-prone. Previous works have assumed that the metadata regarding the quality of data is available, accurate, and unbiased; either published by the data providers themselves or provided by user rankings of the most accurate or reliable data sources.

To complete this overview, the next subsection will present the main techniques of data quality improvement, in particular record linkage and data cleaning techniques with Extraction-Transformation-Loading (ETL) operations.

3.5 Techniques for Record Linkage

The principle of record linkage is simply to compare and bring together records from two (or more) sources that are believed to relate to the same real-world entity and whose (presumably overlapping) descriptions can be matched in such a way that they may then be treated as a single record [38]. The problem of detecting duplicate entities that describe the same real-world object (and purging them) is an important data cleaning task. This task is necessary to improve data quality, in particular in the context of data integration prior to warehousing, where data from distributed and heterogeneous data sources is combined and loaded into a data warehouse system. Deduplication of relational data received considerable attention in database and knowledge discovery and data mining communities as mentioned in the introduction. Domain-dependent solutions were proposed for detecting duplicates for Census datasets [98], medical data [37], genealogical data [16] or bibliographic data [12, 3]. Domain-independent solutions include identity uncertainty [36], entity identification [50], entity resolution [9], or approximate string joins [49]. More generally, most of the methods proposed in the literature for record linkage consist of the five following steps:

i) Pre-processing for coding, formatting and standardizing the data to compare
ii) Selecting a blocking method to reduce the search space by partitioning the datasets into mutually exclusive blocks to compare (e.g., with hashing, sorting keys, sorted nearest neighbors or windowing techniques over one or more keys (or attributes)) [58, 80]
iii) Selecting and computing a comparison function: this step consists of measuring the similarity distance between the pairs of records for string matching [30], e.g., using the simple string distance, or the occurrence frequency-weighted distance, or another distance such as Hamming distance [86], N-grams [43], Q-grams [48], etc. (see Table 2)
iv) Selecting a decision model: this step consists of assigning and classifying pairs of records as matching, non-matching or potentially matching records with a method that can be probabilistic (with or without training datasets), knowledge-based or empirical. Table 3 presents the methods and tools proposed in the literature and presented in [11]
v) Validation of the method and feedback.

The problem of identifying duplicate records in databases was originally identified by Newcombe *et al.* in 1959 [37] as record linkage on medical records for identifying the same individual over different time periods. Fellegi and Sunter [38] developed a formal theory for record linkage and offered statistical methods for estimating matching parameters and error rates. Among the empirical approaches, Hernandez and Stolfo [58] developed the sorted neighborhood method for limiting the number of potential duplicate pairs that require distance computation. Among the knowledge-based approaches, Tejada *et al.* [88] developed a system that employs active learning methods for

Table 2. Main Characteristics of Distances Functions for String Matching

Distance	Main Characteristics
Hamming Distance [86]	Fixed numerical fields (e.g. SSN, Zip Code) without considering added/missing characters
Edit Distance	Compute the minimal cost of transformation using add/drop/exchange of characters
Jaro/Winkler Algorithm [99]	$(C/L_1 + C/L_2 + (2C - T)/2C)/3$ with C the number of common characters and T the number of transposed characters in the two strings of length L_1 and L_2
N-grams Distance [43]	$\sqrt{\sum_{\forall x} \lvert f_a(x) - f_b(x) \rvert}$ extended by Q-grams [48]
Soundex	Based on phonetics and consonants (e.g., the code of "John" and "Jon" is J500) using the first letter of the string and the correspondences of the Soundex code guide: 1 for B,F,P,V; 2 for C,G,J,K,Q,S,X,Z; 3 for D,T; 4 for L; 5 for M,N; 6 for R.

Table 3. Decision Models for Duplicate Identification and Record Linkage

Method *(Tool)*	Authors	Model
Error-based Model	Fellegi and Sunter, 1969 [38]	
Expectation Maximization based Method	Dempster *et al.*, 1977, [7]	Probabilistic
Induction	Bilenko and Mooney, 2003 [12]	
Clustering for Record Linkage *(Tailor)*	Elfeky *et al.*, 2002, [65]	
1-1 matching and Bridging File	Winkler, 2004, [98]	
Sorted-Nearest Neighbors method	Hernandez and Stolfo, 1995, [58]	
XML Object Matching	Weis and Naumann, 2004, [63]	
Hierarchical Structure *(Delphi)*	Ananthakrishna *et al.*, 2002, [8]	Empirical
Matching Prediction *(ClueMaker)*	Buechi *et al.*, 2003, [17]	
Functional Dependencies Inference	Lim *et al.*,1993, [50]	
Transformation function *(Active Atlas)*	Tejada *et al.*,2001, [88]	Knowledge -based
Rules and sorted-nearest neighbors *(Intelliclean)*	Low *et al.*, 2001, [100]	

selecting record pairs that are informative for training the record-level classifier. The classifier combines similarity estimates from multiple fields across different metrics. In all of these approaches, fixed-cost similarity metrics were used to compare the database records. Several academic tools and algorithms are proposed for Record Linkage: IntelliClean [100], Tailor [65], ClueMaker [17] and Febrl [70].

3.6 Extraction, Transformation and Loading for Data Cleaning

Under the general acronym ETL, the Extraction-Transformation-Loading activities cover the most prominent tasks of data preparation before the data warehousing and mining processes [74]. They include: *i)* the identification of relevant information at the source side, *ii)* the extraction of this information, *iii)* the transformation and integration of the information coming from multiple sources into a common format and, *iv)* the cleaning and correction of the integrated dataset. Despite the specialized ETL tools (mainly dedicated to relational data) available on the market, data preparation and cleaning processes remain complex, costly and critical [6]. This area has raised a lot of interest from the research community [23, 74], now focusing on semi-structured data [63]. Several academic tools and algorithms were proposed for

Table 4. Main Data Transformation Operators for ETL

ETL Operator	Definition
Format [94] Map [34]	Applies a function to every value in an attribute column of a relational table (such as regular-expression based substitutions and arithmetic operations or user-defined functions).
Add, Drop, Copy	Allow users to add a new column, or to drop or copy a column.
Merge	Transforms concatenates values in two columns, optionally interposing a constant in the middle, to form a new column.
Split	Splits a column into two or more parts, and is used typically to parse a value into its constituent parts. The split positions can be specified by character positions, regular expressions, or by interactively performing splits on example values.
Divide	Conditionnaly divides a column, sending values into one of two new columns based on a predicate.
Fold	Flattens tables by converting one row into multiple rows, folding a set of columns together into one column and replicating the rest. Conversely Unfold unflattens tables: it takes two columns, collects that have the same values for all the other columns, and unfolds the two chosen columns.

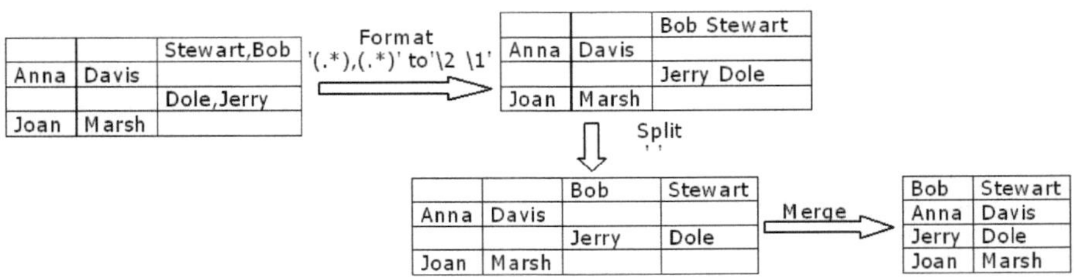

Fig. 4. Example of Data Transformation [94]

data transformation and conciliation: AJAX [34], Potter's Wheel [94], ARKTOS [76], Telcordia [4].

AJAX [34] is an SQL extension for specifying each data transformation (such as matching, merging, mapping, and clustering) for the data cleaning process. These transformations standardize data formats when possible and find pairs of records that most probably refer to the same real-world object. The duplicate elimination step is applied if approximate duplicate records are found, and multi-table matching computes similarity joins between distinct data flows and consolidates them. Table 4 presents the main ETL operators (e.g., *format, add, merge,* etc.) proposed in Potter's Wheel [94] and AJAX [34] systems with, in the first column, the name of the operator and, in the second column, its definition. Figure Fig. 4 gives a simple but illustrative example of data transformations using three operators (*i.e., format, split* and *merge*) of Potter's Wheel system [94] applied to the attribute values of a table. Other propositions concern the definition of declarative language extensions for specifying/querying quality metadata or for applying data transformations necessary to a specific cleaning process. The prototype Q-Data described by Sheth, Wood and Kashyap [5] checks if the existing data are correct and ensures data validation and cleanup by using a logical database language (LDL++).

The system employs data validation constraints and data cleanup rules. Both ETL tools and algorithms operate in a batch and off-line manner but "active data warehousing" (also called "real time warehousing") [2] refers to a new trend where higher levels of data freshness are required for data warehouses that must be updated as frequently as possible with all the performance and overloading issues this raises for ETL tasks based on filters, transformers and binary operations over the multi-source datasets.

4 A Framework of Data Quality Awareness for the KDD Process

Based on this exhaustive and database-oriented overview of data quality management, this section describes a pragmatic framework for data quality awareness preceding and during the knowledge discovery process. Each step may use and combine the previously mentioned approaches, methods and techniques proposed in the literature for characterizing and measuring data quality, record linking, and data cleaning. The grand view of the framework is depicted in Fig. 5. This framework is divided into parts upstream and downstream of the KDD process (respectively left-hand and right-hand sides of Fig. 5). It consists of five steps from U1 to U5 for upstream and seven steps from D1 to D7 for downstream that will be described in the next subsections.

4.1 KDD Upstream Quality-Aware Activities

The first upstream step denoted U1 in Figure 5 consists of: *i)* selecting the data sources from which data will be extracted by automatic and massive import procedures and *ii)* defining a clear, consistent set of data quality requirements for the data warehouse, for the input data sources and the entire KDD workflow and information chain (*i.e.*, qualitative and quantitative descriptions of quality criteria for the information units, KDD processes and sub-processes). In the U2 step, it is necessary to provide the basis for establishing measures and statistical controls of the data quality dimensions previously specified in the first step. Data items do not have the same importance, they may not be equivalent from a "strategic" point of view for a company and thus do not have to be considered in a uniform way for scheduling ETL and mining activities. The data quality measurement step U2 should provide a first characterization of the data quality of the pre-selected data sources (by Exploratory Data Mining techniques, for instance) prior to data loading into the data warehouse. The EDM summaries will be stored as quality metadata characterizing the quality of each data source. With a more technical and system-centered perspective, different levels of periodic measurement and control can be implemented in a data warehouse as shown in Figure 6

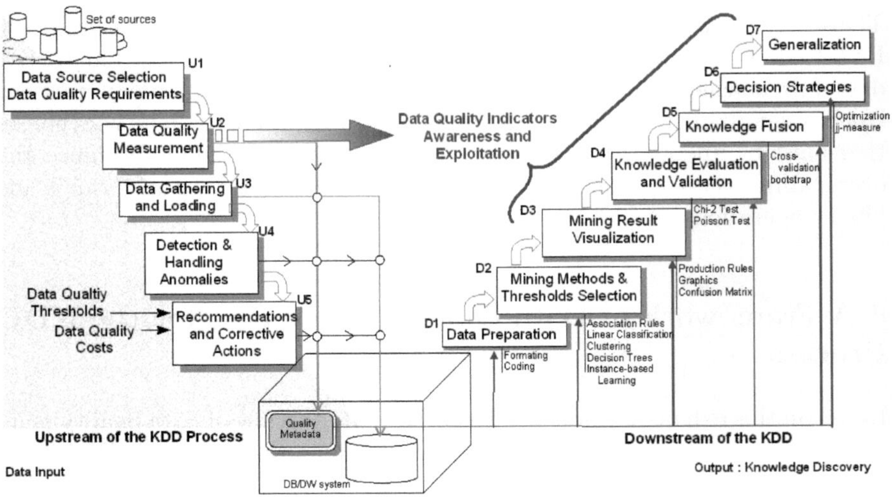

Fig. 5. General Framework of Data Quality Awareness for the KDD Process

(listed from A. to J.). The U2 step consists in computing quality scores and indicators mainly using pre-validation programs (see H. in Figure 6). These programs may characterize some quantitative dimensions of data quality with statistical indicators. The U3 step consists of possibly cleaning, reconciling, aggregating data and loading data into the data warehouse with appropriate ETL tools and record linking strategies. Different ETL operators (mentioned in Table 4) may be used for formatting, mapping or merging the matching records to be stored in the data warehouse. The goal of the U4 step is: *i)* to detect the problems of data quality (*i.e.,* errors, outliers or poor quality data) throughout data processing with post-validation programs applied on the data warehouse system (see I. in Figure 6), *ii)* to refresh the quality metadata and *iii)* to analyze the causes depending on the data quality (non-) tolerance thresholds and data quality costs. The purpose of the U5 step is to propose corrective actions and recommendations of improvements for each of the four previous upstream steps in order to set up quality feedback loops.

4.2 KDD Downstream: Quality Metadata Exploitation

Starting the KDD process, the classical first step is data preparation noted D1 for downstream of the KDD process in Fig. 5. It consists of a succession of tasks such as: *i)* selection of datasets and objects, *ii)* selection and weighting of the variables and features, *iii)* (re-)coding data, *iv)* analyzing missing values and the data sources. Different kinds of missing values are distinguished (e.g., unknown vs. unrecorded vs. irrelevant vs. truncated data), *v)* detection of data anomalies: because outliers are individual data values that are inconsistent with the majority of values in the data collection, it's important to

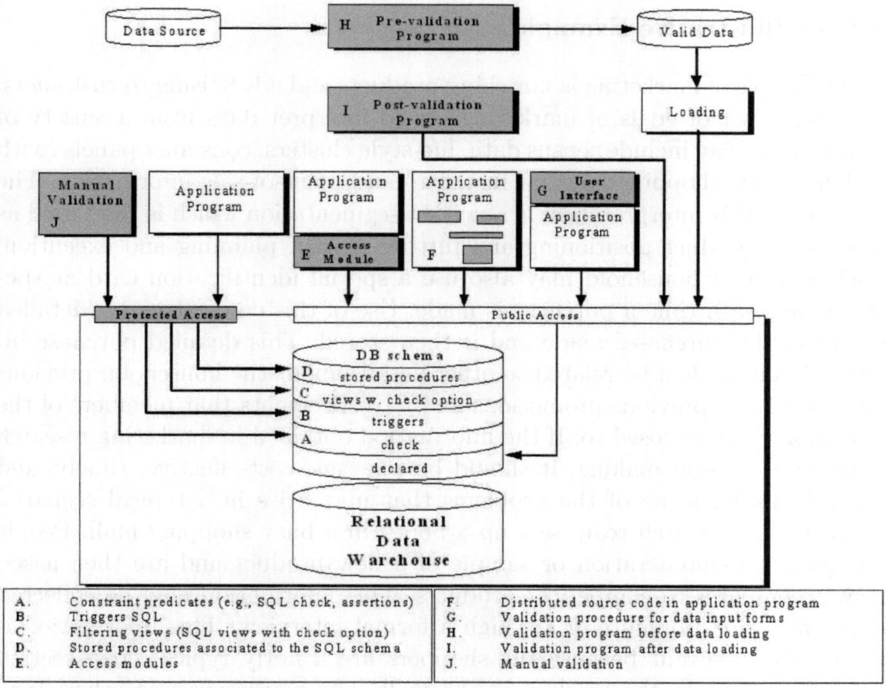

Fig. 6. Different Levels for Controlling and Measuring Data Quality

characterize the side-effects of the strategy of detecting and omitting outliers, *vi)* the homogenization of the data files, *vii)* the discretization of the continuous attributes and *viii)* the possible use of quantization strategies for real variables (e.g., defining quartiles or deciles) or high-dimension reduction techniques. The next steps D2 and D3 in Fig. 5 consist of selecting the mining methods, configuration parameters and thresholds and, therefore the knowledge representation for visualizing the mining results (e.g., decision tables, decision trees, classification rules, association rules, instance-based learning representation and clusters). For quality-awareness, these steps (in particular D3) should include the visualization of data quality indicators previously computed in step U2 and stored in the quality metadata repository of the data warehouse.

The added-value of quality metadata consists in their exploitation as explanation factors for evaluating discovered knowledge and for validating one mining process (step D4) and combining the results of several mining techniques (step D5) and also drive the decision strategies (step D6) and generalization (step D7).

4.3 An Illustrative Example

The principle of marketing is matching products and advertising to customers. The initial set of goals of marketing are to interpret data from a variety of sources that may include census data, life-style clusters, consumer panels (with looking inside shopping bags for instance) and point-of-sale information. The outcome of this interpretation is a market segmentation which is then used as a basis for product positioning and further market planning and execution. Each consumer household may also use a special identification card at specific stores each time a purchase is made. Use of this card triggers a detailed record of the purchases made and is then stored. This detailed purchase information may then be related to other details about the household, previous purchases, and previous promotions and advertisements that members of the household were exposed to. If the information obtained in marketing research is to help decision making, it should be relevant, cost-effective, timely, and valid. Consider some of the problems that may arise in a typical scenario. A marketing research team sets up a booth in a busy shopping mall. People are given a demonstration or sample of a new product and are then asked how it compares to competitor products. This information may be collected either as a questionnaire or through a formal interview. This information is presumably relevant because the shoppers are a fairly typical cross-section of people who buy the product in the mall. The marketing researchers may actually screen people according to criteria such as age, gender, or type of clothing, etc. Responses are entered into a database, tabulated and a report is generated. The first upstream step of our approach denoted U1 in Fig. 5 consists of: *i)* selecting the relevant sources of data (e.g., product and purchase data, point-of-sale data collected by bar code scanning and census data) and *ii)* defining a clear, consistent set of data quality requirements for the data warehouse that stores the integrated data. For the example, the considered data quality criteria for the input datasets may be:

- Data consistency that is defined as the percent of values that satisfy a set of pre-defined integrity constraints on one (or a set of) record(s) (e.g., "name = Smith John, gender = F" does not satisfy the constraint specifying the correspondance between the gender and the first name). The constraints may involve additional knowledge (e.g., a thesaurus or a dictionary of first names)
- Data completeness that is defined as the percent of data fields having non-null values
- Data validity that is defined as the percent of data having values that fall within their respective domain of allowable values. EDM summaries may also be used to characterize the data distribution and detect outliers
- Data timeliness that is defined by the percent of data available within a specified threshold time frame (e.g., days, hours, minutes, seconds).

In the U2 step, the measures, statistical controls or EDM summaries characterizing these data quality dimensions are computed for each data source by pre-validation programs before loading the data. The EDM summaries will be stored as quality metadata characterizing the quality of each data source. The U3 step consists of linking records and aggregating data before loading data into the data warehouse. The following three records:

r1 ["name = Smith John, gender = F, address=4 rue des Champs Elysées, City=Paris, Country=France"]
r2 ["name = J. Smith, gender = M, address=Champs Elyses, City=75000, Country=F"]
r3 ["name = John Smith, gender = ?, address=?, City=Paris, Citizenship=French"]

will be formatted by ETL operators, identified as legitimately matching records and finally merged and stored as a single record in the data warehouse. The goal of the U4 step is to detect the problems of data quality in the data warehouse using post-validation programs (see I. in Fig. 6). Then the quality dimensions such as consistency, completeness, validity and timeliness will be computed on the datasets of the warehouse. In the U5 step, two corrective actions may be to avoid the usage of one of the input data sources or to improve the way the answers of the questionnaire are collected and entered in the database. The approach proposed in the framework is to combine the data quality awareness with the data mining process itself. The quality indicators stored as quality metadata in the data warehouse repository are used and combined to characterize the quality of the discovered association rules, for instance (or decision trees or clusters), resulting from a data mining process. There are as many repeated downstream processes as the number of data mining models. Different mining objectives can lead to different models and different quality measures. Starting the KDD process, the first step of data preparation noted D1 in Figure Fig. 5 consists of selecting the datasets, and variables, (re-)coding and normalizing the data, and analyzing missing values (completeness). For instance, deviations from the standard and from typical correlations are detected (validity). For the next steps D2 and D3 in Figure Fig. 5, consider that association rules discovery has been selected as one of the mining methods for prediction. An example of applied rule-based prediction is the prediction of demand for stock items on the basis of historical marketing data. In this case, the premise of a predictive rule is a set of conditions such as promotional and advertising campaigns undertaken, and the conclusion is the number of orders for a given item within a time period. Here is an example of a rule:

IF $10 \leq CashRebate \leq 15$ AND
$3,000 \leq CurrentItem345Sales \leq 6,000$ AND
$300,000 \leq TVReach \leq 600,000$
THEN $5,000 \leq RebateClaim \leq 8,000$
WITH confidence: 100% AND
applicable months: July 2004, Sept. 2004, Oct. 2004, Nov. 2004, Dec. 2004, Jan. 2005.

The interpretation of this rule is "if an active rebate between \$10 and \$15 is in place and sales are currently between 3,000 and 6,000 and between

300,000 to 600,000 TV viewers are reached then further sales between 5,000 and 8,000 taking advantage of the rebate can be expected". This rule can be discovered from historical sales data for products. By using this rule one can perform better material resource planning, anticipate demand, and know when to run rebate programs. The data quality specific issues which can affect a predictive rule are: *i)* the historical time period from which the rule is discovered (timeliness), *ii)* the lack of completeness or consistency of some of the analyzed data fields, *iii)* the presence of invalid data that may have been (or not) considered in the rule discovery process.

Predictive stability and validity are not just problems for rule-based prediction. It affects all forms of prediction. One strategy for dealing with these problems is to continuously test and update rules. Rules should be scored according to how well they predict new data. Rules that fall below a certain level of performance should be discarded and replaced by new rules that provide a better level of prediction. One way of selecting new rules in this situation is to deliberately discard the data that supported the faulty rule and carry out rule discovery on the same conclusion in order to identify the replacement rule. Thus rule-based prediction requires continuous testing, validation, and refinement in order to be maximally effective. Another more investigative and "quality-aware" way described in [46] is to exploit and aggregate data quality indicators computed on all the item sets composing the premises and the conclusions of the discovered rules in order to understand the consistency, completeness, validity and timeliness of the data the rule has been computed from, and explain why this rule even with good confidence may be faulty and useless for prediction because of low-quality data.

5 Conclusion

This chapter gave an exhaustive overview of data quality management and related techniques that can be employed for improving the data quality awareness of knowledge discovery and data mining techniques. Three application examples have introduced the importance of data quality-awareness for knowledge discovery activities. The chapter also provided a pragmatic step-by-step framework for a quality-driven KDD process illustrated by an example in marketing rule mining. A higher vision of the approach proposed in this chapter is more for helping users and decision makers to improve data mining results interpretations rather than to improve the data mining performance. Black-box approaches (in case of predictive modeling for instance) will perfom equally well even without data quality indicators, but such mathematical models are difficult to interpret and thus difficult to be accepted by domain experts. Being totally aware of the data quality and having it well specified and quantified, the most important features coming out of a data mining process can be trusted and well explained.

References

1. Avenali A, Batini C, Bertolazzi P, and Missier P. A formulation of the data quality optimization problem. In *Proc. of the Intl. CAiSE Workhop on Data and Information Quality (DIQ)*, pages 49–63, Riga, Latvia, 2004.
2. Karakasidis A, Vassiliadis P, and Pitoura E. Etl queues for active data warehousing. In *Proc. of the 2nd ACM SIGMOD Workshop on Information Quality in Information Systems (IQIS) in conjunction with ACM PODS/SIGMOD*, pages 28–39, Baltimore, MD, USA, 2005.
3. McCallum A, Nigam K, and Ungar LH. Efficient clustering of high-dimensional data sets with application to reference matching. In *Proc. of the 6th ACM SIGKDD Conf. on Knowledge Discovery and Data Mining (KDD)*, pages 169–178, Boston, MA, USA, 2000.
4. Monge A. Matching algorithms within a duplicate detection system. *IEEE Data Eng. Bull.*, 23(4):14–20, 2000.
5. Sheth A, Wood C, and Kashyap V. Q-data: Using deductive database technology to improve data quality. In *Proc. of Intl. Workshop on Programming with Logic Databases (ILPS)*, pages 23–56, 1993.
6. Simitsis A, Vassiliadis P, and Sellis TK. Optimizing etl processes in data warehouses. In *Proc. of the 11th Intl. Conf. on Data Engineering (ICDE)*, pages 564–575, Tokyo, Japan, 2005.
7. Dempster AP, Laird NM, and Rubin DB. Maximum likelihood from incomplete data via the em algorithm. *Journal of the Royal Statistical Society*, 39:1–38, 1977.
8. Kahn B, Strong D, and Wang R. Information quality benchmark: Product and service performance. *Com. of the ACM*, 45(4):184–192, 2002.
9. Batini C, Catarci T, and Scannapiceco M. A survey of data quality issues in cooperative information systems. In *Tutorial presented at the 23rd Intl. Conf. on Conceptual Modeling (ER)*, Shanghai, China, 2004.
10. Djeraba C. Association and content-based retrieval. *IEEE Transactions on Knowledge and Data Engineering (TDKE)*, 15(1):118–135, 2003.
11. Fox C, Levitin A, and Redman T. The notion of data and its quality dimensions. *Information Processing and Management*, 30(1), 1994.
12. Ordonez C and Omiecinski E. Discovering association rules based on image content. In *Proc. of IEEE Advances in Digital Libraries Conf. (ADL'99)*, pages 38–49, 1999.
13. Carlson D. Data stewardship in action. *DM Review*, 2002.
14. Loshin D. *Enterprise Knowledge Management: The Data Quality Approach.* . Morgan Kaufmann, 2001.
15. Pyle D. *Data Preparation for Data Mining.* Morgan Kaufmann, 1999.
16. Quass D and Starkey P. Record linkage for genealogical databases. In *Proc. of ACM SIGKDD'03 Workshop on Data Cleaning, Record Linkage and Object Consolidation*, pages 40–42, Washington, DC, USA, 2003.
17. Theodoratos D and Bouzeghoub M. Data currency quality satisfaction in the design of a data warehouse. *Special Issue on Design and Management of Data Warehouses, Intl. Journal of Cooperative Inf. Syst.*, 10(3):299–326, 2001.
18. Paradice DB and Fuerst WL. A mis data quality management strategy based on an optimal methodology. *Journal of Information Systems*, 5(1):48–66, 1991.
19. Ballou DP and Pazer H. Designing information systems to optimize the accuracy-timeliness trade-off. *Information Systems Research*, 6(1), 1995.

20. Ballou DP and Pazer H. Modeling completeness versus consistency trade-offs in information decision contexts. *IEEE Transactions on Knowledge and Data Engineering (TDKE)*, 15(1):240–243, 2002.
21. Guérin E, Marquet G, Burgun A, Loral O, Berti-Équille L, Leser U, and Moussouni F. Integrating and warehousing liver gene expression data and related biomedical resources in gedaw. In *Proc. of the 2nd Intl. Workshop on Data Integration in the Life Science (DILS)*, San Diego, CA, USA, 2005.
22. Knorr E and Ng R. Algorithms for mining distance-based outliers in large datasets. In *Proc. of the 24th Intl. Conf. on Very Large Data Bases (VLDB)*, pages 392–403, New York City, USA, 1998.
23. Rahm E and Do H. Data cleaning: Problems and current approaches. *IEEE Data Eng. Bull.*, 23(4):3–13, 2000.
24. Caruso F, Cochinwala M, Ganapathy U, Lalk G, and Missier P. Telcordia's database reconciliation and data quality analysis tool. In *Proc. of the 26th Intl. Conf. on Very Large Data Bases (VLDB)*, pages 615–618, Cairo, Egypt, September 10–14 2000.
25. Naumann F. *Quality-Driven Query Answering for Integrated Information Systems*, volume 2261 of *LNCS*. Springer, 2002.
26. Naumann F, Leser U, and Freytag JC. Quality-driven integration of heterogeneous information systems. In *Proc. of the 25th Intl. Conf. on Very Large Data Bases (VLDB)*, pages 447–458, Edinburgh, Scotland, 1999.
27. De Giacomo G, Lembo D, Lenzerini M, and Rosati R. Tackling inconsistencies in data integration through source preferences. In *Proc. of the 1rst ACM SIGMOD Workshop on Information Quality in Information Systems (IQIS)*, pages 27–34, Paris, France, 2004.
28. Delen G and Rijsenbrij D. The specification, engineering and measurement of information systems quality. *Journal of Software Systems*, 17:205–217, 1992.
29. Liepins G and Uppuluri V. *Data Quality Control: Theory and Pragmatics*. M. Dekker, 1990.
30. Navarro G. A guided tour to approximate string matching. *ACM Computer Surveys*, 33(1):31–88, 2001.
31. Shankaranarayan G, Wang RY, and Ziad M. Modeling the manufacture of an information product with ip-map. In *Proc. of the 6th Intl. Conf. on Information Quality*, Boston, MA, USA, 2000.
32. Mihaila GA, Raschid L, and Vidal M. Using quality of data metadata for source selection and ranking. In *Proc. of the 3rd Intl. WebDB Workshop*, pages 93–98, Dallas, TX, USA, 2000.
33. Tayi GK and Ballou DP. Examining data quality. *Com. of the ACM*, 41(2):54–57, 1998.
34. Galhardas H, Florescu D, Shasha D, Simon E, and Saita C. Declarative data cleaning: Language, model and algorithms. In *Proc. of the 9th Intl. Conf. on Very Large Data Bases (VLDB)*, pages 371–380, Roma, Italy, 2001.
35. Müller H, Leser U, and Freytag JC. Mining for patterns in contradictory data. In *Proc. of the 1rst ACM SIGMOD Workshop on Information Quality in Information Systems (IQIS) in conjunction with ACM PODS/SIGMOD*, pages 51–58, Paris, France, 2004.
36. Pasula H, Marthi B, Milch B, Russell S, and Shpitser I. Identity uncertainty and citation matching. In *Proc. of the Intl. Conf. Advances in Neural Information Processing Systems (NIPS)*, pages 1401–1408, Vancouver, British Colombia, 2003.

37. Newcombe HB, Kennedy JM, Axford SJ, and James AP. Automatic linkage of vital records. *Science*, 130:954–959, 1959.
38. Fellegi IP and Sunter AB. A theory for record linkage. *Journal of the American Statistical Association*, 64:1183–1210, 1969.
39. Celko J and McDonald J. Don't warehouse dirty data. *Datamation*, 41(18), 1995.
40. Rothenberg J. Metadata to support data quality and longevity. In *Proc. of the 1st IEEE Metadata Conf.*, 1996.
41. Schlimmer J. Learning determinations and checking databases. In *Proc. of AAAI Workshop on Knowledge Discovery in Databases*, 1991.
42. Schafer JL. *Analysis of Incomplete Multivariate Data.* Chapman & Hall, 1997.
43. Ullmann JR. A binary n-gram technique for automatic correction of substitution, deletion, insertion and reversal errors in words. *The Computer Journal*, 20(2):141–147, 1997.
44. Fan K, Lu H, Madnick S, and Cheung D. Discovering and reconciling value conflicts for numerical data integration. *Information Systems*, 26(8):235–656, 2001.
45. Huang K, Lee Y, and Wang R. *Quality Information and Knowledge Management.* Prentice Hall, New Jersey, 1999.
46. Berti-Équille L. Data quality awareness: a case study for cost-optimal association rule mining. *Knowl. Inf. Syst.*, 2006.
47. English L. *Improving Data Warehouse and Business Information Quality.* Wiley, New York, 1998.
48. Gravano L, Ipeirotis PG, Jagadish HV, Koudas N, Muthukrishnan S, Pietarinen L, and Srivastava D. Using Q-grams in a DBMS for Approximate String Processing. *IEEE Data Eng. Bull.*, 24(4), December 2001.
49. Gravano L, Ipeirotis PG, Koudas N, and Srivastava D. Text joins in an rdbms for web data integration. In *Proc. of the 12th Intl. World Wide Web Conf. (WWW)*, pages 90–101, Budapest, Hungary, 2003.
50. Lim L, Srivastava J, Prabhakar S, and Richardson J. Entity identification in database integration. In *Proc. of the 9th Intl. Conf. on Data Engineering (ICDE)*, pages 294–301, Vienna, Austria, 1993.
51. Liu L and Chi L. Evolutionary data quality. In *Proc. of the 7th Intl. Conf. on Information Quality (IQ)*, MIT, Cambridge, USA, 2002.
52. Santis LD, Scannapieco M, and Catarci T. Trusting data quality in cooperative information systems. In *Proc. of the Intl. Conf. on Cooperative Information Systems (CoopIS)*, pages 354–369, Catania, Sicily, Italy, 2003.
53. Bilenko M and Mooney RJ. Adaptive duplicate detection using learnable string similarity measures. In *Proc. of the 9th ACM SIGKDD Conf. on Knowledge Discovery and Data Mining (KDD)*, pages 39–48, Washington, DC, USA, 2003.
54. Bouzeghoub M and Peralta V. A framework for analysis of data freshness. In *Proc. of the 1st ACM SIGMOD Workshop on Information Quality in Information Systems (IQIS)*, pages 59–67, Paris, France, 2004.
55. Breunig M, Kriegel H, Ng R, and Sander J. Lof: Identifying density-based local outliers. In *Proc. of 2000 ACM SIGMOD Conf.*, pages 93–104, Dallas, TX, USA, May 16-18 2000.
56. Buechi M, Borthwick A, Winkel A, and Goldberg A. Cluemaker: a language for approximate record matching. In *Proc. of the 8th Intl. Conf. on Information Quality (IQ)*, MIT, Cambridge, USA, 2003.

57. Goodchild M and Jeansoulin R. *Data Quality in Geographic Information: From Error to Uncertainty.* Hermès, 1998.
58. Hernandez M and Stolfo S. Real-world data is dirty: Data cleansing and the merge/purge problem. *Data Mining and Knowledge Discovery*, 2(1):9–37, 1998.
59. Jarke M, Jeusfeld MA, Quix C, and Vassiliadis P. Architecture and quality in data warehouses. In *Proc. of the 10th Intl. Conf. on Advanced Information Systems Engineering (CAiSE)*, pages 93–113, Pisa, Italy, 1998.
60. Piattini M, Calero C, and Genero M, editors. *Information and Database Quality*, volume 25. Kluwer International Series on Advances in Database Systems, 2002.
61. Piattini M, Genero M, Calero C, Polo C, and Ruiz F. *Chapter 14: Advanced Database Technology and Design*, chapter Database Quality, pages 485–509. Artech House, 2000.
62. Scannapieco M, Pernici B, and Pierce E. *Advances in Management Information Systems - Information Quality Monograph (AMIS-IQ)*, chapter IP-UML: A Methodology for Quality Improvement Based on IP-MAP and UML. Sharpe, 2004.
63. Weis M and Naumann F. Detecting duplicate objects in xml documents. In *Proc. of the 1st Intl. ACM SIGMOD Workshop on Information Quality in Information Systems (IQIS) in conjunction with ACM PODS/SIGMOD*, pages 10–19, Paris, France, 2004.
64. Jeusfeld MA, Quix C, and Jarke M. Design and analysis of quality information for data warehouses. In *Proc. of 17th Intl. Conf. Conceptual Modelling (ER)*, pages 349–362, Singapore, 1998.
65. Elfeky MG, Verykios VS, and Elmagarmid AK. Tailor: A record linkage toolbox. In *Proc. of the 19th Intl. Conf. on Data Engineering (ICDE)*, pages 1–28, San Jose, CA, USA, 2002.
66. Brodie ML. Data quality in information systems. *Information and Management*, 3:245–258, 1980.
67. Lavrač N, Flach PA, and Zupan B. Rule evaluation measures: A unifying view. In *Proc. of the Intl. Workshop on Inductive Logic Programming (ILP)*, pages 174–185, Bled, Slovenia, 1999.
68. Benjelloun O, Garcia-Molina H, Su Q, and Widom J. Swoosh: A generic approach to entity resolution. Technical report, Stanford Database Group., 2005.
69. Zaïane O, Han J, and Zhu H. Mining recurrent items in multimedia with progressive resolution refinement. In *Proc. of the 16th Intl. Conf. on Data Engineering (ICDE), p.461–476*, San Diego, CA, USA, 2000.
70. Christen P, Churches T, and Hegland M. Febrl - a parallel open source data linkage system. In *Proc. of the 8th Pacific Asia Conf. on Advances in Knowledege Discovery and Data Mining (PAKDD)*, pages 638–647, Sydney, Australia, May 26–28 2004.
71. Missier P and Batini C. A multidimensional model for information quality in cis. In *Proc. of the 8th Intl. Conf. on Information Quality (IQ)*, MIT, Cambridge, MA, USA, 2003.
72. Perner P. *Data Mining on Multimedia*, volume LNCS 2558. Springer, 2002.
73. Vassiliadis P. *Data Warehouse Modeling and Quality Issues.* PhD thesis, Technical University of Athens, Greece, 2000.
74. Vassiliadis P, Simitsis A, Georgantas P, and Terrovitis M. A framework for the design of etl scenarios. In *Proc. of the 15th Intl. Conf. on Advanced Information Systems Engineering (CAiSE)*, pages 520–535, Klagenfurt, Austria, 2003.

75. Vassiliadis P, Bouzeghoub M, and Quix C. Towards quality-oriented data warehouse usage and evolution. In *Proc. of the 11th Intl. Conf. on Advanced Information Systems Engineering (CAiSE)*, pages 164–179, Heidelberg, Germany, 1999.
76. Vassiliadis P, Vagena Z, Skiadopoulos S, and Karayannidis N. ARKTOS: A Tool For Data Cleaning and Transformation in Data Warehouse Environments. *IEEE Data Eng. Bull.*, 23(4):42–47, 2000.
77. Tan PN, Kumar V, and Srivastava J. Selecting the right interestingness measure for association patterns. In *Proc. of the 8th ACM SIGKDD Conf. on Knowledge Discovery and Data Mining (KDD)*, pages 32–41, Edmonton, Canada, 2002.
78. Agrawal R, Imielinski T, and Swami AN. Mining association rules between sets of items in large databases. In *Proc. of the 1993 ACM SIGMOD Conf.*, pages 207–216, Washington, DC,USA, 1993.
79. Ananthakrishna R, Chaudhuri S, and Ganti V. Eliminating fuzzy duplicates in datawarehouses. In *Proc. of the 28th Intl. Conf. on Very Large Data Bases (VLDB)*, pages 586–597, Hong-Kong, China, 2002.
80. Baxter R, Christen P, and Churches T. A comparison of fast blocking methods for record linkage. In *Proc. of ACM SIGKDD'03 Workshop on Data Cleaning, Record Linkage and Object Consolidation*, pages 27–29, Washington, DC, USA, 2003.
81. Wang R. A product perspective on total data quality management. *Com. of the ACM*, 41(2):58–65, 1998.
82. Wang R. *Advances in Database Systems*, volume 23, chapter Journey to Data Quality. Kluwer Academic Press, Boston, MA, USA, 2002.
83. Wang R, Storey V, and Firth C. A framework for analysis of data quality research. *IEEE Transactions on Knowledge and Data Engineering (TDKE)*, 7(4):670–677, 1995.
84. Little RJ and Rubin DB. *Statistical Analysis with Missing Data.* Wiley, New-York, 1987.
85. Pearson RK. Data mining in face of contaminated and incomplete records. In *Proc. of SIAM Intl. Conf. Data Mining*, 2002.
86. Hamming RW. Error-detecting and error-correcting codes. *Bell System Technical Journal*, 29(2):147–160, 1950.
87. Chaudhuri S, Ganjam K, Ganti V, and Motwani R. Robust and efficient fuzzy match for online data cleaning. In *Proc. of the 2003 ACM SIGMOD Intl. Conf. on Management of Data*, pages 313–324, San Diego, CA, USA, 2003.
88. Tejada S, Knoblock CA, and Minton S. Learning object identification rules for information integration. *Information Systems*, 26(8), 2001.
89. Ahmed T, Asgari AH, Mehaoua A, Borcoci E, Berti-Équille L, and Kormentzas G. End-to-end quality of service provisioning through an integrated management system for multimedia content delivery. *Special Issue of Computer Communications on Emerging Middleware for Next Generation Networks*, 2005.
90. Dasu T and Johnson T. *Exploratory Data Mining and Data Cleaning.* Wiley, New York, 2003.
91. Dasu T, Johnson T, Muthukrishnan S, and Shkapenyuk V. Mining database structure or how to build a data quality browser. In *Proc. of the 2002 ACM SIGMOD Intl. Conf.*, pages 240–251, Madison, WI, USA, 2002.

92. Johnson T and Dasu T. Comparing massive high-dimensional data sets. In *Proc. of the 4th Intl. Conf. KDD*, pages 229–233, New York City, New York, USA, 1998.
93. Redman T. *Data Quality: The Field Guide*. Digital Press, Elsevier, 2001.
94. Raman V and Hellerstein JM. Potter's wheel: an interactive data cleaning system. In *Proc. of the 26th Intl. Conf. on Very Large Data Bases (VLDB)*, pages 381–390, Roma, Italy, 2001.
95. DuMouchel W, Volinsky C, Johnson T, Cortez C, and Pregibon D. Squashing flat files flatter. In *Proc. of the 5th ACM SIGKDD Conf. on Knowledge Discovery and Data Mining (KDD)*, pages 6–16, San Diego, CA, USA, 1999.
96. Madnick SE Wang R, Kon HB. Data quality requirements analysis and modeling. In *Proc. of the 9th Intl. Conf. on Data Engineering (ICDE)*, pages 670–677, Vienna, Austria, 1993.
97. Hou WC and Zhang Z. Enhancing database correctness: A statistical approach. In *Proc. of the 1995 ACM SIGMOD Intl. Conf. on Management of Data*, San Jose, CA, USA, 1995.
98. Winkler WE. Methods for evaluating and creating data quality. *Information Systems*, 29(7), 2004.
99. Winkler WE and Thibaudeau Y. An application of the fellegi-sunter model of record linkage to the 1990 u.s. decennial census. Technical Report Statistical Research Report Series RR91/09, U.S. Bureau of the Census, Washington, DC, USA, 1991.
100. Low WL, Lee ML, and Ling TW. A knowledge-based approach for duplicate elimination in data cleaning. *Information System*, 26(8), 2001.
101. Cui Y and Widom J. Lineage tracing for general data warehouse transformation. In *Proc. of the 27th Intl. Conf. on Very Large Data Bases (VLDB)*, pages 471–480, Roma, Italy, September 11–14 2001.
102. Zhu Y and Shasha D. Statstream: Statistical monitoring of thousands of data streams in real time. In *Proc. of the 10th Intl. Conf. on Very Large Data Bases (VLDB)*, pages 358–369, Hong-Kong, China, 2002.

Quality and Complexity Measures for Data Linkage and Deduplication

Peter Christen and Karl Goiser

Department of Computer Science, The Australian National University,
Canberra ACT 0200, Australia {peter.christen,karl.goiser}@anu.edu.au

Summary. Deduplicating one data set or linking several data sets are increasingly important tasks in the data preparation steps of many data mining projects. The aim of such linkages is to match all records relating to the same entity. Research interest in this area has increased in recent years, with techniques originating from statistics, machine learning, information retrieval, and database research being combined and applied to improve the linkage quality, as well as to increase performance and efficiency when linking or deduplicating very large data sets. Different measures have been used to characterise the quality and complexity of data linkage algorithms, and several new metrics have been proposed. An overview of the issues involved in measuring data linkage and deduplication quality and complexity is presented in this chapter. It is shown that measures in the space of record pair comparisons can produce deceptive quality results. Various measures are discussed and recommendations are given on how to assess data linkage and deduplication quality and complexity.

Key words: data or record linkage, data integration and matching, deduplication, data mining pre-processing, quality and complexity measures

1 Introduction

With many businesses, government organisations and research projects collecting massive amounts of data, the techniques collectively known as *data mining* have in recent years attracted interest from both industry and academia. While there is much ongoing research in data mining algorithms and techniques, it is well known that a large proportion of the time and effort in real-world data mining projects is spent understanding the data to be analysed, as well as in the data preparation and preprocessing steps (which may dominate the actual data mining activity) [38]. It is generally accepted [43] that about 20 % to 30 % of the time and effort in a data mining project is used for data understanding, and about 50 % to 70 % for data preparation.

Peter Christen and Karl Goiser: *Quality and Complexity Measures for Data Linkage and Deduplication*, Studies in Computational Intelligence (SCI) **43**, 127–151 (2007)
www.springerlink.com

An increasingly important task in the data preprocessing step of many data mining projects is detecting and removing duplicate records that relate to the same entity within one data set. Similarly, linking or matching records relating to the same entity from several data sets is often required as information from multiple sources needs to be integrated, combined or linked in order to allow more detailed data analysis or mining. The aim of such linkages is to match and aggregate all records relating to the same entity, such as a patient, a customer, a business, a product description, or a genome sequence.

Data linkage and deduplication can be used to improve data quality and integrity, to allow re-use of existing data sources for new studies, and to reduce costs and efforts in data acquisition. They can also help to enrich data that is used for pattern detection in data mining systems. In the health sector, for example, linked data might contain information that is needed to improve health policies [2, 8, 28], and which traditionally has been collected with time consuming and expensive survey methods. Statistical agencies routinely link census data for further analysis [22, 49], and businesses often deduplicate and link their data sets to compile mailing lists. Within taxation offices and departments of social security, data linkage and deduplication can be used to identify people who register for assistance multiple times or who work and collect unemployment benefits. Another application of current interest is the use of data linkage in crime and terror detection. Security agencies and crime investigators increasingly rely on the ability to quickly access files for a particular individual, which may help to prevent crimes by early intervention.

The problem of finding similar entities doesn't only apply to records which refer to persons. In bioinformatics, data linkage can help find genome sequences in a large data collection that are similar to a new, unknown sequence at hand. Increasingly important is the removal of duplicates in the results returned by Web search engines and automatic text indexing systems, where copies of documents – for example bibliographic citations – have to be identified and filtered out before being presented to the user. Finding and comparing consumer products from different online stores is another application of growing interest. As product descriptions are often slightly different, linking them becomes difficult.

If unique entity identifiers (or keys) are available in all the data sets to be linked, then the problem of linking at the entity level becomes trivial: a simple database *join* is all that is required. However, in most cases no unique keys are shared by all the data sets, and more sophisticated linkage techniques need to be applied. These techniques can be broadly classified into *deterministic*, *probabilistic*, and modern approaches, as discussed in Sect. 2. The notation and problem analysis are presented in Sect. 3, and an overview of the various quality measures used to assess data linkage and deduplication techniques is given in Sect. 4. When linking large data sets, it is normally not feasible to compare all possible record pairs due to the resulting computational complexity, so special *blocking* techniques have to be applied. Several recently proposed complexity measures, and the influence of blocking upon quality measurements, are discussed in Sect. 5. A real-world example is used in Sect. 6 to illustrate the effects of different quality and complexity measures. The issues involved in measuring quality in data linkage and deduplication are discussed and a series of recommendations is given in Sect. 7. Finally, the chapter is concluded with a short summary in Sect. 8.

2 Data Linkage Techniques

Data linkage and deduplication techniques have traditionally been used in the health sector for cleaning and compiling data sets for longitudinal or other epidemiological studies [2, 8, 28], and in statistics for linking census and related data [22, 49]. Computer-assisted data linkage goes back as far as the 1950s. At that time, most linkage projects were based on *ad hoc* heuristic methods. The basic ideas of probabilistic data linkage were introduced by Newcombe and Kennedy [35] in 1962, and the mathematical foundation was provided by Fellegi and Sunter [20] in 1969.

Similar techniques were independently developed by computer scientists in the area of document indexing and retrieval [17]. However, until recently few cross-references could be found between the statistical and the computer science community. While statisticians and epidemiologists speak of *record* or *data linkage* [20], the computer science and database communities often refer to the same process as *data* or *field matching*, *data scrubbing*, *data cleaning* [21, 39], *data cleansing* [30], *preprocessing*, *duplicate detection* [7], *entity uncertainty*, or as the *object identity problem*. In commercial processing of customer databases or business mailing lists, data linkage is sometimes called *merge/purge processing* [26], *data integration* [14], *list washing* or *ETL* (extraction, transformation and loading).

2.1 The Data Linkage Process

A general schematic outline of the data linkage process is given in Fig. 1. As most real-world data collections contain noisy, incomplete and incorrectly formatted information, data cleaning and standardisation are important preprocessing steps for successful data linkage, or before data can be loaded into data warehouses or used for further analysis [39]. Data may be recorded or captured in various, possibly obsolete formats and data items may be missing, out of date, or contain errors. The cleaning and standardisation of names and addresses is especially important to make sure that no misleading or redundant information is introduced (e.g. duplicate records). Names are often reported differently by the same person depending upon the organisation they are in contact with, resulting in missing middle names, initials-only, or even swapped name parts. Additionally, while for many regular words there is only one correct spelling, there are often different written forms of proper names,

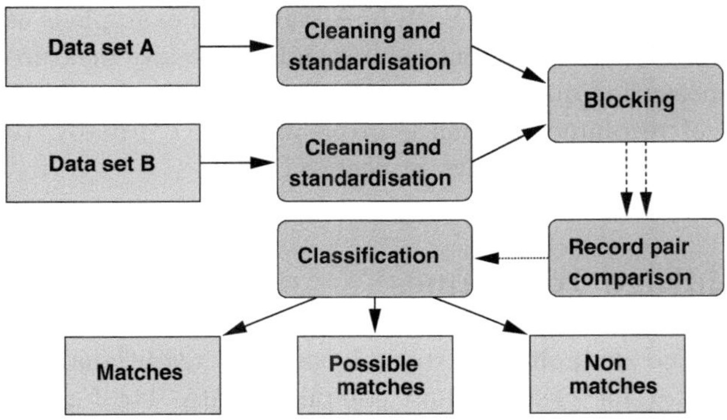

Fig. 1. General linkage process. The output of the blocking step are record pairs, and the output of the comparison step are vectors with numerical matching weights

for example *'Gail'* and *'Gayle'*. The main task of data cleaning and standardisation is the conversion of the raw input data into well defined, consistent forms, as well as the resolution of inconsistencies in the way information is represented and encoded [12, 13].

If two data sets, **A** and **B**, are to be linked, potentially each record from **A** has to be compared with all records from **B**. The number of possible record pair comparisons thus equals the product of the size of the two data sets, $|\mathbf{A}| \times |\mathbf{B}|$. Similarly, when deduplicating one data set, **A**, the number of possible record pairs is $|\mathbf{A}| \times (|\mathbf{A}| - 1)/2$. The performance bottleneck in a data linkage or deduplication system is usually the expensive detailed comparison of fields (or attributes) between pairs of records [3], making it unfeasible to compare all pairs when the data sets are large. For example, linking two data sets with 100,000 records each would result in 10^{10} (ten billion) record pair comparisons. On the other hand, the maximum number of true matches that are possible corresponds to the number of records in the smaller data set (assuming a record in **A** can only be linked to a maximum of one record in **B**, and vice versa). Therefore, the number of potential matches increases linearly when linking larger data sets, while the computational efforts increase quadratically. The situation is the same for deduplication, where the number of duplicate records is always less than the number of records in a data set.

To reduce the large amount of possible record pair comparisons, traditional data linkage techniques [20, 49] employ *blocking*, i.e. they use one or a combination of record attributes (called the *blocking variable*) to split the data sets into blocks. All records having the same value in the blocking variable will be put into the same block, and only records within a block will be compared. This technique becomes problematic if a value in the blocking variable is recorded wrongly, as a potentially matching record may be inserted into a different block, prohibiting the possibility of a match. To overcome this problem, several passes (iterations) with different blocking variables are normally performed.

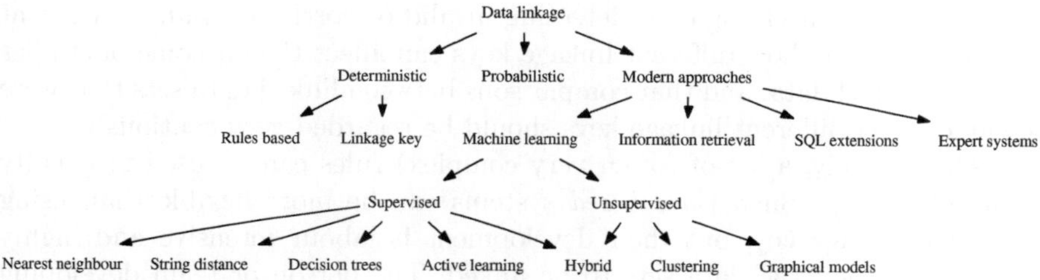

Fig. 2. Taxonomy of data linkage techniques, with a focus on modern approaches

While the aim of blocking is to reduce the number of record pair comparisons made as much as possible (by eliminating pairs of records that obviously are not matches), it is important that no potential match is overlooked because of the blocking process. An alternative to standard blocking is the *sorted neighbourhood* [27] approach, where records are sorted according to the values of the blocking variable, then a sliding window is moved over the sorted records, and comparisons are performed between the records within the window. Newer experimental approaches based on approximate q-gram indices [3, 10] or high-dimensional overlapping clustering [32] are current research topics. The effects of blocking upon the quality and complexity of the data linkage process are discussed in Sect. 5.

The record pairs not removed by the blocking process are compared by applying a variety of comparison functions to one or more – or a combination of – attributes of the records. These functions can be as simple as a numerical or an exact string comparison, can take into account typographical errors [37], or be as complex as a distance comparison based on look-up tables of geographic locations (longitude and latitude). Each comparison returns a numerical value, often positive for agreeing values and negative for disagreeing values. For each compared record pair a vector is formed containing all the values calculated by the different comparison functions. These vectors are then used to classify record pairs into *matches*, *non-matches*, and *possible matches* (depending upon the decision model used). Figure 2 shows a taxonomy of the various techniques employed for data linkage. They are discussed in more detail in the following sections.

2.2 Deterministic Linkage

Deterministic linkage techniques can be applied if unique entity identifiers (or keys) are available in all the data sets to be linked, or a combination of attributes can be used to create a *linkage key* [2] which is then employed to match records that have the same key value. Such linkage systems can be developed based on standard *SQL* queries. However, they only achieve good linkage results if the entity identifiers or linkage keys are of high quality. This means they have to be precise, stable over time, highly available, and robust with regard to errors. Extra robustness for identifiers can be obtained by including a check digit for detecting invalid or corrupted values. A recent study [2] showed how different linkage keys can affect the outcome of studies that use linked data, and that comparisons between linked data sets that were created using different linkage keys should be regarded very cautiously.

Alternatively, a set of (often very complex) rules can be used to classify pairs of records. Such *rules based* systems can be more flexible than using a simple linkage key, but their development is labour intensive and highly dependent upon the data sets to be linked. The person or team developing such rules not only needs to be proficient with the data to be deduplicated or linked, but also with the rules system. In practise, therefore, deterministic rules based systems are limited to ad-hoc linkages of smaller data sets. In a recent study [23], an iterative deterministic linkage system was compared with the commercial probabilistic system *AutoMatch* [31], and the presented results showed that the probabilistic approach achieved better linkage quality.

2.3 Probabilistic Linkage

As common, unique entity identifiers are rarely available in all data sets to be linked, the linkage process must be based on the existing common attributes. These normally include person identifiers (like names and dates of birth), demographic information (like addresses), and other data specific information (like medical details, or customer information). These attributes can contain typographical errors, they can be coded differently, parts can be out-of-date or swapped, or they can be missing.

In the traditional probabilistic linkage approach [20, 49], pairs of records are classified as matches if their common attributes predominantly agree, or as non-matches if they predominantly disagree. If two data sets (or files) $\mathbf{A}$ and $\mathbf{B}$ are to be linked, the set of record pairs

$$\mathbf{A} \times \mathbf{B} = \{(a,b);\ a \in \mathbf{A},\ b \in \mathbf{B}\}$$

is the union of the two disjoint sets

$$M = \{(a,b);\ a = b,\ a \in \mathbf{A},\ b \in \mathbf{B}\} \tag{1}$$

of true matches, and

$$U = \{(a,b);\ a \neq b,\ a \in \mathbf{A},\ b \in \mathbf{B}\} \tag{2}$$

of true non-matches. Fellegi and Sunter [20] considered ratios of probabilities of the form

$$R = \frac{P(\gamma \in \Gamma | M)}{P(\gamma \in \Gamma | U)} \tag{3}$$

where γ is an arbitrary agreement pattern in a comparison space Γ. For example, Γ might consist of six patterns representing simple agreement or disagreement on given name, surname, date of birth, street address, locality and postcode. Alternatively, some of the γ might additionally consider typographical errors [37], or account for the relative frequency with which specific values occur. For example, a surname value *'Miller'* is much more common in many western countries than a value *'Dijkstra'*, resulting in a smaller agreement value for *'Miller'*. The ratio R, or any monotonically increasing function of it (such as its logarithm) is referred to as a *matching weight.* A decision rule is then given by

if $R > t_{upper}$, then	designate a record pair as *match*,
if $t_{lower} \leq R \leq t_{upper}$, then	designate a record pair as *possible match*,
if $R < t_{lower}$, then	designate a record pair as *non-match.*

The thresholds t_{lower} and t_{upper} are determined by a-priori error bounds on false matches and false non-matches. If $\gamma \in \Gamma$ for a certain record pair mainly consists of agreements, then the ratio R would be large and thus the pair would more likely be designated as a match. On the other hand, for a $\gamma \in \Gamma$ that primarily consists of disagreements the ratio R would be small.

The class of possible matches are those record pairs for which human oversight, also known as *clerical review,* is needed to decide their final linkage status. In theory, it is assumed that the person undertaking this clerical review has access to additional data (or may be able to seek it out) which enables her or him to resolve the linkage status. In practice, however, often no additional data is available and the clerical review process becomes one of applying experience, common sense or human intuition to make the decision. As shown in an early study [44] comparing a computer-based probabilistic linkage system with a fully manual linkage of health records, the computer based approach resulted in more reliable, consistent and more cost effective results.

In the past, generally only small data sets were linked (for example for epidemiological survey studies), and clerical review was manageable in a reasonable amount of time. However, with today's large administrative data collections with millions of records, this process becomes impossible. In these cases, even a very small percentage being passed for clerical review will result in hundreds of thousands of record pairs. Clearly, what is needed are more accurate and automated decision models that will reduce – or even eliminate – the amount of clerical review needed, while keeping a high linkage quality. Developments towards this ideal are presented in the following section.

2.4 Modern Approaches

Improvements [48] upon the classical probabilistic linkage [20] approach include the application of the expectation-maximisation (EM) algorithm for improved parameter estimation [46], the use of approximate string comparisons [37] to calculate partial agreement weights when attribute values have typographical errors, and the application of Bayesian networks [47]. A system that is capable of extracting probable matches from very large data sets with hundreds of millions of records is presented in [50]. It is based on special sorting, preprocessing and indexing techniques and assumes that the smaller of two data sets fits into the main memory of a large computing server.

In recent years, researchers have started to explore the use of techniques originating in machine learning, data mining, information retrieval and database research to improve the linkage process. A taxonomy is shown in Fig. 2. Many of these approaches are based on supervised learning techniques and assume that training data is available (i.e. record pairs with known linkage or deduplication status).

An *information retrieval* based approach is to represent records as document vectors and compute the *cosine distance* [14] between such vectors. Another possibility is to use an *SQL* like language [21] that allows approximate joins and cluster building of similar records, as well as decision functions that determine if two records represent the same entity. A generic knowledge-based framework based on rules and an *expert system* is presented in [29]. The authors also describe the precision-recall trade-off (which will be discussed in Sect. 4), where choosing a higher recall results in lower precision (more non-matches being classified as matches), and vice versa.

A popular approach [6, 10, 15, 34, 51, 52] is to learn distance measures that are used for approximate string comparisons. The authors of [6] present a framework for improving duplicate detection using trainable measures of textual similarity. They argue that both at the character and word level there are differences in importance of certain character or word modifications (like inserts, deletes, substitutions, and transpositions), and accurate similarity computations require adapting string similarity metrics with respect to the particular data domain. They present two learnable string similarity measures, the first based on edit distance (and better suitable for shorter strings) and the second based on a support vector machine (more appropriate for attributes that contain longer strings). Their results on various data sets show that learned edit distance resulted in improved precision and recall results. Similar approaches are presented in [10, 51, 52]. [34] uses support vector machines for of classifying record pairs. As shown in [15], combining different learned string comparison methods can result in improved linkage classification.

The authors of [42] use *active learning* to address the problem of lack of training data. Their approach involves repeatedly (i) selecting an example that a vote of classifiers disagree on the most, (ii) manually classifying it, then (iii) adding it to the training data and (iv) re-training the classifiers. The key idea is to use human input only where the classifiers could not provide a clear result. It was found that less than 100 examples selected in this manner provide better results than the random selection of 7,000 examples. A similar approach is presented in [45], where a committee of *decision trees* is used to learn mapping rules (i.e. rules describing linkages).

A *hybrid system* is described in [18] which utilises both supervised and unsupervised machine learning techniques in the data linkage process, and introduces metrics for determining the quality of these techniques. The authors find that machine learning techniques outperform probabilistic techniques, and provide a lower proportion of possible matching pairs. In order to overcome the problem of the lack of availability of training data in real-world data sets, they propose a hybrid technique where class assignments are made to a sample of the data through unsupervised clustering, and the resulting data is then used as a training set for a supervised classifier (specifically, a decision tree or an instance-based classifier).

High-dimensional overlapping *clustering* is used in [32] as an alternative to traditional blocking in order to reduce the number of record pair comparisons to be made, while in [25] the use of simple k-means clustering together with a user-tunable fuzzy region for the class of possible matches is explored, thus allowing control over the trade-off between accuracy and the amount of clerical review needed. Methods based on *nearest neighbours* are explored in [11], with the idea being to capture local structural properties instead of a single global distance approach. *Graphical models* [40] are another unsupervised technique not requiring training data. This approach aims to use the structural information available in the data to build hierarchical probabilistic graphical models. Results are presented that are better than those achieved by supervised techniques.

An overview of other methods (including statistical outlier identification, clustering, pattern matching, and association rules) is given in [30].

Different measures for the quality of the achieved linkages and the complexity of the presented algorithms have been used in many recent publications. An overview of these measures is given in Sects. 4 and 5.

3 Notation and Problem Analysis

The notation used in this chapter follows the traditional data linkage literature [20, 49, 48]. The number of elements in a set $\mathbf{X}$ is denoted $|\mathbf{X}|$. A general linkage situation is assumed, where the aim is to link two sets of entities. For example, the first set could be patients of a hospital, and the second set people who had a car accident. Some of the car accidents have resulted in people being admitted into the hospital. The two sets of entities are denoted as $\mathbf{A}_e$ and $\mathbf{B}_e$. $\mathbf{M}_e = \mathbf{A}_e \cap \mathbf{B}_e$ is the intersection set of matched entities that appear in both $\mathbf{A}_e$ and $\mathbf{B}_e$, and $\mathbf{U}_e = (\mathbf{A}_e \cup \mathbf{B}_e) \backslash \mathbf{M}_e$ is the set of non-matched entities that appear in either $\mathbf{A}_e$ or $\mathbf{B}_e$, but not in both. The space described by the above is illustrated in Fig. 3 and termed *entity space*.

The maximum possible number of matched entities corresponds to the size of the smaller set of $\mathbf{A}_e$ or $\mathbf{B}_e$. This is the situation when the smaller set is a proper subset of the larger one, which also results in the minimum number of non-matched entities. The minimum number of matched entities is zero, which is the situation when no entities appear in both sets. In this situation the number of non-matched entities corresponds to the sum of the entities in both sets. The following equations show this in a formal way:

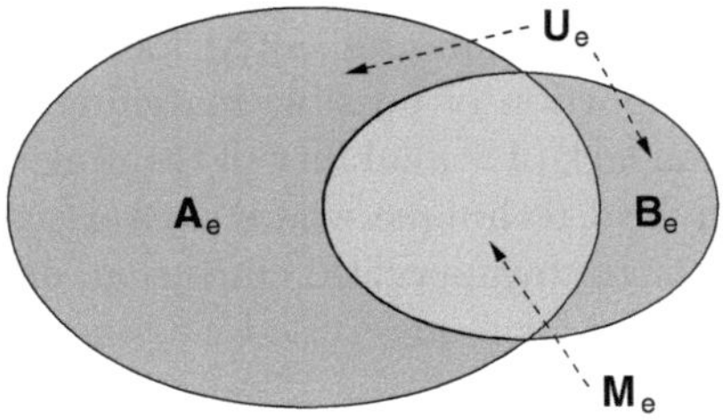

Fig. 3. General linkage situation with two sets of entities $\mathbf{A}_e$ and $\mathbf{B}_e$, their intersection $\mathbf{M}_e$ (entities that appear in both sets), and the set $\mathbf{U}_e$ (entities that appear in either $\mathbf{A}_e$ or $\mathbf{B}_e$, but not in both)

$$0 \leq |\mathbf{M}_e| \leq min(|\mathbf{A}_e|, |\mathbf{B}_e|) \tag{4}$$

$$abs(|\mathbf{A}_e| - |\mathbf{B}_e|) \leq |\mathbf{U}_e| \leq |\mathbf{A}_e| + |\mathbf{B}_e| \,. \tag{5}$$

Example 1. Assume the set $\mathbf{A}_e$ contains 5 million entities (e.g. hospital patients), and set $\mathbf{B}_e$ contains 1 million entities (e.g. people involved in car accidents), with 700,000 entities present in both sets (i.e. $|\mathbf{M}_e| = 700,000$). The number of non-matched entities in this situation is $|\mathbf{U}_e| = 4,600,000$, which is the sum of the entities in both sets (6 million) minus twice the number of matched entities (as they appear in both sets $\mathbf{A}_e$ and $\mathbf{B}_e$).

Records which refer to the entities in $\mathbf{A}_e$ and $\mathbf{B}_e$ are now stored in two data sets (or databases or files), denoted by $\mathbf{A}$ and $\mathbf{B}$, such that there is exactly one record in $\mathbf{A}$ for each entity in $\mathbf{A}_e$ (i.e. the data set contains no duplicate records), and each record in $\mathbf{A}$ corresponds to an entity in $\mathbf{A}_e$. The same holds for $\mathbf{B}_e$ and $\mathbf{B}$. The aim of a data linkage process is to classify pairs of records as matches or non-matches in the product space $\mathbf{A} \times \mathbf{B} = M \cup U$ of true matches and true non-matches [20, 49], as defined in (1) and (2).

It is assumed that no blocking or indexing (as discussed in Sect. 2.1) is applied, and that all pairs of records are compared. The total number of comparisons equals $|\mathbf{A}| \times |\mathbf{B}|$, which is much larger than the number of entities available in $\mathbf{A}_e$ and $\mathbf{B}_e$ together. In the case of the deduplication of a single data set $\mathbf{A}$, the number of record pair comparisons equals $|\mathbf{A}| \times (|\mathbf{A}| - 1)/2$, as each record in the data set will be compared to all others, but not to itself. The space of record pair comparisons is illustrated in Fig. 4 and called the *comparison space*.

Example 2. For Example 1 given above, the comparison space consists of $|\mathbf{A}| \times |\mathbf{B}| = 5,000,000 \times 1,000,000 = 5 \times 10^{12}$ record pairs, with $|M| = 700,000$ and $|U| = 5 \times 10^{12} - 700,000 = 4.9999993 \times 10^{12}$ record pairs.

A linkage algorithm compares record pairs and classifies them into $\tilde{M}$ (record pairs considered to be a match by the algorithm) and $\tilde{U}$ (record pairs considered to be a non-match). To keep this analysis simple, it is assumed here that the linkage algorithm does not classify record pairs as possible matches

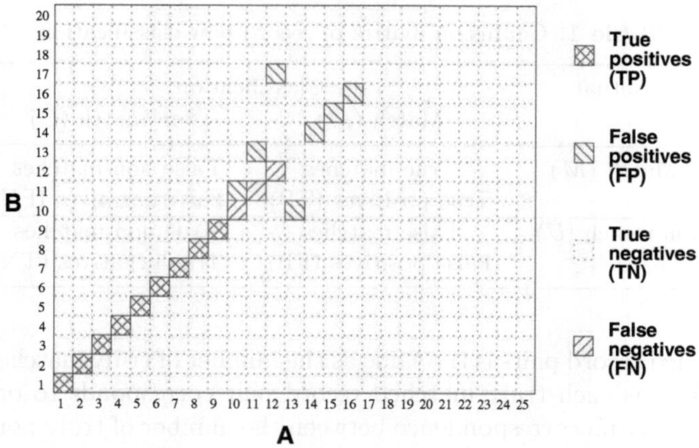

Fig. 4. Record pair comparison space with 25 records in data set **A** arbitrarily arranged on the horizontal axis and 20 records in data set **B** arbitrarily arranged on the vertical axis. The full rectangular area corresponds to all possible record pair comparisons. Assume that record pairs $(A1, B1)$, $(A2, B2)$ up to $(A12, B12)$ are true matches. The linkage algorithm has wrongly classified $(A10, B11)$, $(A11, B13)$, $(A12, B17)$, $(A13, B10)$, $(A14, B14)$, $(A15, B15)$, and $(A16, B16)$ as matches (false positives), but missed $(A10, B10)$, $(A11, B11)$, and $(A12, B12)$ (false negatives)

(as discussed in Sect. 2.3). Where a record pair comparison in $\tilde{M}$ is actually a match (a truly matched record pair), both of its records will refer to the same entity in $\mathbf{M}_e$. Records in un-matched record pairs, on the other hand, correspond to different entities in $\mathbf{A}_e$ and $\mathbf{B}_e$, with the possibility of both records of such a pair corresponding to different entities in $\mathbf{M}_e$. As each record relates to exactly one entity, and it is assumed there are no duplicates in the data sets, a record in data set **A** can only be matched to a maximum of one record in data set **B**, and vice versa.

Given the binary classification into $\tilde{M}$ and $\tilde{U}$, and knowing the true classification of a record pair comparison, an assignment to one of four categories can be made [19]. This is illustrated in the confusion matrix in Table 1. Truly matched record pairs from M that are classified as matches (into $\tilde{M}$) are called *true positives* (TP). Truly non-matched record pairs from U that are classified as non-matches (into $\tilde{U}$) are called *true negatives* (TN). Truly matched record pairs from M that are classified as non-matches (into $\tilde{U}$) are called *false negatives* (FN), and truly non-matched record pairs from U that are classified as matches (into $\tilde{M}$) are called *false positives* (FP). As illustrated, $M = TP + FN$, $U = TN + FP$, $\tilde{M} = TP + FP$, and $\tilde{U} = TN + FN$.

When assessing the quality of a linkage algorithm, the general interest is in how many truly matched entities and how many truly non-matched entities have been classified correctly as matches and non-matches, respectively. However, as the record pair comparisons occur in the comparison space, the results of measurements are also bound to this space. While the number of

Table 1. Confusion matrix of record pair classification

Actual	Classification	
	Match ($\tilde{M}$)	Non-match ($\tilde{U}$)
Match (M)	True matches True positives (TP)	False non-matches False negatives (FN)
Non-match (U)	False matches False positives (FP)	True non-matches True negatives (TN)

truly matched record pairs is the same as the number of truly matched entities, $|M| = |\mathbf{M}_e|$ (as each truly matched record pair corresponds to one entity), there is however no correspondence between the number of truly non-matched record pairs and non-matched entities. Each non-matched pair contains two records that correspond to two different entities, and each un-matched entity can be part of many record pairs. It is thus more difficult than it would first seem to decide on a proper value for the number of non-matched entities.

If no duplicates are assumed in the data sets $\mathbf{A}$ and $\mathbf{B}$, then the maximum number of truly matched entities is given by (4). From this follows the maximum number of record pairs a linkage algorithm should classify as matches is $|\tilde{M}| \leq |\mathbf{M}_e| \leq min(|\mathbf{A}_e|, |\mathbf{B}_e|)$. As the number of classified matches $|\tilde{M}| = (|TP| + |FP|)$, it follows that $(|TP| + |FP|) \leq |\mathbf{M}_e|$. With $|M| = (|TP| + |FN|)$, it also follows that both the numbers of FP and FN will be small compared to the number of TN, and they will not be influenced by the quadratic increase between the entity and the comparison space. The number of TN will dominate (as illustrated in Fig. 4), because in the comparison space the following equation holds:

$$|TN| = |\mathbf{A}| \times |\mathbf{B}| - |TP| - |FN| - |FP|.$$

Therefore (assuming no duplicates in the data sets) any quality measure used in data linkage or deduplication that uses the number of TN will give deceptive results, as will be shown in Sects. 4 and 6.

In reality, data sets are known to contain duplicate records, in which case a *one-to-one* assignment restriction [5] can be applied if there is only interest in the best match for each record. On the other hand, *one-to-many* and *many-to-many* linkages or deduplications are also possible. Examples include longitudinal studies of administrative health data where several records might correspond to a certain patient over time, or business mailing lists where several records can relate to the same customer (this happens when data sets have not been properly deduplicated). In such cases, a linkage algorithm may classify more record pairs as matches than there are entities (or records in a data set). The inequality $|\tilde{M}| \leq |\mathbf{M}_e|$ is not valid anymore in this context. The number of matches for a single record, however, will be small compared to the total number of record pair comparisons, as in practise often only a small number of best matches for each record are of interest. While a simple analysis as done above would not be possible, the issue of having a very large number of TN still holds in one-to-many and many-to-many linkage situations.

In the following section the different quality measures that have been used for assessing data linkage algorithms [4, 6, 11, 18, 32, 42, 45, 52] are presented. Various publications have used measures that include the number of TN, which leads to deceptive results.

4 Quality Measures

Given that data linkage and deduplication are classification problems, various quality measures are available to the data linkage researcher and practitioner [19]. With many recent approaches being based on supervised learning, no clerical review process (i.e. no possible matches) is often assumed and the problem becomes a binary classification, with record pairs being classified as either matches or non-matches, as shown in Table 1. One issue with many algorithms is the setting of a threshold value which determines the classifier performance. In order to select a threshold for a particular problem, comparative evaluations must be sourced or conducted. An obvious, much used, and strongly underpinned methodology for doing this involves the use of statistical techniques. In [41] this issue is described in terms of data mining and the use of machine learning algorithms. Several pitfalls are pointed out which can lead to misleading results, and a solution to overcome them is offered. This issue of classifier comparison is discussed in more detail first, before the different quality measures are presented in Sect. 4.2.

4.1 On Comparing Classifiers

When different classifiers are compared on the same problem class, care has to be taken to make sure that the achieved quality results are statistically valid and not just an artifact of the comparison procedure. One pitfall in particular, the *multiplicity effect* [41], means that, when comparing algorithms on the same data, because of the lack of independence of the data, the chances of erroneously achieving significance on a single test increases. So the level below which significance of the statistical p-value is accepted must be adjusted down (a conservative correction used in the statistics community known as the *Bonferroni* adjustment). In an example [41], if 154 variations (i.e. combinations of parameter settings) of a test algorithm are used, there is a 99.96 % chance that one of the variations will be incorrectly significant at the 0.05 level. Multiple independent researchers using the same data sets (e.g. community repositories like the UCI machine learning repository [36]) can suffer from this problem as well. Tuning – the process of adjusting an algorithm's parameters in an attempt to increase the quality of the classification – is subject to the same issue if the data for tuning and testing are the same.

A recommended solution [41] for the above is to use k-fold cross validation (k-times hold out one k'th of the data for testing), and to also hold out a portion of the training data for tuning. Also, since the lack of independence rules out the use of the t-test, it is suggested in [41] to use the binomial test or the analysis of variance (ANOVA) of distinct random samples.

While the aim of this chapter is not to compare the performance of classifiers for data linkage, it is nevertheless important for both researchers and practitioners working in this area to be aware of the issues discussed.

4.2 Quality Measures used for Data Linkage and Deduplication

In this section, different measures [19] that have been used for assessing the quality of data linkage algorithms [7] are presented. Using the simple example from Sect. 3, it is shown how the calculated results can be deceptive for some measures. The assumption is that a data linkage technique is used that classifies record pairs as matches and non-matches, and that the true matches and true non-matches are known, resulting in a confusion matrix of classified record pairs as shown in Table 1. The linkage classifier is assumed to have a single threshold parameter t (with no possible matches: $t_{lower} = t_{upper}$), which determines the cut-off between classifying record pairs as matches (with matching weight $R \geq t$) or as non-matches ($R < t$). Increasing the value of t can result in an increased number of TN and FN and in a reduction in the number of TP and FP, while lowering t can reduce the number of TN and FN and increase the number of TP and FP. Most of the quality measures presented here can be calculated for different values of such a threshold (often only the quality measure values for an optimal threshold are reported in empirical studies). Alternatively, quality measures can be visualised in a graph over a range of threshold values, as illustrated by the example in Sect. 6. The following list presents the commonly used quality measures, as well as a number of other popular measures used for binary classification problems (citations given refer to data linkage or deduplication publications that have used these measures in recent years).

- **Accuracy** [18, 25, 42, 45, 53] is measured as $acc = \frac{|TP|+|TN|}{|TP|+|FP|+|TN|+|FN|}$. It is a widely used measure and mainly suitable for balanced classification problems. As this measure includes the number of TN, it is affected by their large number when used in the comparison space (i.e. $|TN|$ will dominate the formula). The calculated accuracy values will be too high. For example, erroneously classifying all matches as non-matches will still result in a very high accuracy value. Accuracy is therefore not a good quality measure for data linkage and deduplication, and should not be used.
- **Precision** [4, 14, 32] is measured as $prec = \frac{|TP|}{|TP|+|FP|}$ and is also called *positive predictive value* [8]. It is the proportion of classified matches that are true matches, and is widely used in information retrieval [1] in combination with the *recall* measure for visualisation in *precision-recall graphs*.

- **Recall** [25, 32, 53] is measured as $rec = \frac{|TP|}{|TP|+|FN|}$ (true positive rate). Also known as *sensitivity* (commonly used in epidemiological studies [53]), it is the proportion of actual matches that have been classified correctly.
- **Precision-recall graph** [6, 11, 16, 33] is created by plotting precision values on the vertical and recall values on the horizontal axis. In information retrieval [1], the graph is normally plotted for eleven standardised recall values at $0.0, 0.1, \ldots, 1.0$, and is interpolated if a certain recall value is not available. In data linkage, a varying threshold can be used. There is a trade-off between precision and recall, in that high precision can normally only be achieved at the cost of lower recall values, and vice versa [29].
- **Precision-recall break-even point** is the value where precision becomes equal to recall, i.e. $\frac{|TP|}{|TP|+|FP|} = \frac{|TP|}{|TP|+|FN|}$. At this point, positive and negative misclassifications are made at the same rate, i.e. $|FP| = |FN|$. This measure is a single number.
- **F-measure** [16, 32] (or *F-score*) is the harmonic mean of precision and recall and is calculated as $f\text{-}meas = 2(\frac{prec \times rec}{prec+rec})$. It will have a high value only when both precision and recall have high values, and can be seen as a way to find the best compromise between precision and recall [1].
- **Maximum F-measure** is the maximum value of the F-measure over a varying threshold. This measure is a single number.
- **Specificity** [53] (which is the *true negative rate*) is calculated as $spec = \frac{|TN|}{|TN|+|FP|}$. This measure is used frequently in epidemiological studies [53]. As it includes the number of TN, it suffers from the same problem as accuracy, and should not be used for data linkage and deduplication.
- **False positive rate** [4, 27] is measured as $fpr = \frac{|FP|}{|TN|+|FP|}$. Note that $fpr = (1 - spec)$. As this measure includes the number of TN, it suffers from the same problem as accuracy and specificity, and should not be used.
- **ROC curve** (Receiver operating characteristic curve) is plotted as the true positive rate (which is the recall) on the vertical axis against the false positive rate on the horizontal axis for a varying threshold. While ROC curves are being promoted to be robust against skewed class distributions [19], the problem when using them in data linkage is the number of TN, which only appears in the false positive rate. This rate will be calculated too low, resulting in too optimistic ROC curves.
- **AUC** (Area under ROC curve) is a single numerical measure between 0.5 and 1 (as the ROC curve is always plotted in the unit square, with a random classifier having an AUC value of 0.5), with larger values indicating better classifier performance. The AUC has the statistical property of being equivalent to the statistical *Wilcoxon* test [19], and is also closely related to the *Gini* coefficient.

Example 3. Continuing the example from Sect. 3, assume that for a given threshold a linkage algorithm has classified $|\tilde{M}| = 900{,}000$ record pairs as matches and the rest ($|\tilde{U}| = 5 \times 10^{12} - 900{,}000$) as non-matches. Of these

Table 2. Quality measure results for Example 3

Measure	Entity space	Comparison space
Accuracy	94.340 %	99.999994 %
Precision	72.222 %	72.222000 %
Recall	92.857 %	92.857000 %
F-measure	81.250 %	81.250000 %
Specificity	94.565 %	99.999995 %
False positive rate	5.435 %	0.000005 %

900,000 classified matches 650,000 were true matches (TP), and 250,000 were false matches (FP). The number of falsely non-matched record pairs (FN) was 50,000, and the number of truly non-matched record pairs (TN) was $5 \times 10^{12} - 950,000$. When looking at the entity space, the number of non-matched entities is $4,600,000 - 250,000 = 4,350,000$. Table 2 shows the resulting quality measures for this example in both the comparison and the entity spaces. As can be seen, the results for accuracy, specificity and the false positive rate all show misleading results when based on record pairs (i.e. measured in the comparison space). This issue will be illustrated and discussed further in Sects. 6 and 7.

The authors of a recent publication [7] discuss the issue of evaluating data linkage and deduplication systems. They advocate the use of precision-recall graphs over the use of single number measures like accuracy or maximum F-measure, on the grounds that such single number measures assume that an optimal threshold value has been found. A single number can also hide the fact that one classifier might perform better for lower threshold values, while another has improved performance for higher thresholds.

In [8] a method is described which aims at estimating the positive predictive value (precision) under the assumption that there can only be one-to-one matches (i.e. a record can only be involved in one match). Using combinatorial probabilities the number of FP is estimated, allowing quantification of the linkage quality without training data or a *gold standard* data set.

While all quality measures presented so far assume a binary classification without clerical review, a new measure has been proposed recently [25] that aims to quantify the proportion of possible matches within a traditional probabilistic linkage system (which classifies record pairs into matches, non-matches and possible matches, as discussed in Sect. 2.3). The measure $pp = \frac{N_{P,M}+N_{P,U}}{|TP|+|FP|+|TN|+|FN|}$ is proposed, where $N_{P,M}$ is the number of true matches that have been classified as possible matches, and $N_{P,U}$ is the number of true non-matches that have been classified as possible matches. This measure quantifies the proportion of record pairs that are classified as possible matches, and therefore needing manual clerical review. Low *pp* values are desirable, as they correspond to less manual clerical review.

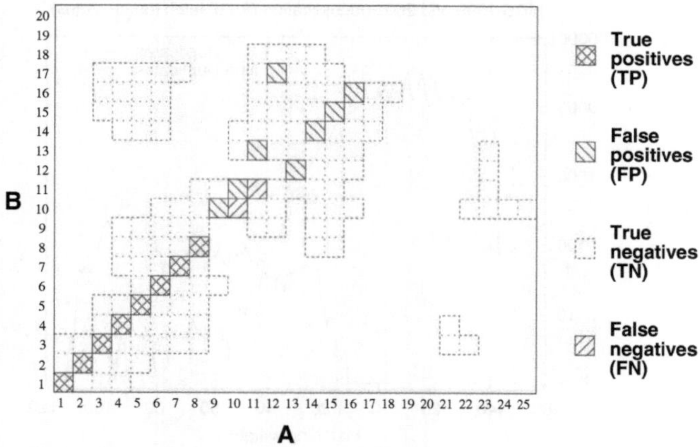

Fig. 5. Version of Fig. 4 in a blocked comparison space. The empty space are the record pairs which were removed by blocking. Besides many non-matches, the blocking process has also removed the truly matched record pairs $(A9, B9)$ and $(A12, B12)$, and then the linkage algorithm has wrongly classified the pairs $(A9, B10)$ and $(A12, B17)$ as matches

5 Blocking and Complexity Measures

An assumption in the analysis and discussion of quality measures given so far has been that all record pairs are compared. The number of comparisons in this situation equals $|\mathbf{A}| \times |\mathbf{B}|$, which is computationally feasible only for small data sets. In practise, blocking [3, 20, 49], sorting [27], filtering [24], clustering [32], or indexing [3, 10] techniques are used to reduce the number of record pair comparisons (as discussed in Sect. 2.1). Collectively known as *blocking*, these techniques aim at cheaply removing as many record pairs as possible from the set of non-matches U that are obvious non-matches, without removing any pairs from the set of matches M. Two complexity measures that quantify the efficiency and quality of such blocking methods have recently been proposed [18] (citations given refer to data linkage or deduplication publications that have used these measures):

- **Reduction ratio** [3, 18, 24] is measured as $rr = 1 - \frac{N_b}{|\mathbf{A}| \times |\mathbf{B}|}$, with $N_b \leq (|\mathbf{A}| \times |\mathbf{B}|)$ being the number of record pairs produced by a blocking algorithm (i.e. the number of record pairs not removed by blocking). The reduction ratio measures the relative reduction of the comparison space, without taking into account the quality of the reduction, i.e. how many record pairs from U and how many from M are removed by blocking.
- **Pairs completeness** [3, 18, 24] is measured as $pc = \frac{N_m}{|M|}$ with $N_m \leq |M|$ being the number of correctly classified truly matched record pairs in the blocked comparison space, and $|M|$ the total number of true matches as defined in Sect. 3. Pairs completeness can be seen as being analogous to recall.

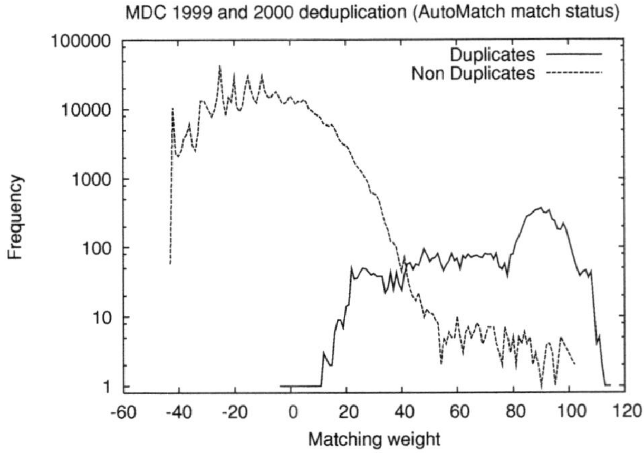

Fig. 6. The histogram plot of the matching weights for a real-world administrative health data set. This plot is based on record pair comparisons in a blocked comparison space. The lowest matching weight is -43 (disagreement on all comparisons), and the highest 115 (agreement on all comparisons). Note that the vertical axis with frequency counts is on a logarithmic scale

There is a trade-off between the reduction ratio and pairs completeness [3] (i.e. between number of removed record pairs and the number of missed true matches). As no blocking algorithm is perfect and will thus remove record pairs from M, the blocking process will affect both true matches and true non-matches. All quality measures presented in Sect. 4 will therefore be influenced by blocking.

6 Illustrative Example

In this section the previously discussed issues of quality and complexity measures are illustrated using a real-world administrative health data set, the *New South Wales Midwives Data Collection* (MDC) [9]. 175,211 records from the years 1999 and 2000 were extracted, containing names, addresses and dates of birth of mothers giving birth in these two years. This data set has previously been deduplicated (and manually clerically reviewed) using the commercial probabilistic linkage system *AutoMatch* [31]. According to this deduplication, the data set contains 166,555 unique mothers, with 158,081 having one, 8,295 having two, 176 having three, and 3 having four records in this data set. Of these last three mothers, two gave birth to twins twice in the two years 1999 and 2000, while one mother had a triplet and a single birth. The *AutoMatch* deduplication decision was used as the true match (or deduplication) status.

A deduplication was then performed using the *Febrl* (Freely extensible biomedical record linkage) [12] data linkage system. Fourteen attributes in the MDC were compared using various comparison functions (like exact and approximate string, and date of birth comparisons), and the resulting numerical values were summed into a matching weight R (as discussed in Sect. 2.3) ranging from -43 (disagreement on all fourteen comparisons) to 115 (agreement on all comparisons). As can be seen in Fig. 6, almost all true matches (record pairs classified as true duplicates) have positive matching weights, while the majority of non-matches have negative weights. There are, however, non-matches with rather large positive matching weights, which is due to the differences in calculating the weights between *AutoMatch* and *Febrl*.

The full comparison space for this data set with 175,211 records would result in $175,211 \times 175,210/2 = 15,349,359,655$ record pairs, which is infeasible to process even with today's powerful computers. Standard blocking was used to reduce the number of comparisons, resulting in 759,773 record pair comparisons (corresponding to each record being compared to around 4 other records). The reduction ratio in this case was therefore

$$rr = 1.0 - \frac{759,773}{15,349,359,655} = 1.0 - 4.9499 \times 10^{-5} = 0.99995.$$

This corresponds to only around 0.005 % of all record pairs in the full comparison space. The total number of truly classified matches (duplicates) was 8,841 (for all the duplicates as described above), with 8,808 of the 759,773 record pairs in the blocked comparison space corresponding to true duplicates. The resulting pairs completeness value therefore was

$$pc = \frac{8,808}{8,841} = 0.99626,$$

which corresponds to more than 99.6 % of all the true duplicates being included in the blocked comparison space and classified as duplicates by both *AutoMatch* and *Febrl*.

The quality measures discussed in Sect. 4 applied to this real-world deduplication are shown in Fig. 7 for a varying threshold $-43 \leq t \leq 115$. The aim of this figure is to illustrate how the different measures look for a deduplication example taken from the real world. The measurements were done in the blocked comparison space as described above. The full comparison space (15,349,359,655 record pairs) was simulated by assuming that blocking removed mainly record pairs with negative comparison weights (normally distributed between -43 and -10). This resulted in different numbers of TN between the blocked and the (simulated) full comparison spaces.

As can be seen, the precision-recall graph is not affected by the blocking process, and the F-measure graph differs only slightly. All other measures, however, resulted in graphs of different shape. The large number of TN compared to the number of TP resulted in the specificity measure being very similar to the accuracy measure. Interestingly, the ROC curve, being promoted as robust with regard to skewed classification problems [19], resulted in the least illustrative graph, especially for the full comparison space, making it not very useful for data linkage and deduplication.

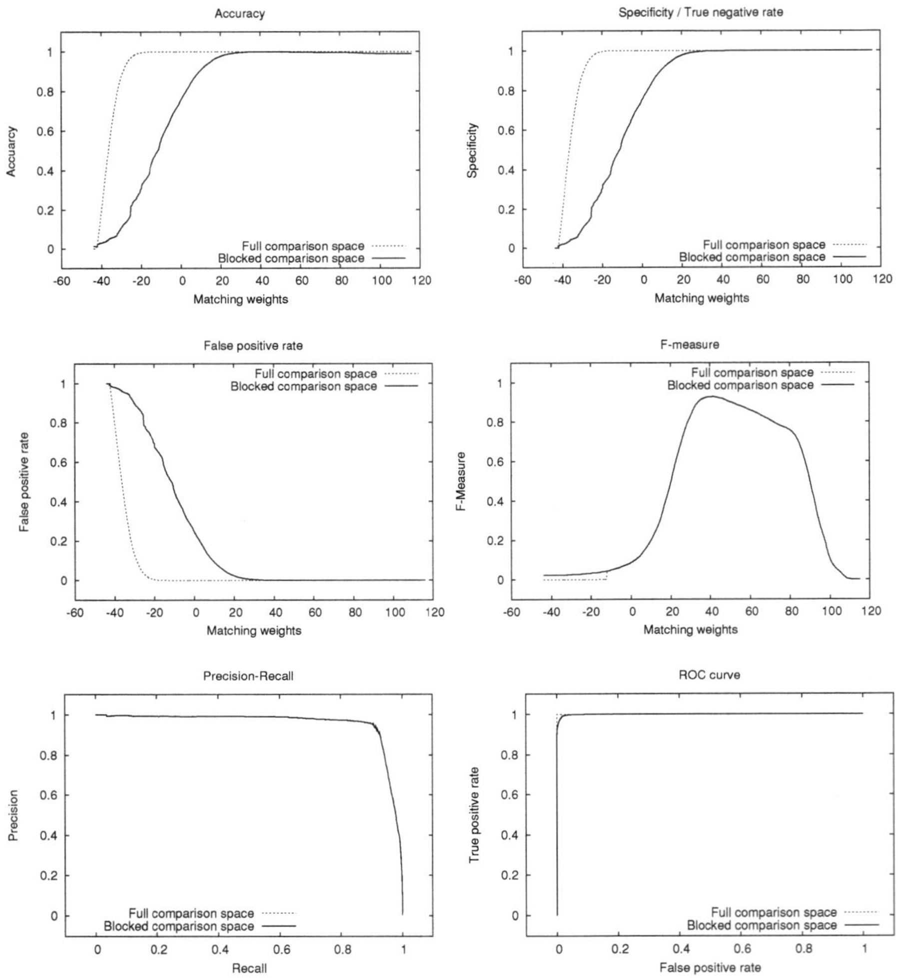

Fig. 7. Quality measurements of a real-world administrative health data set. The full comparison space (15,349,359,655 record pairs) was simulated by assuming that the record pairs removed by blocking were normally distributed with matching weights between −43 and −10. Note that the precision-recall graph does not change at all, and the change in the F-measure graph is only slight. Accuracy and specificity are almost the same, as both are dominated by the large number of true negatives. The ROC curve is the least illustrative graph, which is again due to the large number of true negatives

7 Discussion and Recommendations

Primarily, the measurement of quality in data linkage and deduplication involves either absolute or relative results (for example, "either technique X had an accuracy of 93 %", or "technique X performed better than technique

Y on all data examined"). In order for a practitioner or researcher to make informed choices, the results of experiments must be comparable, or the techniques must be repeatable so comparisons between techniques can be made.

It is known, however, that the quality of techniques vary depending on the nature of the data sets the techniques are applied to [6, 41]. Whether producing absolute or comparable results, it is necessary for the experiments to be conducted using the same data. Therefore, results should be produced from data sets which are available to researchers and practitioners in the field. However, this does not preclude research on private data sets. The applicability of a technique to a type of data set may be of interest, but the results produced are not beneficial for evaluating relative quality of techniques.

Of course, for researchers to compare techniques against earlier ones, either absolute results must be available, or the earlier techniques must be repeatable for comparison. Ultimately, and ideally, a suite of data sets should be collected and made publicly available for this process, and they should encapsulate as much variation in types of data as feasible.

Recommendations for the various steps of a data linkage process are given in the following sections. Their aim is to provide both the researcher and practitioner with guidelines on how to perform empirical studies on different linkage algorithms or production linkage projects, as well as on how to properly assess and describe the outcome of such linkages or deduplications.

7.1 Record Pair Classification

Due to the problem of the number of true negatives in any comparison, quality measures which use that number (for example accuracy, specificity, false positive rate, and thus ROC curve) should not be used. The variation in the quality of a technique against particular types of data means that results should be reported for particular data sets. Also, given that the nature of some data sets may not be known in advance, the average quality across all data sets used in a certain study should also be reported. When comparing techniques, precision-versus-recall or F-measure graphs provide an additional dimension to the results. For example, if a small number of highly accurate links is required, the technique with higher precision for low recall would be chosen [7].

7.2 Blocking

The aim of blocking is to cheaply remove obvious non-matches before the more detailed, expensive record pair comparisons are made. Working perfectly, blocking would only remove record pairs that are true non-matches, thus affecting the number of true negatives, and possibly the number of false positives. To the extent that, in reality, blocking also removes record pairs from the set of true matches (resulting in a pairs completeness $pc < 1$), it will also affect the number of true positives and false negatives. Blocking can thus be seen to be a *confounding* factor in quality measurement – the types of blocking procedures and the parameters chosen will potentially affect the results obtained for a given linkage procedure.

If computationally feasible, for example in an empirical study using small data sets, it is strongly recommended that all quality measurement results be obtained without the use of blocking. It is recognised that it may not be possible to do this with larger data sets. A compromise, then, would be to publish the blocking measures, reduction ratio and pairs completeness, and to make the *blocked* data set available for analysis and comparison by other researchers. At the very least, the blocking procedure and parameters should be specified in a form that can enable other researchers to repeat it.[1]

7.3 Complexity

The overall complexity of a linkage technique is fundamentally important due to the potential size of the data sets it could be applied to: when sizes are in the millions or even billions, techniques which are $O(n^2)$ become problematic and those of higher complexity cannot even be contemplated. While blocking can provide improvements, complexity is still important. For example, if linkage is attempted on a real-time data stream, a complex algorithm may require faster hardware, more optimisation, or replacement. As data linkage, being an important step in the data mining process, is a field rooted in practice, the practicality of a technique's implementation and use on very large data sets should be indicated. Thus, at least, the reporting of the complexity of a technique in $O()$ terms should always be made. The reporting of other usage, such as disk space and memory size, could also be beneficial.

8 Conclusions

Data linkage and deduplication are important steps in the pre-processing phase of many data mining projects, and also important for improving data quality before data is loaded into data warehouses. An overview of data linkage techniques has been presented in this chapter, and the issues involved in measuring both the quality and complexity of linkage algorithms have been discussed. It is recommended that the quality be measured using the precision-recall or F-measure graphs (over a varying threshold) rather than single numerical values, and that quality measures that include the number of true negative matches should not be used due to their large number in the space of record pair comparisons. When publishing empirical studies researchers should aim to use non-blocked data sets if possible, or otherwise at least report measures that quantify the effects of the blocking process.

[1] Note that the example given in Sect. 6 doesn't follow the recommendations presented here. The aim of the section was to illustrate the presented issues, not the actual results of the deduplication.

Acknowledgements

This work is supported by an Australian Research Council (ARC) Linkage Grant LP0453463 and partially funded by the NSW Department of Health. The authors would like to thank Markus Hegland for insightful discussions.

References

1. Baeza-Yates RA, Ribeiro-Neto B. *Modern information retrieval.* Addison-Wesley Longman Publishing Co., Boston, 1999.
2. Bass J. Statistical linkage keys: How effective are they? In *Symposium on Health Data Linkage*, Sydney, 2002. Available online at: `http://www.publichealth.gov.au/symposium.html`.
3. Baxter R, Christen P, Churches T. A comparison of fast blocking methods for record linkage. In *Proceedings of ACM SIGKDD workshop on Data Cleaning, Record Linkage and Object Consolidation*, pages 25–27, Washington DC, 2003.
4. Bertolazzi P, De Santis L, Scannapieco M. Automated record matching in cooperative information systems. In *Proceedings of the international workshop on data quality in cooperative information systems*, Siena, Italy, 2003.
5. Bertsekas DP. Auction algorithms for network flow problems: A tutorial introduction. *Computational Optimization and Applications*, 1:7–66, 1992.
6. Bilenko M, Mooney RJ. Adaptive duplicate detection using learnable string similarity measures. In *Proceedings of ACM SIGKDD*, pages 39–48, Washington DC, 2003.
7. Bilenko M, Mooney RJ. On evaluation and training-set construction for duplicate detection. In *Proceedings of ACM SIGKDD workshop on Data Cleaning, Record Linkage and Object Consolidation*, pages 7–12, Washington DC, 2003.
8. Blakely T, Salmond C. Probabilistic record linkage and a method to calculate the positive predictive value. *International Journal of Epidemiology*, 31:6:1246–1252, 2002.
9. Centre for Epidemiology and Research, NSW Department of Health. New South Wales mothers and babies 2001. *NSW Public Health Bull*, 13:S-4, 2001.
10. Chaudhuri S, Ganjam K, Ganti V, Motwani R. Robust and efficient fuzzy match for online data cleaning. In *Proceedings of ACM SIGMOD*, pages 313–324, San Diego, 2003.
11. Chaudhuri S, Ganti V, Motwani R. Robust identification of fuzzy duplicates. In *Proceedings of the 21st international conference on data engineering (ICDE'05)*, pages 865–876, Tokyo, 2005.
12. Christen P, Churches T, Hegland M. Febrl – a parallel open source data linkage system. In *Proceedings of the 8th PAKDD, Springer LNAI 3056*, pages 638–647, Sydney, 2004.
13. Churches T, Christen P, Lim K, Zhu JX. Preparation of name and address data for record linkage using hidden markov models. *BioMed Central Medical Informatics and Decision Making*, 2(9), 2002. Available online at: `http://www.biomedcentral.com/1472-6947/2/9/`.
14. Cohen WW. Integration of heterogeneous databases without common domains using queries based on textual similarity. In *Proceedings of ACM SIGMOD*, pages 201–212, Seattle, 1998.

15. Cohen WW, Ravikumar P, Fienberg SE. A comparison of string distance metrics for name-matching tasks. In *Proceedings of IJCAI-03 workshop on information integration on the Web (IIWeb-03)*, pages 73–78, Acapulco, 2003.
16. Cohen WW, Richman J. Learning to match and cluster large high-dimensional data sets for data integration. In *Proceedings of ACM SIGKDD*, pages 475–480, Edmonton, 2002.
17. Cooper WS, Maron ME. Foundations of probabilistic and utility-theoretic indexing. *Journal of the ACM*, 25(1):67–80, 1978.
18. Elfeky MG, Verykios VS, Elmagarmid AK. TAILOR: A record linkage toolbox. In *Proceedings of ICDE*, pages 17–28, San Jose, 2002.
19. Fawcett T. ROC Graphs: Notes and practical considerations for researchers. Technical Report HPL-2003-4, HP Laboratories, Palo Alto, 2004.
20. Fellegi I, Sunter A. A theory for record linkage. *Journal of the American Statistical Society*, 64(328):1183–1210, 1969.
21. Galhardas H, Florescu D, Shasha D, Simon E. An extensible framework for data cleaning. In *Proceedings of ICDE*, page 312, 2000.
22. Gill L. Methods for automatic record matching and linking and their use in national statistics. Technical Report National Statistics Methodology Series, no 25, National Statistics, London, 2001.
23. Gomatam S, Carter R, Ariet M, Mitchell G. An empirical comparison of record linkage procedures. *Statistics in Medicine*, 21(10):1485–1496, 2002.
24. Gu L, Baxter R. Adaptive filtering for efficient record linkage. In *SIAM international conference on data mining*, Orlando, 2004.
25. Gu L, Baxter R. Decision models for record linkage. In *Proceedings of the 3rd Australasian data mining conference*, pages 241–254, Cairns, 2004.
26. Hernandez MA, Stolfo SJ. The merge/purge problem for large databases. In *Proceedings of ACM SIGMOD*, pages 127–138, San Jose, 1995.
27. Hernandez MA, Stolfo SJ. Real-world data is dirty: Data cleansing and the merge/purge problem. *Data Mining and Knowledge Discovery*, 2(1):9–37, 1998.
28. Kelman CW, Bass AJ, Holman CD. Research use of linked health data – a best practice protocol. *Aust NZ Journal of Public Health*, 26:251–255, 2002.
29. Lee ML, Ling TW, Low WL. IntelliClean: a knowledge-based intelligent data cleaner. In *Proceedings of ACM SIGKDD*, pages 290–294, Boston, 2000.
30. Maletic JI, Marcus A. Data cleansing: beyond integrity analysis. In *Proceedings of the Conference on Information Quality (IQ2000)*, pages 200–209, Boston, 2000.
31. MatchWare Technologies. *AutoStan and AutoMatch, User's Manuals*. Kennebunk, Maine, 1998.
32. McCallum A, Nigam K, Ungar LH. Efficient clustering of high-dimensional data sets with application to reference matching. In *Proceedings of ACM SIGKDD*, pages 169–178, Boston, 2000.
33. Monge A, Elkan C. The field-matching problem: Algorithm and applications. In *Proceedings of ACM SIGKDD*, pages 267–270, Portland, 1996.
34. Nahm UY, Bilenko M, Mooney RJ. Two approaches to handling noisy variation in text mining. In *Proceedings of the ICML-2002 workshop on text learning (TextML'2002)*, pages 18–27, Sydney, 2002.
35. Newcombe HB, Kennedy JM. Record linkage: making maximum use of the discriminating power of identifying information. *Communications of the ACM*, 5(11):563–566, 1962.

36. Newman DJ, Hettich S, Blake CL, Merz CJ. UCI repository of machine learning databases, 1998.
URL: http://www.ics.uci.edu/~mlearn/MLRepository.html.
37. Porter E, Winkler WE. Approximate string comparison and its effect on an advanced record linkage system. Technical Report RR97/02, US Bureau of the Census, 1997.
38. Pyle D. *Data preparation for data mining.* Morgan Kaufmann Publishers, San Francisco, 1999.
39. Rahm E, Do HH. Data cleaning: problems and current approaches. *IEEE Data Engineering Bulletin*, 23(4):3–13, 2000.
40. Ravikumar P, Cohen WW. A hierarchical graphical model for record linkage. In *Proceedings of the 20th conference on uncertainty in artificial intelligence*, pages 454–461, Banff, Canada, 2004.
41. Salzberg S. On comparing classifiers: pitfalls to avoid and a recommended approach. *Data Mining and Knowledge Discovery*, 1(3):317–328, 1997.
42. Sarawagi S, Bhamidipaty A. Interactive deduplication using active learning. In *Proceedings of ACM SIGKDD*, pages 269–278, Edmonton, 2002.
43. Shearer C. The CRISP-DM model: The new blueprint for data mining. *Journal of Data Warehousing*, 5(4):13–22, 2000.
44. Smith ME, Newcombe HB. Accuracies of computer versus manual linkages of routine health records. *Methods of Information in Medicine*, 18(2):89–97, 1979.
45. Tejada S, Knoblock CA, Minton S. Learning domain-independent string transformation weights for high accuracy object identification. In *Proceedings of ACM SIGKDD*, pages 350–359, Edmonton, 2002.
46. Winkler WE. Using the EM algorithm for weight computation in the Fellegi-Sunter model of record linkage. Technical Report RR00/05, US Bureau of the Census, 2000.
47. Winkler WE. Methods for record linkage and Bayesian networks. Technical Report RR2002/05, US Bureau of the Census, 2002.
48. Winkler WE. Overview of record linkage and current research directions. Technical Report RR2006/02, US Bureau of the Census, 2006.
49. Winkler WE, Thibaudeau Y. An application of the Fellegi-Sunter model of record linkage to the 1990 U.S. decennial census. Technical Report RR91/09, US Bureau of the Census, 1991.
50. Yancey WE. BigMatch: a program for extracting probable matches from a large file for record linkage. Technical Report RRC2002/01, US Bureau of the Census, 2002.
51. Yancey WE. An adaptive string comparator for record linkage. Technical Report RR2004/02, US Bureau of the Census, 2004.
52. Zhu JJ, Ungar LH. String edit analysis for merging databases. In *KDD workshop on text mining, held at ACM SIGKDD*, Boston, 2000.
53. Zingmond DS, Ye Z, Ettner SL, Liu H. Linking hospital discharge and death records – accuracy and sources of bias. *Journal of Clinical Epidemiology*, 57:21–29, 2004.

Statistical Methodologies for Mining Potentially Interesting Contrast Sets

Robert J. Hilderman, Terry Peckham

Department of Computer Science, University of Regina, Regina, Saskatchewan, Canada S4S 0A2, {hilder, peckham}@cs.uregina.ca

Summary. One of the fundamental tasks of data analysis in many disciplines is to identify the significant differences between classes or groups. Contrast sets have previously been proposed as a useful tool for describing these differences. A contrast set is a set of association rules for which the antecedents describe distinct groups, a common consequent is shared by all the rules, and support for the rules is significantly different between groups. The intuition is that comparing the support between groups may provide some insight into the fundamental differences between the groups. In this chapter, we compare two contrast set mining methodologies that rely on different statistical philosophies: the well-known STUCCO approach and CIGAR, our proposed alternative approach. Following a brief introduction to general issues and problems related to statistical hypothesis testing in data mining, we survey and discuss the statistical measures underlying the two methods using an informal tutorial approach. Experimental results show that both methodologies are statistically sound, representing valid alternative solutions to the problem of identifying potentially interesting contrast sets.

Key words: Statistics, association rules, contrast sets.

1 Introduction

One of the fundamental tasks of data analysis in many disciplines is to identify the significant differences between classes or groups. For example, an epidemiological study of self-reported levels of stress experienced by health care providers could be used to characterize the differences between those who work in rural and urban communities. The differences could be conveniently described using pairs of contrasting conditional probabilities, such as `P(Stress=high ∧ Income=low | Location=rural) = 32%` and `P(Stress=high ∧ Income=low | Location= urban) = 25%`. The conditional probabilities shown here are equivalent to rules of the form `Location= rural ⇒ Stress=high ∧ Income=low (32%)` and `Location = urban ⇒`

Robert J. Hilderman and Terry Peckham: *Statistical Methodologies for Mining Potentially Interesting Contrast Sets*, Studies in Computational Intelligence (SCI) **43**, 153–177 (2007)
www.springerlink.com

`Stress=high ∧ Income=low (25%)`, known as *association rules* [1]. A set of association rules, where the *antecedents* describe distinct groups (i.e., `Location=rural` and `Location=urban`), a common *consequent* is shared by all groups (i.e., `Stress=high ∧ Income=low`), and the percentages represent the number of examples in each group for which the association rule is true (called *support*), is called a *contrast set* [3]. The intuition is that comparing the support for the rules between groups may provide some insight into explaining the fundamental differences between the groups. (At first glance, the term support, as used and intended in this paragraph and throughout the remainder of this chapter, seems to actually describe confidence. However, the reader should note the following peculiarity. That is, when generating contrast sets, the groups are treated as if they are contained in separate and unique datasets. The support calculation for each group is then based upon only those rules that are found within each group. When the groups are viewed in this way, support is the appropriate term. However, we acknowledge that if the groups are treated as if they are contained in one dataset, the traditional term used in association rule mining is confidence. For historical reasons and to be consistent with previous work, we use the term support.)

Contrast set mining is an association rule-based discovery technique that was originally introduced as emerging pattern mining [8]. The problem of emerging pattern mining focuses on datasets of a temporal nature and is essentially a special case of the more general contrast set mining problem. There has since been extensive analysis of, and refinements to, emerging pattern mining techniques. Some of the most recent work is reported in [10], where an excellent bibliography on emerging pattern mining can be found.

More generally applicable work in contrast set mining can be found in [3], [4], and [14]. In [3], contrast set mining is studied within the context of an association rule-based technique called STUCCO (Searching and Testing for Understandable Consistent COntrasts). For an extensive description and evaluation of STUCCO, see [4]. STUCCO is an effective technique for generating contrast sets that models the search space as a set-enumeration tree [4] to improve search efficiency. The fundamental characteristic of this approach is that it utilizes a canonical ordering of nodes in the search space, such that any node that cannot be pruned is visited only once. STUCCO also utilizes χ^2 testing of two-dimensional contingency tables to determine whether differences between rules in a contrast set are statistically significant. In addition, to correct for multiple comparisons during significance testing, a variant of the widely applied Bonferroni method [11] is used to control Type I error.

Group differences are also studied in [14] within the context of an association rule-like technique called Magnum Opus, a commercial exploratory rule discovery tool. There, a different approach to mining contrast sets is used that restricts the consequent of the rules generated to only those attribute values that define group membership. To determine whether differences between rules are statistically significant, a binomial sign test is used, but no mechanisms are utilized to correct for multiple comparisons. However, the statistical reasoning used by Magnum Opus is flawed, as it actually performs a within-groups comparison rather than a between-groups comparison [12]. Consequently it only finds a subset of the contrast sets generated by STUCCO, so we do not discuss it further in this work.

In this chapter, we discuss STUCCO in detail, and we introduce CIGAR (ContrastIng, Grouped Association Rules), a contrast set mining technique that relies on an alternative statistical philosophy to the discovery of statistically significant contrast sets, yet still adheres to sound and accepted practices. CIGAR not only considers whether the difference in support between groups is significant, it also utilizes a fine-grained approach that specifically identifies which pairs of groups are significantly different and considers whether the attributes in a contrast set are correlated. In contrast to the aggressive Type I error control approach used by STUCCO, CIGAR seeks to control Type II error through increasing the significance level for the significance tests and by not correcting for multiple comparisons. The primary reason for not correcting for multiple comparisons is that we want to avoid over-pruning (i.e., pruning contrast sets containing significant differences). A correlational pruning technique that compares r-values between parents and children in the search space is also utilized to reduce the size of the search space.

Our objective in this chapter is to focus on relevant statistical methods for identifying potentially interesting differences between groups and statistical methods for pruning the search space. Thus, our focus is not on algorithms, per se. We introduce and describe, using a tutorial approach, the statistical techniques used by STUCCO and CIGAR for determining the statistical validity of contrast sets. We also take advantage of statistical properties of association rules and contrast sets to prune the search space whenever it is prudent to do so. However, while we cannot say for certain that the contrast sets generated are the most "interesting" as might be determined by a domain expert, a fundamental assumption is that they likely provide the domain expert with a reasonable starting point for further analysis and evaluation of discovered knowledge.

The remainder of the chapter is organized as follows. In the next section, we briefly describe the contrast set mining problem and show how it can be viewed as an extension to the common association rule mining problem. In Section 3, we discuss general issues and problems related to statistical hypothesis testing in contrast set mining. In Section 4, we survey and discuss the statistical techniques utilized by STUCCO and CIGAR, and provide some illustrative examples. In Section 5, we present experimental results that demonstrate the differences between the results generated by STUCCO and CIGAR. In Section 6, we conclude with a summary of our work.

2 The Contrast Set Mining Framework

Mining contrast sets is an exploratory rule discovery technique that seeks to find qualitative rules describing relationships between groups. In this section, we describe the classic exploratory rule discovery problem called association rule mining, and then describe the contrast set mining problem, a generalization of the association rule mining problem from binomial or transactional data types to multinomial, grouped categorical data.

2.1 The Association Rule Mining Problem

Mining contrast sets is based upon the problem of mining association rules [1]. The problem of association rule mining is typically studied within the context of discovering buying patterns from retail sales transactions (i.e., market basket data), and is formally defined as follows. Let $A = \{\mathtt{A}_1, \mathtt{A}_2, \ldots, \mathtt{A}_\mathtt{m}\}$ be a set of attributes called *items*. Let D be a set of *transactions*, where each transaction T is described by a vector of m attribute-value pairs $\mathtt{A}_1 = \mathtt{V}_1, \mathtt{A}_2 = \mathtt{V}_2, \ldots, \mathtt{A}_\mathtt{m} = \mathtt{V}_\mathtt{m}$, and each $\mathtt{V}_\mathtt{j}$ is selected from the set $\{1, 0\}$ (i.e., $\mathtt{V}_\mathtt{j} = 1$ ($\mathtt{V}_\mathtt{j} = 0$) indicates that item $\mathtt{A}_\mathtt{j}$ was purchased (not purchased)). The collection of purchased items contained in transaction T is an *itemset*. Transaction T *contains* $\mathtt{X}$, a set of purchased items, if $\mathtt{X} \subseteq T$. An *association rule* is an implication of the form $\mathtt{X} \Rightarrow \mathtt{Y}$, where $\mathtt{X} \subset A$, $\mathtt{Y} \subset A$, and $\mathtt{X} \cap \mathtt{Y} = \emptyset$. *Confidence* in $\mathtt{X} \Rightarrow \mathtt{Y}$ is the percentage of transactions in D containing $\mathtt{X}$ that also contain $\mathtt{Y}$. *Support* for $\mathtt{X} \Rightarrow \mathtt{Y}$ is the percentage of transactions in D containing $\mathtt{X} \cup \mathtt{Y}$. It is important to observe that implication in association rules does not correspond to the logical notion of implication. Instead, the confidence in an association rule $\mathtt{X} \Rightarrow \mathtt{Y}$ actually measures the conditional probability of $\mathtt{Y}$ given $\mathtt{X}$, denoted $\mathtt{P(Y|X)}$. The goal of association rule mining is to identify all itemsets whose support and confidence exceed some user-defined thresholds.

2.2 The Contrast Set Mining Problem

In the problem of contrast set mining [3], [4] transaction set D is generalized to a set of multinomial *examples*, where each example E is described by a vector of m attribute-value pairs $\mathtt{A}_1 = \mathtt{V}_1, \mathtt{A}_2 = \mathtt{V}_2, \ldots, \mathtt{A}_\mathtt{m} = \mathtt{V}_\mathtt{m}$, and each $\mathtt{V}_\mathtt{i}$ is selected from the finite set of discrete domain values in the set $\{\mathtt{V}_{\mathtt{i}_1}, \mathtt{V}_{\mathtt{i}_2}, \ldots, \mathtt{V}_{\mathtt{i}_\mathtt{n}}\}$ associated with $\mathtt{A}_\mathtt{i}$. One attribute $\mathtt{A}_\mathtt{j}$ in D is a distinguished attribute whose value $\mathtt{V}_{\mathtt{j}_\mathtt{k}}$ in example E is used to assign E into one of $\mathtt{n}$ mutually exclusive groups $G_1, G_2, \ldots, G_\mathtt{n}$. A *contrast set* is a set of association rules consisting of conjunctions of attribute-value pairs defined on $G_1, G_2, \ldots, G_\mathtt{n}$, such that no $\mathtt{A}_\mathtt{i}$ occurs more than once. From these conjunctions, we obtain rules of the form $\mathtt{A}_\mathtt{j} = \mathtt{V}_{\mathtt{j}_\mathtt{k}} \Rightarrow \mathtt{X}$, where the antecedent contains the distinguished attribute and determines group membership, and the consequent is a common conjunction of attribute-value pairs shared by all groups. *Support* for association rule

$A_j = V_{j_k} \Rightarrow X$ is the percentage of examples in G_j containing X. The goal of contrast set mining is to identify all contrast sets for which the support is significantly different between groups.

2.3 The Search Space

The problem of mining contrast sets can be modeled as a tree search problem, where canonical ordering of the nodes can be utilized to improve the efficiency of traversing the search space [4], [4]. That is, either a node is visited only once, or it is pruned and not visited at all. For example, given the attributes A_1, A_2, and A_3, whose domain values are $\{V_{11}, V_{12}\}$, $\{V_{21}, V_{22}\}$, and $\{V_{31}, V_{32}\}$, respectively, and where A_1 is the distinguished attribute, the search space of all possible rules that can be generated is shown in Figure 1. The root at depth zero corresponds to the empty rule. The search space is built in two phases, as follows. In the first phase, at depth one, all possible single attribute-value pairs from the example set are enumerated. Thus, given the three attributes A_1, A_2, and A_3, each having two possible values, six rules are generated at depth one, as shown. The second phase generates all other rules in the search space.

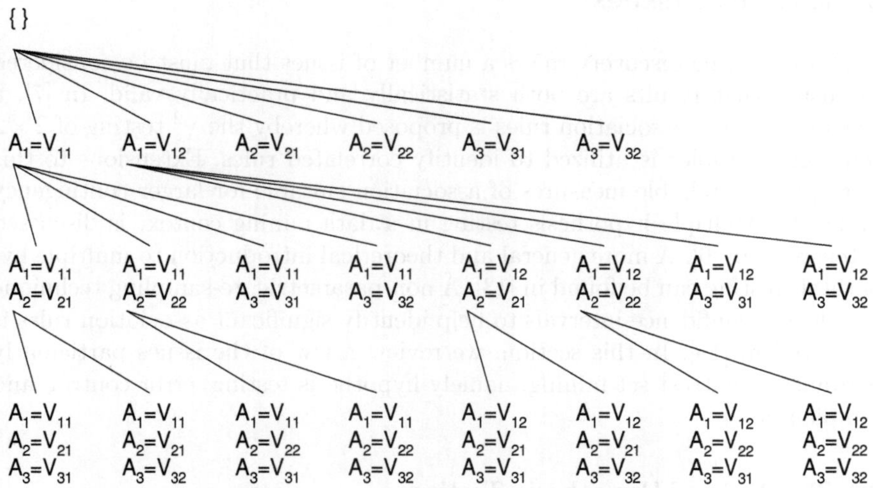

Fig. 1. The search space for attributes A_1, A_2, and A_3

In phase two, the rules at depth $k > 0$ are used to generate the rules at depth $k + 1$. For example, in Figure 1, the rules at depth two are generated by forming all possible combinations of conjunctions from the attribute-value pairs at depth one associated with A_2 and A_3, in turn with those of A_1, the distinguished attribute. Essentially, children are formed by appending attribute-value pairs that follow all existing attribute-value-pairs in the initial order.

For example, taking $A_2 = V_{21}$, $A_2 = V_{22}$, $A_3 = V_{31}$, and $A_3 = V_{32}$, in turn with $A_1 = V_{11}$, we generate the first four rules at depth two of the search space. We do not form conjunctions of attribute-value pairs at depth one associated with A_1, in turn with those of A_2 and A_3 because this results in rules that have already been generated. For example, the rule $A_1 = V_{11} \wedge A_2 = V_{21}$ is equivalent to $A_2 = V_{21} \wedge A_1 = V_{11}$. We also do not form conjunctions of attribute-value pairs at depth one associated with A_3, in turn with those of A_2 because these conjunctions do not contain the distinguished attribute.

Attribute-value pairs at a node related to attribute-value pairs at other nodes form a contrast set. For example, in Figure 1, the rules $A_1 = V_{11} \Rightarrow A_2 = V_{21}$ and $A_1 = V_{12} \Rightarrow A_2 = V_{21}$ (i.e., the first and fifth pairs at depth two) are related because the antecedents describe group membership and the consequents are common to both rules. Thus, four contrast sets can be formed from the rules at level two. In this example, the contrast set consists of just two rules because only two groups are defined for A_1, the distinguished attribute. However, the size of the contrast set depends on the number of unique values defined for the distinguished attribute, so the number of rules in a contrast set can be arbitrarily large.

3 Statistical Issues

Exploratory rule discovery raises a number of issues that must be considered to ensure that results are both statistically and practically valid. In [7], a generalization of association rules is proposed whereby the χ^2 testing of 2×2 contingency tables is utilized to identify correlated rules. Extensions to this work propose reliable measures of association suitable for larger contingency tables [2]. Multiple hypothesis testing in a data mining context is discussed extensively in [6]. A more general and theoretical introduction to multiple hypothesis testing can be found in [13]. A non-parametric, re-sampling technique that learns confidence intervals to help identify significant association rules is presented in [15]. In this section, we review a few of the issues particularly relevant to contrast set mining, namely hypothesis testing, error control, and sample size.

3.1 The Role of Hypothesis Testing

To determine whether support for rules is significantly different between groups, two-dimensional contingency table analysis is used. For example, if we assume from Figure 1 that A_1 corresponds to `Location`, V_{11} to `urban`, V_{12} to `rural`, A_2 to `Stress`, and V_{21} to `high`, then we can construct the two-dimensional contingency table in Table 1 showing the incidence of stress for urban and rural health care providers. Each cell in this contingency table contains the number of examples for which the corresponding rule is true. For example, 194 examples contain the rule `Location=urban` $\Rightarrow$ `Stress=high`.

Table 1. An example contingency table containing dependent attributes

	`Location=urban`	`Location=rural`	$\sum$ *Row*
`Stress=high`	194	355	549
`¬ (Stress=high)`	360	511	871
$\sum$ *Column*	554	866	1420

The χ^2 statistic is used to compare the observed number of examples (i.e., the *observed frequencies*) with the expected number of examples (i.e., the *expected frequencies*). More formally, the χ^2 statistic tests the null hypothesis that row and column totals are not related (i.e., are *independent*), and is given by

$$\chi^2 = \sum_{i=1}^{2}\sum_{j=1}^{2}\frac{(O_{ij} - E_{ij})^2}{E_{ij}},$$

where O_{ij} is the observed frequency at the intersection of row i and column j, $E_{ij} = (O_{i.} \times O_{.j})/O_{..}$, $O_{i.}$ is the total in row i, $O_{.j}$ is the total in column j, and $O_{..}$ is the sum of the row and column totals (i.e., the total number of examples). The number of *degrees of freedom* for χ^2 is given by $df = (r-1) \times (c-1)$, where r and c are the number of rows and columns, respectively. A sufficiently large χ^2 value will cause rejection of the null hypothesis that row and column totals are not related. For example, $\chi^2 = 5.08$ and $df = 1$ for the rules in Table 1. At the 95% significance level (i.e., $\alpha = 0.05$), $\chi^2_\alpha = 3.84$. Since $5.08 > 3.84$, we reject the null hypothesis. That is, we conclude that `Location` and `Stress` are dependent (i.e., a relationship exists between the two attributes).

In contrast set mining, the size of the search space (i.e., the number of rules generated) depends upon the number of attributes and the number of unique values contained within each attribute. Since many rules are generated, multiple hypothesis tests are required to identify valid relationships. However, as we discuss in the next section, multiple hypothesis tests are not without problems.

3.2 The Problem with Multiple Hypothesis Tests

Performing multiple tests of significance is a common practice in data mining applications and leads to a fundamental problem in inferential statistics. The problem is described, as follows. Say $\alpha = 0.05$ and a null hypothesis is true. Then, we can expect that the null hypothesis will be rejected by chance once in 20 tests. That is, we report a significant result when one does not exist (known as Type I error). Actually, when doing multiple tests, the expected number of errors is much higher. Specifically, as the number of tests increases, the true α level is inflated according to the approximation $\alpha_{true} = 1 - (1-\alpha)^t$, where α is the level you think you are using and t is the number of tests required. So, for example, when $t = 20$, $\alpha_{true} = 1 - (1 - 0.5)^{20} = 0.64$. That is, more likely than not, we will report a significant result when one does not exist. Consequently, methods are required to aid in controlling error.

3.3 Understanding Error

There are two kinds of error that can occur during hypothesis testing [11], as shown in Table 2. *Type I error* occurs when the null hypothesis is rejected when it is, in fact, true (also known as a *false positive*). *Type II error* occurs when the null hypothesis is not rejected when it is, in fact, false (equivalently, the alternative hypothesis is rejected when it is, in fact, true and also known as a *false negative*).

Table 2. Possible outcomes from a hypothesis test

Decision	H_0 *True*	H_0 *False*
Reject H_0	Type I Error	Correct
$\neg$ Reject H_0	Correct	Type II Error

Naturally, we hope to minimize both kinds of error. Unfortunately, though, we cannot decrease Type I error without increasing Type II error (the converse is also true). Type II error results in a lost opportunity to reject the null hypothesis. However, not rejecting the null hypothesis does not mean that it is accepted. That is, when the null hypothesis is not rejected, no conclusion can be drawn. In contrast, Type I error leads to a conclusion being drawn (i.e., that a relationship exists between the attributes being tested). Consequently, a Type I error is generally considered to be more of a problem than a Type II error, but the seriousness or impact of either kind of error is domain dependent. For example, in failing to identify a relationship between two attributes in a disease dataset (i.e., Type II error), a physician may not provide adequate treatment to patients. In a sports statistic dataset, a bettor could fail to pick the best player in the office pool (and one would hope and assume this is not life threatening).

3.4 Statistical Significance Versus Practical Significance

In short, statistical significance implies the rejection of the null hypothesis. The ability of a test of significance to detect differences that lead to a rejection of the null hypothesis can depend on the size of the dataset being tested. For example, the contingency table shown in Table 3 will be used to demonstrate this phenomenon. As in Table 1, this contingency table also shows the incidence of stress for urban and rural health care providers. Table 3 was derived from Table 1 by multiplying the cells of Table 1 by 0.1. Now $\chi^2 = 0.67$ for the rules in Table 3. So although the proportions of each cell to the row and column totals is unchanged from Table 1, at the 95% significance level, where

$\alpha = 0.05$ and $\chi^2_\alpha = 3.84$, we do not reject the null hypothesis that row and column totals are not related because $0.67 < 3.84$. The only difference between the two tests is that the number of examples (i.e., the dataset size) is smaller in Table 3.

Table 3. An example contingency table containing independent attributes

	Location=urban	Location=rural	$\sum$ *Row*
Stress=high	19	36	55
¬ (Stress=high)	36	51	87
$\sum$ *Column*	55	87	142

Thus, with a large dataset, the statistical test may cause us to reject the null hypothesis, implying that the attributes are related. However, in practice, the implied relationship between two attributes could be small to the point of having no practical significance. Consequently, it is important that the domain expert does not just blindly apply tests and draw weak or false conclusions, the domain expert should also combine theoretical judgment with statistical analysis.

4 Statistical Measures in Practice

In this section, we adopt a tutorial format to describe the methodologies utilized in STUCCO and CIGAR. While both methodologies are statistically sound, fundamentally they represent alternative philosophies to the problem of identifying interesting contrast sets.

4.1 STUCCO

The objective of STUCCO is to find contrast sets from grouped categorical data, where a dataset D can be divided into n mutually exclusive groups, such that for groups G_i and G_j, $G_i \cap G_j = \emptyset$, for all $i \neq j$. Specifically, we want to identify all contrast sets, such that the conditions

$$\exists ij \mathtt{P}(\mathtt{X}|G_i) \neq \mathtt{P}(\mathtt{X}|G_j)$$

and

$$max_{ij}|\mathtt{support}(\mathtt{X},G_i) - \mathtt{support}(\mathtt{X},G_j)| \geq \delta$$

are satisfied, where X is a conjunction of attribute-value pairs, G_k is a group, and δ is the user-defined minimum support difference (i.e., the minimum support difference between two groups). Contrast sets satisfying the first condition are called *significant*, those satisfying the second condition are called *large*, and those satisfying both conditions are called *deviations*.

Finding Deviations

As mentioned in Section 2.3, the search space consists of a canonical ordering of nodes, where all possible combinations of attribute-value pairs are enumerated. A contrast set formed from related nodes is called a *candidate set* until it has been determined that it meets the criteria required to be called a contrast set (i.e., it is significant and large).

Determining Support for a Candidate Set. The search for contrast sets follows a breadth-first search strategy, and is based upon support for a candidate set. For example, as shown in Table 1, `Support(Location=urban` $\wedge$ `Stress= high)` $= 194/554 = 0.35$ (or 35%) and `Support(Location=rural` $\wedge$ `Stress= high)` $= 355/866 = 0.41$ (or 41%).

Determining Whether a Candidate Set is Large. As mentioned above, two rules whose support difference exceeds some user-defined threshold are called large rules. For example, `|Support(Location=urban` $\wedge$ `Stress=high) - Support(Location=rural` $\wedge$ `Stress=high)|` $=$ `|0.35 - 0.41|` $= 0.06$ (or 6%). If we assume that $\delta = 0.05$ (or 5%), then the candidate set `Location=urban` $\wedge$ `Stress=high` and `Location=rural` $\wedge$ `Stress=high` is large.

Determining Whether a Candidate Set is Significant. To test the hypothesis that the difference in support for a candidate set is significantly different across groups, the χ^2 statistic is used. A $2 \times n$ contingency table is constructed, where the rows represent the truth of the consequent of the rules in the candidate set and the columns represent the groups in the antecedents. The procedure followed for the χ^2 test is described in Section 3.1, where it was determined that `Location` and `Stress` in the rules `Location=urban` $\wedge$ `Stress=high` and `Location=rural` $\wedge$ `Stress=high` are dependent.

However, we have failed to consider the effects of multiple hypothesis tests. As described in Section 3.2, the α level is used to control the maximum probability of falsely rejecting the null hypothesis in a single χ^2 test (i.e., Type I error). In the example of Section 3.1, we used $\alpha = 0.05$. But since STUCCO performs multiple hypothesis tests, a modified Bonferroni statistic is employed to limit the total Type I error rate for all χ^2 tests to α. The modified Bonferroni statistic uses a different α for contrast sets being tested at different levels of the search space. That is, at level i in the search space,

$$\alpha_i = min((\alpha/2^i)/|C_i|, \alpha_{i-1}),$$

where $|C_i|$ is the number of candidates at level i. The net effect, then, is that as we descend through the search space, α_i is half that of α_{i-1}, so a significant difference is increasingly restrictive as we descend. In the case of the example in Section 3.1, the rules being tested are found at level two of the search space. If we assume that 10 tests will be conducted at level two, then $i = 2$ and $\alpha_2 = ((0.05/2^2)/|10|) = 0.00125$. For $\alpha = 0.00125$, $\chi^2_\alpha \approx 10.83$. Since $5.08 < 10.83$, we accept the null hypothesis. That is, we conclude that `Location` and `Stress` are independent (i.e., no relationship exists between the two attributes). And since the rules are not significantly different, they are not significant. Finally, since the rules are not both large and significant, they are not deviations, and therefore, do not constitute a contrast set.

Pruning the Search Space

As with all exploratory data mining techniques, STUCCO has the potential to generate many results. Consequently, effective pruning strategies are required to reduce the number of results that need to be considered by a domain expert. Conceptually, the basic pruning strategy is simple: a node in the search space can be pruned whenever it fails to be significant and large.

Effect Size Pruning. When the maximum support difference, δ_{max}, between all possible pairs of groups has been considered and $\delta_{max} < \delta$, then the corresponding nodes can be pruned from the search space. This ensures that the effect size is large enough to be considered important by the domain expert.

Statistical Significance Pruning. The accuracy of the χ^2 test depends on the expected frequencies in the contingency table. When an expected frequency is too small, the validity of the χ^2 test may be questioned. However, there is no universal agreement on what is appropriate, so frequencies ranging anywhere from 1 (liberal) to 5 (conservative) are considered acceptable. Thus, nodes are pruned whenever an expected frequency is considered unacceptable (i.e., the test is invalid).

Maximum χ^2 Pruning. As we descend through the search space, the number of attribute-value pairs in a contrast set increases (i.e., itemsets are larger as the rules become more specific). And at each successive lower level in the search space, the support for a contrast set at level i is bounded by the parent at level $i - 1$. For example, given the rule `Location=rural` $\Rightarrow$ `Stress=high (54%)`, then the rule `Location=rural` $\Rightarrow$ `Stress=high` $\wedge$ `Income=low (65%)` cannot possibly be true. That is, since 54% of the transactions satisfy `Stress=high`, then a specialization of this rule (i.e., `Stress=high` $\wedge$ `Income=low`) cannot be any more than 54%. Consequently, the support for the parent rule `Stress=high` becomes an upper bound for all descendants in the search space. Similarly, as we ascend through the search space, the support for a contrast set at level i is bounded by the child at level $i + 1$. That is, the support for the child rule becomes a lower bound for all ancestors in the search space. Within the context of a contingency table, the observed frequencies in the upper (lower) row decrease (increase) as the contrast set becomes more specialized.

Since the support is bounded, the maximum possible χ^2 value for all specializations at the next level can be determined and used to prune the specialization if it cannot meet the χ^2 cutoff for α at that level. Let u_i and l_i represent the upper and lower bounds, respectively, of the observed values in position i of row one across all specializations. For example, if we have three specializations, say, and the observed values at position $i = 2$ of the three specializations are 4, 2, and 5, then $u_2 = 5$ and $l_2 = 2$. The maximum χ^2 value possible for any specialization of a rule is given by

$$\chi^2_{max} = max_{o_i \in \{u_i, l_i\}} \chi^2(o_1, o_2, \ldots, o_n),$$

where $\chi^2(o_1, o_2, \ldots, o_n)$ is the value for a contingency table with $\{o_1, o_2, \ldots, o_n\}$ as the observed values in the first row. The rows that we use to determine the maximum χ^2 value are based upon the n upper and lower bounds from the specializations. For example, if the first rows of our three specializations are {5, 4, 9}, {3, 2, 10}, and {8, 5, 6}, then the upper and lower bounds are {8, 5, 10} and {3, 2, 6}, respectively. We generate all 2^n possible first rows from combinations of the values in the upper and lower bounds. For example, from the upper and lower bounds given previously, the 2^3 unique first rows that we can generate are {8, 5, 10}, {3, 5, 10}, {8, 2, 10}, {3, 2, 10}, {8, 5, 6}, {3, 5, 6}, {8, 2, 6}, and {3, 2, 6}. These rows actually correspond to the extreme points (i.e., corners) of a feasible region where the maximum χ^2 value can be found. Since the values in the second row of each contigency table are determined by the values in the first row (since the column totals are fixed), then each contingency table is unique. For example, if the column totals are {15, 7, 13}, then the second row corresponding to {8, 5, 10} is {7, 2, 3}. We then simply determine the χ^2 value for each of the generated contingency tables and take the maximum. If χ^2_{max} exceeds the α cutoff, then none of the specializations can be pruned.

Interest Based Pruning. Specializations with support identical to the parent are not considered interesting by STUCCO. That is, when an additional attribute-value pair is added to a rule, if the resulting rule has the same support as the original rule, then the node containing the new rule is pruned from the search space, as it contains no new information. Similarly, when the support for one group is much higher than other groups, it will sometimes remain much higher regardless of the nature of any additional attribute-value pairs that are added to the rule. Specializations of the rule are pruned from the search space.

Statistical Surprise Pruning. When the observed frequencies are statististically different from expected frequencies (i.e., statistically surprising), a contrast set is considered interesting. For cases involving two variables, the expected frequency can be determined by multiplying the respective observed frequencies. For example, if `P(Stress=high | Location=rural)` `= 40%` and `P(Stress=high | Income=low) = 65%`, then `P(Stress=high` $\wedge$

`Income=low | Location=rural) = 26%`. If the product is within some threshold range, the contrast set is considered uninteresting and pruned from the search space. For more complicated cases, iterative proportional fitting can be used [9].

4.2 CIGAR

CIGAR relies on the same general approach as STUCCO for mining contrast sets from grouped categorical data. However, whereas STUCCO answers the question whether a difference exists between contrast sets in two or more groups through the analysis of $2 \times n$ contingency tables, CIGAR seeks a more fine grained approach by breaking the $2 \times n$ contingency tables down into a series of 2×2 contingency tables to try to explain where these differences actually occur. So, while we still want to identify all contrast sets such that the conditions

$$\exists ij \texttt{P}(\texttt{X}|G_i) \neq \texttt{P}(\texttt{X}|G_j)$$

and

$$max_{ij}|\texttt{support}(\texttt{X},G_i) - \texttt{support}(\texttt{X},G_j)| \geq \delta$$

used by STUCCO are satisfied (i.e., to find the significant and large contrast sets), CIGAR also utilizes three additional constraints. That is, we also want to identify all contrast sets such that the conditions

$$\texttt{support}(\texttt{X},G_i) \geq \beta,$$

$$\texttt{correlation}(\texttt{X},G_i,G_j) \geq \lambda,$$

and

$$|\texttt{correlation}(\texttt{X},G_i,G_j) - \texttt{correlation}(\texttt{child}(\texttt{X},G_i,G_j))| \geq \gamma$$

are satisfied, where X is a conjunction of attribute-value pairs, G_k is a group, β is the user-defined minimum support threshold, λ is the user-defined minimum correlation threshold, and γ is the user-defined minimum correlation difference. Contrast sets satisfying the third condition are called *frequent*. We believe a support threshold can aid in identifying outliers. Since outliers can dramatically affect the correlation value, the minimum support threshold provides an effective tool for removing them. Contrast sets satisfying the fourth condition are called *strong*. This measures the strength of any linear relationship between the contrast set and group membership. Contrast sets satisfying the first four conditions are called *deviations*. Those deviations that fail to satisfy the last condition are called *spurious* and pruned from the search space.

Finding Deviations

CIGAR generates candidate sets by traversing the search space in the same way as STUCCO. However, with CIGAR, before a candidate set can become a contrast set it must meet more restrictive criteria than STUCCO (i.e., it must not only be significant and large, it must also be frequent and strong).

Determining Support for a Candidate Set. CIGAR determines support for a candidate set in the same way as STUCCO, but CIGAR also uses support in ways that STUCCO does not. That is, while STUCCO determines the support for candidate sets, it does not utilize a minimum support threshold. However, the minimum support threshold used in CIGAR is useful for two reasons. First, the domain expert may not be interested in low support rules. Consequently, the nodes for these rules and all the descendant nodes can be pruned, reducing the size of the search space. Second, if the rule support is 0% or 100%, then the rule is pruned since a conjunction of this rule with any other does not create any new information.

Determining Whether a Candidate Set is Large and/or Significant. CIGAR determines whether a candidate set is large and/or significant in the same way as STUCCO.

Determining Whether a Candidate Set is Correlated. In CIGAR, correlation is calculated using the Phi correlation coefficient. The Phi correlation coefficient is a measure of the degree of association between two dichotomous variables, such as those contained in a 2×2 contingency table, and is conveniently expressed in terms of the observed frequencies. For example, given the generic 2×2 contingency table shown in Table 4, the Phi correlation coefficient is given by

$$r = \frac{O_{11}O_{22} - O_{12}O_{21}}{\sqrt{(O_{11} + O_{21})(O_{12} + O_{22})(O_{11} + O_{12})(O_{21} + O_{22})}}.$$

Table 4. A generic contingency table

	G_1	G_2	$\sum$ *Row*
`Contrast Set`	O_{11}	O_{12}	$O_{11} + O_{12}$
`¬ (Contrast Set)`	O_{21}	O_{22}	$O_{21} + O_{22}$
$\sum$ *Column*	$O_{11} + O_{21}$	$O_{12} + O_{22}$	$O_{11} + O_{12} + O_{21} + O_{22}$

The Phi correlation coefficient compares the diagonal cells (i.e., O_{11} and O_{22}) to the off-diagonal cells (i.e., O_{21} and O_{12}). The variables are considered positively associated if the data is concentrated along the diagonal, and negatively associated if the data is concentrated off the diagonal. To represent this association, the denominator ensures that the Phi correlation coefficient takes values between 1 and -1, where zero represents no relationship. However, the calculation for the Phi correlation coefficient can be expressed in terms of the χ^2 value (which we have to calculate anyway), and is given by

$$r = \sqrt{\chi^2/N},$$

where $N = O_{11} + O_{12} + O_{21} + O_{22}$. So, for the example in Section 3.1, and using $\chi^2 = 5.08$ and $\alpha = 0.05$, we have $r = \sqrt{5.08/1420} = 0.06$. Now a general rule of thumb is that $0.0 \leq r \leq 0.29$ represents little or no association, $0.3 \leq r \leq 0.69$ represents a weak positive association, and $0.7 \leq r \leq 1.0$ represents a strong positive association. Consequently, although we previously determined that a significant relationship exists between `Location` and `Stress` in Section 3.1 (i.e., prior to considering the effects of multiple hypothesis tests), at $r = 0.06$, this relationship is very weak.

Pruning the Search Space

CIGAR provides a powerful alternative strategy for reducing the number of results that must be considered by a domain expert. Conceptually, the basic pruning strategy is that a node in the search space is pruned whenever it fails to be significant, large, frequent, and strong.

Look-Ahead χ^2 Pruning. The χ^2 look-ahead approach calculates the χ^2 value for each specialization of a rule. If no specialization is found to be significant, then all the specializations are pruned from the search space. If at least one specialization is found to be significant, all the specializations are considered candidate sets at the next level of the search tree.

Statistical Significance Pruning. As mentioned in Section 4.1, the validity of the χ^2 test may be questioned when the expected frequencies are too small. To address this problem, Yates' correction for continuity has been suggested. Although there is no universal agreement on whether this adjustment should be used at all, there does seem to be some consensus that indicates the correction for continuity should be applied to all 2×2 contingency tables and/or when at least one expected frequency is less than five (liberal) or 10 (conservative) [9]. Either way, Yates' correction provides a more conservative estimate of the χ^2 value that is, hopefully, a more accurate estimate of the significance level. Yates' correction for continuity is given by

$$\chi^2 = \sum_{i=1}^{2} \sum_{j=1}^{2} \frac{(|O_{ij} - E_{ij}| - 0.5)^2}{E_{ij}}.$$

For example, Yates' $\chi^2 = 4.84$ for the contingency table in Table 1.

Minimum Support Pruning. The minimum support threshold utilized by CIGAR is the first line pruning strategy. For example, when determining correlation between `Location` and `Stress`, if one group happens to have very low support, it will likely affect the correlation for all pairwise group comparisons. Consequently, a domain expert may decide to exclude the low support group.

Minimum Correlation Pruning. The Phi correlation coefficient provides a basis for determining whether a contrast set is worth further consideration. For example, the lower the correlation, the higher the likelihood that no relationship actually exists between a rule and the group. That is, even if a rule is considered significant, if the correlation is zero, then the probability that the rules is simply a statistical artifact is high. Therefore, the removal of rules that do not meet the minimum correlation criteria eliminates the likelihood of reporting statistical artifacts. For example, we determined in the previous section that the relationship between `Location` and `Stress` is very weak at $r = 0.06$ and the candidate set should be removed from further consideration.

When a high minimum correlation threshold is used, many significant rules may be pruned, resulting in an increase in Type II error. Similarly, when a low minimum correlation threshold is used, many spurious rules may not be pruned. This is analogous to the problem of setting support thresholds in the classic association rule mining problem.

CIGAR is different from STUCCO in that it tends to report more specialized contrast sets rather than generalized contrast sets. The assumption behind approaches that report more generalized rules is that more general rules are better for prediction. However, the complex relationships between groups can often be better explained with more specialized rules.

Minimum Correlation Difference Pruning. CIGAR calculates the difference between the correlation of a rule and the correlations of specializations of that rule. If the difference in correlation between a rule and a specialization is less than the minimum correlation difference threshold, the specialization is pruned from the search space. That is, if the addition of a new attribute-value pair to a contrast set does not add any new information that directly speaks to the strength of the relationship, then the contrast set is spurious. For example, assume that $r = 0.7$ for the rule `Location=rural` $\wedge$ `Income=low`. If $\gamma = 0.05$, and $r = 0.72$ for the specialization `Location=rural` $\wedge$ `Income=low` $\wedge$ `Stress=high`, then the specialization is pruned from the search space because $|0.70 - 0.72| = 0.02$ and $0.02 < 0.05$.

Generally, as we descend through the search space, the support for rules at lower levels decreases. As a result, the χ^2 value and r generally decrease, as well. The decision on whether to prune a rule from the search space is then a fairly easy one. However, it is possible that as we descend through the search space the χ^2 value and/or r can increase, as the previous example showed (i.e., it did not exceed the minimum correlation difference threshold). It is also possible the χ^2 value and/or r can decrease and then increase again. If the correlation difference between a rule and one of its specializations is less than the minimum correlation difference, regardless of whether the difference represents a decrease or an increase, the domain expert has to be careful when deciding whether to prune the specialization. That is, pruning a specialization that fails to meet the minimum correlation difference criteria at level i could result in the loss of a specialization at level $i + 1$ that does meet the minimum correlation difference criteria. So, some statistical judgment may be required on the part of the domain expert to ensure that only unproductive and spurious rules can be pruned.

5 Experimental Results

In this section, we present the results of our experimental evaluation and comparison of STUCCO and CIGAR. STUCCO was supplied by the original authors [3], [4]. STUCCO, implemented in C++ and compiled using gcc (version 2.7.2.1), was run on a Sun Microsystems Enterprise 250 Model 1400 with two UltraSparc-II 400 MHz processors and 1 GB of memory. CIGAR was implemented by the authors of this paper in Java 1.4.1 and was run under Windows XP on an IBM compatible PC with a 2.4 GHz AMD Athlon processor and 1 GB of memory. The performance of the two software tools was compared by generating contrast sets from publicly available datasets.

5.1 The Datasets

Discovery tasks were run on three datasets: Mushroom, GSS Social, and Adult Census. The Mushroom dataset, available from the UCI Machine Learning Repository, describes characteristics of gilled mushrooms. The GSS Social dataset is a survey dataset from Statistics Canada that contains the responses to the General Social Survey of Canada (1986 - Cycle 2): Social Activities and Language Use. The Adult Census dataset is a subset of the Adult Census Data: Census Income (1994/1995) dataset, a survey dataset from the U.S. Census Bureau.

The characteristics of the three datasets are shown in Table 5. In Table 5, the *Tuples* column describes the number of tuples in the dataset, the *Attributes* column describes the number of attributes, the *Values* column describes the number of unique values contained in the attributes, and the *Groups* column describes the number of distinct groups defined by the number of unique values in the distinguished attribute.

Table 5. Characteristics of the three datasets

Dataset	*Tuples*	*Attributes*	*Values*	*Groups*
Mushroom	8,142	23	130	2
GSS Social	179,148	16	2,026	7
Adult Census	826	13	129	2

5.2 The Parameters

CIGAR requires that four parameters be set to constrain the results generated. The values used by CIGAR for the minimum support (β), minimum support difference (δ), minimum correlation (λ), and minimum correlation difference (γ) thresholds are shown in Table 6. For example, the results from the Mushroom dataset were generated using $\beta = 0.02$, $\delta = 0.04$, $\lambda = 0.0$, and $\gamma = 0.0$. STUCCO requires that just one parameter be set, the minimum support difference (δ). For each of the four datasets, STUCCO used the same value for δ as CIGAR.

Table 6. Parameters used to generate the reported results

Dataset	β	δ	λ	γ
Mushroom	0.02	0.04	0.0	0.0
GSS Social	0.01	0.01	0.0	0.0
Adult Census	0.01	0.01	0.0	0.0

5.3 Issues Affecting a Direct Comparison of STUCCO and CIGAR

A number of issues affect the direct comparison of STUCCO and CIGAR. First, STUCCO and CIGAR were developed in different languages and run on different platforms. This affects our ability to do any kind of meaningful evaluation of the computational performance of the two approaches with respect to time. Second, on the one hand, as you will see, CIGAR generates considerably more candidate sets than STUCCO, so one would expect STUCCO to generate results faster than CIGAR. Third, on the other hand, STUCCO uses a maximum χ^2 calculation that is exponential in the number of groups represented in the distinguished attribute. It may be possible that this could give CIGAR some advantage. Finally, STUCCO generates the whole search space prior to pruning (necessary for the modified Bonferroni calculations), whereas CIGAR utilizes a generate-and-test approach. These are just a few of the more obvious issues that will affect computational performance. There are likely many, perhaps more subtle, computational issues.

One obvious procedural issue affecting a direct comparison is that STUCCO and CIGAR differ in the number and variety of parameters required to generate results, and the underlying philosophy regarding statistical error control is different. In an attempt to address this procedural issue and "force" the results to be somewhat comparable, the minimum correlation and minimum correlation difference thresholds used by CIGAR were set to zero. However, later on, when we demonstrate the effect of correlational pruning, these thresholds will be set to a non-zero value. When this is the case, the thresholds used will be specified.

It is important to remember, when studying the results that follow, that we are not trying to show that CIGAR is better or worse than STUCCO, or that CIGAR is right and STUCCO is wrong. Rather, we are merely trying to compare different philisohical and methodological approaches. They both generate statistically valid results.

5.4 The Effect of 2 × 2 Contingency Tables

The number of groups contained in the distinguished attribute affect the size of the contingency tables required by STUCCO because STUCCO gathers the related rules into a $2 \times n$ contingency table, where n is the number of groups. For example, a 2×7 contingency table for the contrast set `Activity Code = shops daily` generated by STUCCO for the seven groups in the GSS Social dataset is shown in Table 7. The χ^2 value and degrees of freedom calculated for this table are $\chi^2 = 386.38$ and $df = 6$, respectively. From this information, STUCCO reports that a significant difference exists between groups and generates the rule `All Groups` ⇒ `Activity Code = shops daily`. But other than pointing out that the relationship between the contrast set `Activity Code = shops daily` and group is not randomly causal, it does not provide any details as to where the differences actually occur. That is, it does not provide any details as to which groups are different.

Table 7. 2×7 contingency table for `Activity Code = shops daily`

	G_1	G_2	G_3	G_4	G_5	G_6	G_7	$\sum$ *Row*
`Activity Code = shops daily`	164	555	558	650	481	619	718	3,745
`¬ (Activity Code = shops daily)`	17,278	32,655	31,627	31,815	20,078	20,685	21,264	175,402
$\sum$ *Column*	17,442	33,210	32,185	32,465	20,559	21,304	21,982	179,147

CIGAR breaks down the $2 \times n$ contingency table required by STUCCO into a series of 2×2 contingency tables, one for each possible combination of group pairs. For example, from the contingency table in Table 7, CIGAR generates a series of 21 2×2 contingency tables, like the one shown in Table 8 for G_1 and G_2. The 2×2 contingency tables used by CIGAR have the potential to provide more information about the differences between groups than the 2×7 contingency table used by STUCCO. That is, CIGAR is able to provide details as to which groups are different. For example, from these contingency tables, CIGAR determines that for the shared consequent `Activity Code = shops daily`, there are significant differences between G_2 and G_6, G_2 and G_7, G_3 and G_6, G_3 and G_7, and G_4 and G_7. The other sixteen combinations of group pairs failed to meet the minimum support and minimum support difference thresholds. Consequently, not only do we know that a significant difference exists between some groups, we have a fine grained breakdown of the groups involved.

Table 8. 2×2 contingency table for `Activity Code = shops daily`

	G_1	G_2	$\sum$ *Row*
`Activity Code = shops daily`	164	555	719
$\neg$ `(Activity Code = shops daily)`	17,278	32,655	49,933
$\sum$ *Column*	17,442	33,210	50,652

5.5 The Effect of Error Control

Recall that STUCCO seeks to control Type I error (false positives), whereas CIGAR seeks to control Type II error (false negatives). In this section, we compare the error control philosophies of STUCCO and CIGAR to evaluate the impact on the number of candidate sets and contrast sets generated.

The number of candidate sets generated from the Mushroom, GSS Social, and Adult Census datasets by STUCCO and CIGAR is shown in Table 9. The number of contrast sets shown in the STUCCO columns reflect the result of pruning using the maximum χ^2 technique described in Section 4.1. The number shown in the CIGAR columns reflect the result of pruning using the minimum support threshold. Table 9 shows for the Mushroom, GSS Social, and Adult Census datasets that CIGAR generated 9.1, 1.3, and 2.8 times more candidate sets, respectively, than STUCCO. For example, for the Mushroom dataset, CIGAR generated 128,717 candidate sets to a depth of 14 in the search space, while STUCCO generated 14,089 candidate sets to a depth of eight.

Table 9. Summary of candidate sets generated

	Mushroom		*GSS Social*		*Adult Census*	
Depth	*STUCCO*	*CIGAR*	*STUCCO*	*CIGAR*	*STUCCO*	*CIGAR*
2	103	53	11,965	3,009	97	44
3	951	694	13,994	5,980	877	419
4	3,470	3,912	6,670	8,620	2,011	1,680
5	6,025	10,496	4,897	13,168	3,033	3,545
6	3,054	21,006	792	10,298	826	4,806
7	485	28,427	117	5,356	36	4,357
8	1	27,995	6	1,524	0	2,755
9	0	20,189	0	236	0	1,184
10	0	10,545	0	20	0	342
11	0	3,870	0	9	0	60
12	0	939	0	0	0	5
13	0	133	0	0	0	0
14	0	8	0	0	0	0
Total	14,089	128,717	38,411	48,220	6,880	19,197

STUCCO generates 1.4 times more candidate sets than CIGAR to depth three. The primary reason for this is that when STUCCO is determining whether a candidate set is large, it includes groups for which the support is zero. CIGAR uses the minimum support threshold to remove rules with low support from further consideration. At depth four, the significance level of the modified Bonferroni statistic in the maximum χ^2 calculation used in STUCCO starts to become more restrictive than the significance level used in CIGAR. It is at this point that STUCCO, while controlling for Type I error, starts to realize more frequent Type II error. For example, 128,717 candidate sets were generated from the Mushroom dataset with no correlational pruning (i.e., $\lambda = 0.00$). At a conservative 99% significance level, this will result in CIGAR generating approximately 1,287 false positive contrast sets. However, when correlation pruning is applied (i.e., $\lambda > 0.00$), this number decreases. Of course, the amount that it decreases is dependent on the value of the correlation threshold. At the 95% significance level used by STUCCO, 704 false positive contrast sets are generated. Consequently, the actual difference between the number of false positives generated by STUCCO and CIGAR can be relatively small. However, when we consider that CIGAR generates 128,717 candidate sets compared to the 14,089 generated by STUCCO, the false positive rate will be significantly lower for CIGAR. For some domain experts, the rate of Type I error in STUCCO may not outweigh the loss of information resulting from the increase in Type II error.

The number of contrast sets generated from the Mushroom, GSS Social, and Adult Census datasets by STUCCO and CIGAR is shown in Table 10. In Table 10, the number of contrast sets shown in the STUCCO columns is based upon those candidate sets from Table 9 that are both significant and large, while those shown in the CIGAR columns are significant, large, frequent, and strong (however, recall that $\lambda = 0.0$ and $\gamma = 0.0$, so these contrast sets are frequent and strong by default). Table 10 shows that CIGAR generated significantly more contrast sets than STUCCO. For the datasets that contain only two groups in the distinguished attribute (i.e., Mushroom and Adult Census), the number of contrast sets generated is somewhat similar until a depth of four in the search space. However, at a depth of five and beyond, the effects of the modified Bonferroni statistic in controlling Type I error are quite apparent.

Table 10. Summary of contrast sets generated

	Mushroom		*GSS Social*		*Adult Census*	
Depth	*STUCCO*	*CIGAR*	*STUCCO*	*CIGAR*	*STUCCO*	*CIGAR*
2	71	46	83	566	22	23
3	686	548	466	3,081	139	202
4	2,236	2,721	1,292	7,645	353	843
5	2,531	7,577	1,155	10,930	341	1,972
6	714	13,899	199	9,368	64	2,929
7	102	18,293	22	4,852	0	2,920
8	0	17,915	0	1,504	0	2,011
9	0	13,124	0	249	0	943
10	0	7,077	0	20	0	286
11	0	2,715	0	11	0	53
12	0	697	0	0	0	5
13	0	106	0	0	0	0
14	0	7	0	0	0	0
Total	6,340	84,725	3,217	38,226	919	12,187

5.6 The Effect of a Minimum Support Threshold

Recall that one of the constraints utilized by CIGAR in contrast set mining, and not utilized by STUCCO, is a minimum support threshold. To aid in making this discussion clear, we discuss the results generated by STUCCO and CIGAR at depth two in the search space for the Mushroom and Adult Census datasets. These results are shown in Table 11. In Table 11, the *Zero Itemsets* row describes the number of contrast sets that were generated where at least one of the groups had zero support. The *Below Minimum Support* row describes the number of contrast sets where at least one of the groups had support below the minimum support threshold. The *Unmatched Contrast Sets* row describes the number of contrast sets that are found by STUCCO (CIGAR) but not by CIGAR (STUCCO). The *Matched Contrast Sets* row describes the number of contrast sets generated that are common to both STUCCO and CIGAR.

Table 11. Summary of results at depth 2

	Mushroom		*Adult Census*	
Criteria	*STUCCO*	*CIGAR*	*STUCCO*	*CIGAR*
Zero Itemsets	15	0	2	0
Below Minimum Support	10	0	1	0
Unmatched Contrast Sets	0	0	0	4
Matched Contrast Sets	46	46	19	19

Table 11 shows that for the Mushroom and Adult Census datasets, STUCCO generates 25 and 3 contrast sets, respectively, whose support is below the minimum support threshold. These contrast sets represent 35% and 14%, respectively, of the total number of contrast sets generated. On the Mushroom dataset, this represents 100% of the difference between the contrast sets generated by STUCCO and CIGAR. On the Adult Census dataset, four (or 17%) of the contrast sets generated by CIGAR did not have a corresponding contrast set in those generated by STUCCO. These four contrast sets were pruned by STUCCO because they did not meet the significance level cutoff of the modified Bonferroni statistic.

5.7 The Effect of Correlational Pruning

The minimum correlation and minimum correlation difference thresholds utilized by CIGAR can significantly reduce the number of contrast sets that need to be considered by a domain expert, focusing his or her attention on only those contrast sets where the relationship between variables is strong and the addition of terms to a rule is statistically meaningful. The number of significant and large contrast sets generated by CIGAR from the Mushroom dataset for various minimum correlation thresholds is shown in Table 12. In

Table 12, the $\geq \lambda$ columns describe the number of contrast sets whose r-value exceeds the minimum correlation threshold for each of the specified minimum correlation threshold values (i.e., $\lambda = 0.00$ to $\lambda = 0.70$). The $\geq \gamma$ columns describe the number of contrast sets remaining after correlational difference pruning. For all the results reported in this table, the minimum correlation difference threshold was set at 2% (i.e., $\gamma = 0.02$).

Table 12. Contrast sets exceeding the correlational pruning thresholds

	$\lambda = 0.00$		$\lambda = 0.25$		$\lambda = 0.50$		$\lambda = 0.60$		$\lambda = 0.70$	
Depth	$\geq \lambda$	$\geq \gamma$	$\geq \lambda$	$\geq \gamma$	$\geq \lambda$	$\geq \gamma$	$\geq \lambda$	$\geq \gamma$	$\geq \lambda$	$\geq \gamma$
2	46	46	21	21	9	9	0	0	0	0
3	548	531	226	226	53	50	11	11	7	7
4	2,721	2,506	949	882	188	148	36	35	17	14
5	7,577	6,290	2,377	1,956	394	257	53	36	21	4
6	13,899	10,183	4,104	2,838	536	332	35	15	15	0
7	18,293	11,897	5,359	3,063	508	318	10	2	6	0
8	17,915	10,305	5,433	2,562	345	208	0	0	0	0
9	13,124	6,531	4,232	1,671	167	86	0	0	0	0
10	7,077	2,964	2,466	835	55	20	0	0	0	0
11	2,715	931	1035	309	11	2	0	0	0	0
12	697	191	295	80	0	0	0	0	0	0
13	106	23	51	13	0	0	0	0	0	0
14	7	0	4	0	0	0	0	0	0	0
Total	84,725	52,398	26,552	14,456	2,266	1,430	145	99	66	25

Clearly, the choice of a minimum correlation threshold can affect the quantity and validity (i.e., quality) of the contrast sets generated. For example, when $\lambda = 0.25$, 0.50, 0.60, and 0.70, the number of contrast sets generated is 31.3%, 2.6%, 0.002%, and 0.0008% of the number generated when $\lambda = 0.00$ (i.e., when there is no correlational pruning). Thus, while CIGAR generates more contrast sets than STUCCO, correlational pruning provides a powerful facility to constrain the contrast sets generated to only those for which some statistically strong relationship actually exists.

Table 12 also demonstrates that correlational difference pruning can be used to further constrain the number of contrast sets generated. Recall that the correlational difference threshold is used to ensure that adding terms to a rule results in statistically meaningful information being added to the rule. For example, from Figure 1, if we add the attribute-value pair $\texttt{A}_3 = \texttt{V}_{31}$ to the rule $\texttt{A}_1 = \texttt{V}_{11} \Rightarrow \texttt{A}_2 = \texttt{V}_{21}$ yielding $\texttt{A}_1 = \texttt{V}_{11} \Rightarrow \texttt{A}_2 = \texttt{V}_{21} \wedge \texttt{A}_3 = \texttt{V}_{31}$, we want to be sure the correlation difference is large (i.e., meaningful) enough that we would consider the rule $\texttt{A}_1 = \texttt{V}_{11} \Rightarrow \texttt{A}_2 = \texttt{V}_{21} \wedge \texttt{A}_3 = \texttt{V}_{31}$ over $\texttt{A}_1 = \texttt{V}_{11} \Rightarrow \texttt{A}_2 = \texttt{V}_{21}$. That is, we can prune $\texttt{A}_1 = \texttt{V}_{11} \Rightarrow \texttt{A}_2 = \texttt{V}_{21}$. For example, when $\lambda = 0.00$, 0.25, 0.50, 0.60, and 0.70, the number of contrast sets generated when $\gamma = 0.02$ (i.e., the values in the $\geq \gamma$ column) is 61.8%, 54.4%, 63.1%, 68.2%, and 37.9% of the number generated, respectively, with no correlational pruning (i.e., the number in the $\geq \lambda$ column).

Table 12 does not demonstrate the situation where contrast sets at level $i + 1$ in the search space have higher correlation than those at level i, a situation that is possible, as was described in Section 4.2.2. However, it is worth mentioning here anyway. The situation occurs frequently in practice. For example, the rules `Bruise=no (r=0.501)`, `Bruise=no` $\wedge$ `Gill Space=close (r=0.735)`, and `Bruise=no` $\wedge$ `Gill Space=close` $\wedge$ `Veil Color=white (r=0.787)` were generated from the Mushroom dataset. Recall that according to the general rule of thumb previously described, a Phi correlation coefficient in the range $0.7 \leq r \leq 1.0$ represents a strong positive association. If we set the minumim correlation threshold to $\lambda = 0.7$, then the more general rule `Bruise=no (r=0.501)` would have been pruned and the two specializations never would have been generated. This highlights a problem in setting the minimum correlation threshold and shows how it can affect results.

6 Conclusion

We have discussed and demonstrated two alternative approaches to the contrast set mining problem: STUCCO and CIGAR. The two approaches rely on differing statistical philosophies: STUCCO seeks to control Type I error, while CIGAR seeks to control Type II error. A tutorial approach showed how various statistical measures can be applied as heuristics when mining contrast sets. With various constraints on the contrast sets, such as the significance and strength, both approaches generated potentially interesting contrast sets. However, due to the differing underlying statistical assumptions used, experimental results showed that the number of contrast sets generated by the two approaches was considerably different. While both methods do generate some of the same contrast sets, many contrast sets generated by one approach do not have a corresponding contrast set in the other approach. In addition, CIGAR tends to generate more specific rules than STUCCO at lower levels in the search space because STUCCO uses more restrictive significance criteria. We also showed that the correlational pruning technique used by CIGAR can significantly reduce the number of contrasts sets that must be considered.

References

1. R. Agrawal, T. Imielinski, and A. Swami. Mining association rules between sets of items in large databases. In *Proceedings of the ACM SIGMOD International Conference on the Management of Data (SIGMOD'93)*, pages 207–216, Washington, D.C., U.S.A., May 1993.
2. K.M. Ahmed, N.M. El-Makky, and Y. Taha. A note on "Beyond Market Baskets: Generalizing Association Rules to Correlations". *SIGKDD Explorations*, 1(2):46–48, 2000.
3. S.D. Bay and M.J. Pazzani. Detecting change in categorical data: Mining contrast sets. In *Proceedings of the Fifth ACM SIGKDD International Conference on Knowledge Discovery and Data Mining (KDD'99)*, pages 302–306, San Diego, U.S.A., August 1999.

4. S.D. Bay and M.J. Pazzani. Detecting group differences: Mining contrast sets. *Data Mining and Knowledge Discovery*, 5(3):213–246, 2001.
5. R.J. Bayardo. Efficiently mining long patterns from databases. In *Proceedings of the 1998 ACM SIGMOD International Conference on Management of Data (SIGMOD'98)*, pages 85–93, Seattle, U.S.A., June 1998.
6. R.J. Bolton, D.J. Hand, and N.M. Adams. Determining hit rate in pattern search. In *Proceedings of the ESF Exploratory Workshop on Pattern Detection and Discovery*, pages 36–48, London, U.K., September 2002.
7. S. Brin, R. Motwani, and C. Silverstein. Beyond market baskets: Generalizing association rules to correlations. In *Proceedings of the ACM SIGMOD International Conference on Management of Data (SIGMOD'97)*, pages 265–276, May 1997.
8. G. Dong and J. Li. Efficient mining of emerging patterns: Discovering trends and differences. In *Proceedings of the Fifth ACM SIGKDD International Conference on Knowledge Discovery and Data Mining (KDD'99)*, pages 43–52, San Diego, U.S.A., August 1999.
9. B.S. Everitt. *The Analysis of Contingency Tables.* Chapman and Hall, 1992.
10. J. Li, T. Manoukian, G. Dong, and K. Ramamohanarao. Incremental maintenance on the border of the space of emerging patterns. *Data Mining and Knowledge Discovery*, 9(1):89–116, 2004.
11. R.G. Miller. *Simultaneous Statistical Inference, Second Edition.* Springer Verlag, 1981.
12. T. Peckham. Contrasting interesting grouped association rules. Master's thesis, University of Regina, 2005.
13. J.P. Shaffer. Multiple hypothesis testing. *Annual Review of Psychology*, 46:561–584, 1995.
14. G.I. Webb, S. Butler, and D. Newlands. On detecting differences between groups. In *Proceedings of the Ninth ACM SIGKDD International Conference on Knowledge Discovery and Data Mining (KDD'03)*, pages 256–265, Washington, D.C., U.S.A., August 2003.
15. H. Zhong, B. Padmanabhan, and A. Tuzhilin. On the discovery of significant statistical quantitative rules. In *Proceedings of the Tenth ACM SIGKDD International Conference on Knowledge Discovery and Data Mining (KDD'04)*, pages 374–383, Seattle, U.S.A., 2004.

Understandability of Association Rules: A Heuristic Measure to Enhance Rule Quality

Rajesh Natarajan[1] and B. Shekar[2]

[1] Cognizant Technology Solutions India Pvt. Ltd., Techno Complex, 5/535, Old Mahabalipuram Road, Thoraipakkam, Chennai - 600 096, Tamil Nadu, India. rajesh.natarajan@cognizant.com

[2] Quantitative Methods and Information Systems (QMIS) Area, Indian Institute of Management Bangalore (IIMB), Bannerghatta Road, Bangalore - 560 076, Karnataka, India. shek@iimb.ernet.in

Summary. Association Rule (AR) mining has been plagued by the problem of rule immensity. The sheer numbers of discovered rules render comprehension difficult, if not impossible. In addition, due to their representing commonplace facts, most of these rules do not add to the knowledge base of the user examining them. Clustering is a possible approach that mitigates this rule immensity problem. Clustering ARs into groups of 'similar' rules is advantageous in two ways. Related rules being placed together in the same cluster facilitates easy exploration of connections among the rules. Secondly, a user needs to examine only those rules in 'relevant' clusters.

The notion of *weakness* of an AR is introduced. *Weakness* reveals the extent of an AR's inability to explain the presence of its constituents in the database of transactions. We elaborate on its usefulness and relevance in the context of a retail market. After providing the intuition, a distance-function on the basis of weakness is developed. This distance-function forms the basis of clustering ARs. Average linkage method is used to cluster ARs obtained from a small artificial data set. Clusters thus obtained are compared with those obtained by applying a commonly used method (from recent data mining literature) to the same data set.

Key words: Association Rules, Heuristic, Weakness, Clustering.

1 Introduction

In a database of transactions, *Association Rules* (ARs) bring out co-occurrence based affinity between items (attributes). In the retail market-basket context, they reveal items that are likely to be purchased together in the same transaction. Due to the completeness and automated nature of AR mining algorithms, a large number of rules are discovered - numbers far exceeding human comprehensibility. In addition, most of the discovered rules being obvious, they do not add to the knowledge of the domain user. Hence,

Rajesh Natarajan and B. Shekar: *Understandability of Association Rules: A Heuristic Measure to Enhance Rule Quality*, Studies in Computational Intelligence (SCI) **43**, 179–203 (2007)
www.springerlink.com

there is a need for automated methods that select the most appropriate rules for further examination.

Researchers have addressed this rule immensity problem to various extents through diverse techniques. Incorporation of additional constraints [11], rule summarisation [5], and rule ranking [6] using interestingness measures are a few typical techniques. Cluster analysis techniques aim at uncovering the hidden implicit structure in data. The objective is to group objects (or rules) such that objects within a cluster have a high degree of 'natural association,' with the clusters themselves remaining 'relatively distinct' from each other [15]. The underlying structure present in any cluster could aid in the analysis of its constituent rules. In addition clustering is advantageous (as compared to a rank list), since it reveals connections between various sets of rules. Further, the user may also be able to obtain an overview of the domain, thus reducing the intensity of the rule immensity problem.

Although clustering techniques have been well studied in literature, clustering of ARs is a relatively unexplored area. The next section places our work in the context of related work done in AR mining. The notion of *weakness* of an AR is introduced subsequently. *Weakness* is the extent to which an AR is unable to explain the presence of its constituent items from the dataset. Some properties of *weakness* are then examined. The explanatory power of an AR is closely related to its predictive ability. The relationship between *weakness* and *confidence* is then explored leading to an alternative way of interpreting *weakness*.

A distance measure based on *weakness* is constructed. The rationale for this distance measure is elucidated with the help of illustrations. A method for grouping ARs is then outlined. This is based on the classical agglomerative hierarchical clustering scheme [18]. Finally, the proposed method is demonstrated and compared to another method from recent literature thus bringing out the effectiveness of the proposed *weakness*-directed clustering scheme.

2 Related Work

Grouping rules according to some useful criteria helps in understanding the mined rules. This is because the structure imposed on groups of rules guides the user in relating them and extracting pertinent information. One straightforward approach groups ARs on the basis of extraneous characteristics of items such as economic assessment, profit margin and period of purchase [3]. Although these criteria may not be directly reflected in transactions, users may consider them important from financial and other perspectives. Clustering as an important data mining technique has received wide attention in recent times [12, 15, 17]. In data mining, clustering is frequently used to gain understanding of data distribution. However, clustering of ARs is a relatively unexplored area. We cite some important studies and their relevance to our work.

Lent, Swami and Widom [4] introduced the notion of 'clustered' ARs. Here, clustering is defined as merging of ARs with adjacent attribute values, or bins (intervals) of attribute values to form one consolidated rule that can represent the group. Their study limits attributes to two numeric attributes in the antecedent and one categorical attribute in the consequent of the rule. Wang, Tay and Liu [16] on the other hand allowed any combination of numeric and categorical attributes in the antecedent along with one or more categorical attributes in the consequent. Both approaches merge values in a bottom-up fashion. The proposed methodology differs from these studies due to the introduction of *weakness* - an inherent property of an AR.

Adomavicius and Tuzhilin [8] have adopted a similarity-based rule grouping that is based on attribute hierarchies. Rules are grouped by selecting a specific "cut" in the attribute hierarchy. All rules that have the same "aggregated rule" are brought together in a group. However the basis of grouping does not have any inherent semantics, and is more expert-driven. Another approach closer to classical clustering is proposed by Toivonen and associates [10]. Distance between two rules is defined as the number of transactions in which their identical consequents differ. Gupta and others [14] have proposed a new normalized distance function called Conditional Market-Basket probability ($CMPB$) distance. This distance function tends to group those rules that 'cover' the same set of transactions. One of the limitations of both schemes is the arbitrariness of distance measures used for clustering [8]. The proposed distance function is based on the notion of *weakness* of an AR.

Dong and Li [9] have introduced interestingness by considering unexpectedness in terms of neighborhood-based parameters. They define distance between rules in a syntactic fashion. This is then used to construct the 'neighbourhood' of a rule. The distance metric proposed by Dong and Li [9] is aimed at detecting unexpected rules in a defined neighbourhood, and not at clustering. Moreover, their distance function does not capture the actual conceptual distance between rules because of its syntactic nature (based on item matching).

Sahar [19] has proposed a new metric called d_{SC} that combined the approach of Dong and Li [9] and that of Toivonen, et al. [10]. This resulted in a measure that used the five defining characteristics of an AR, namely antecedent set, consequent set, rule support, antecedent support and consequent support. This approach utilized both syntactic matching of item sets and rule coverage of the data. However, this approach does not consider individual items constituting a rule. In addition, intrinsic properties of ARs are not considered. Jorge [1] studied hierarchical clustering in the context of thematic browsing and summarisation of large sets of ARs. Although rules reflecting the same theme are brought together, this approach does not account for other properties intrinsic to ARs. Current research has concentrated either on syntactic (item-matching based) comparison of ARs [8, 9, 1] or on transaction set coverage [14, 19, 10]. Although useful, these approaches do not use certain intrinsic properties characteristic to ARs. We bring out the *weakness* (an intrinsic property of ARs)-based specificity/generality of the AR in describing the presence of its constituents in the database.

We cite some studies in related areas though not directly relevant to the theme of this chapter. A different but related study conducted by Aggarwal and associates [7] used category clustering algorithms as a tool for finding associations in small segments of data. Clustering [13] has also been employed as a technique for partitioning intervals over numeric data domains, which then formed inputs to mining algorithms. Kosters, Marchiori and Oerlemans [2] proposed the use of ARs for extracting clusters from a population of customers.

3 *Weakness* of an Association Rule

Consider a retail store scenario. Let I represent the set of all items available for sale ($I = \{a_1, a_2, a_3, ..., a_n, ..., a_l\}$). Each transaction t (in database D that pertains to a specified time period) consists of a set of items purchased by a customer on a single buying-instance ($t \subseteq I$). Consider an AR, $R : a_1a_2..a_m \rightarrow a_{m+1}a_{m+2}...a_n$, $(1 \leq m < n)$, having support S_R and confidence C_R. *'minsup'* and *'minconf'* are the respective support and confidence thresholds. Each transaction in the $S_R\%$ of D referred by rule R contains all the n items $\{a_1, a_2, ..., a_n\}$ of R. In addition, some of these transactions may also contain other items from I. None of the other transactions in D contain the complete set $\{a_1, a_2, ..., a_n\}$. A rule covers a transaction if all the items of the rule are present in that transaction. Thus, R covers transaction t if and only if $a_i \in R \Rightarrow a_i \in t$.

Let the support of an individual item $a_i \in R$ with respect to D be S_{a_i} ($S_{a_i} \geq S_R$). Although the percentage of transactions containing a_i is S_{a_i}, rule R accounts for only $S_R\%$ of transactions in the database. This means rule R covers or explains only $\frac{S_R}{S_{a_i}}$ percentage of the transactions containing a_i. Stated simply, an AR may not completely explain its constituent's presence in the database. Specifically, for an item a_i, R does not explain the portion (of D) containing $(1 - \frac{S_R}{S_{a_i}})\%$ of the transactions containing a_i. This is because R does not cover those transactions. Since R cannot explain anything about a_i's presence in the rest of the $(1 - \frac{S_R}{S_{a_i}})\%$ of database transactions containing a_i, this fraction may be viewed as representing the *weakness* of R. In addition, an AR does not bring out all features of the transactions it covers. This is because the transactions that gave rise to a particular AR may contain items other than those contained in the AR. This is another aspect of *weakness* (of an AR) with respect to the transactions covered by the rule. However, we are mainly concerned about an AR's inability to explain parts of the presence of its constituents in the database.

We define *weakness* (of an AR) with respect to its constituent a_i as:

$$w_{a_i} = 1 - \frac{S_R}{S_{a_i}} \tag{1}$$

In general, an AR consists of many items. In order to quantify the *weakness* of an AR with respect to all constituents, we compute the mean of the rule's *weakness* over all items. Thus, the *weakness* of an association rule R is given by,

$$w_R = \frac{1}{n} \sum (1 - \frac{S_R}{S_{a_i}}); a_i \in \{a_1, a_2, ..., a_n\} \tag{2}$$

Consider an AR:$\{bread, butter\} \rightarrow \{milk\}$ with support of 0.20. Let the total number of transactions in the database be 100 and the support values of *bread*, *butter* and *milk* be 0.25, 0.35 and 0.6 respectively. The rule covers 80% of *bread*'s presence in the database thus explaining only a fraction of *bread*'s presence. The AR covers only 20 transactions of the 25 transactions containing *bread*. Similarly, the rule explains 57.14% and 33.33% of the transactions containing *butter* and *milk* respectively. The rule does not cover or explain the presence of all its constituent items to an equal extent. *Bread* is covered to a larger extent, as compared to *butter* and *milk*. The rule has a w-value of 0.43176. On average, the rule is unable to explain 43.176 % of the presence of its constituents in the database.

A low w_R value does not completely explain the rule's constituents' behaviour. A transaction t covered by R may contain additional items not belonging to R. Hence, R reveals only some of the relationships present in the transactions it covers. Other ARs (consisting of the same items or R's subset/superset) originating from the same set of transactions, may also get mined. These ARs reveal relationships that are different from those revealed by R. The 'w-value' of a rule is specific to a particular rule and not to its constituents. In fact, any w-value brings out the strength of relationship between an AR and its constituents. A low w-value indicates strong characterization of its constituent items. This is because most of the transactions containing R's constituent items exhibit the behaviour captured by the rule. In addition, a low w-value signifies generality (indicating wider coverage in the database) of the relationship described by the rule. However, a high w-value indicates more specificity in the relationships revealed by the rule and explains a small portion of the database that contains its constituent items. Thus, an increase in the w-value increases its specificity. The generality/specificity revealed by the *weakness* measure is computed only with respect to transactions containing its constituent items, and not with respect to the entire database.

The generality/specificity revealed by the *weakness* measure could help in identifying non-routine purchasing patterns. Certain items could be purchased as sets only on particular occasions. For example, Christmas purchases bring together items that are not normally related during the rest of the year. These purchases could give rise to ARs that have high values of *support* and *confidence*. When a user analyses the entire set of mined association rules, it could be difficult to identify seasonal patterns from normal patterns. However, their *weakness* values could throw some light on their seasonal nature. Rules arising from seasonal or non-routine purchasing behaviour are likely to have high *weakness* values. This is because these rules will be able to explain only a small portion of their constituent item's presence in the database. Purchases characteristic to a certain period are not valid during the rest of the year. On the contrary, rules having low *weakness* values reflect normal purchasing behaviour that may happen throughout the year. Thus, *weakness* measure could help a retail outlet manager in identifying the nature of the sales pattern revealed by an AR leading to designing appropriate retail marketing strategies.

3.1 Range of *Weakness*

Weakness reveals the extent to which the behaviour captured by an AR is specific/general with respect to its constituent items. The lower bound of *weakness* is 0. This happens when $\frac{S_R}{S_{a_i}} = 1 \forall a_i \in R$. w_R being 0 means transaction set $T_R (\subseteq D)$ completely covers the presence of all items in R. The following holds for every transaction in the database and for every item present. If $(a_i \in t_x \wedge a_i \in R)$ then $t_x \in T_R$. It may seem from Equation 1 that w_{a_i} can assume a maximum value of 1. However, since $S_R \geq minsup > 0$ and $S_{a_i} \geq S_R$, the value of ratio $\frac{S_R}{S_{a_i}}$ can never be 0. It may easily be shown that the range of w_R is $[0, \frac{n-1}{n}(1 - minsup)]$. A proof of this is given in theorem 1 of the Annexe. A rule attains its maximum w_R value only when one of its constituent items has a support value of *minsup* with the rest having a support value of 1. In other words, the rule should explain the presence of one item completely and other items of the rule in the database are explained to the minimum possible extent.

3.2 Interpretation of *Weakness* in terms of *Confidence*

Confidence signifies the predictive ability of an AR and is given by:

$$Confidence(a \rightarrow b) = \frac{|a \cap b|}{|b|} = P(b/a),$$

where $|x|$ is the cardinality of the transaction set containing item x and $P(x)$ is the probability of item x. *Confidence* reveals the extent to which one can predict the purchase of item b assuming the customer to have already purchased item a in the same transaction. *Weakness* of an AR can be expressed in terms of *confidence* values of a number of related ARs. This is to be expected since the *weakness* measure captures the inability to explain the presence of an AR's constituent items in the transaction database. In some sense, *weakness* expresses inability to predict.

With $\frac{S_R}{S_{a_i}}$ being the confidence for the AR: $a_i \rightarrow a_1 a_2 \ldots a_{i-1} a_{i+1} \ldots a_n$, Equation 2 may be rewritten as:

$$w_R = \frac{1}{n}\sum_{a_i}(1 - \frac{S_R}{S_{a_i}}) = \frac{1}{n}\sum_{a_i}(1 - conf(a_i \rightarrow a_1 a_2 ... a_{i-1} a_{i+1} ... a_n)) \quad (3)$$

Each term in the summation conveys the extent to which AR ($a_i \rightarrow a_1 a_2 ... a_{i-1} a_{i+1} ... a_n$) is unable to predict the purchase of set $\{a_1, a_2, ..., a_{i-1}, a_{i+1}, ..., a_n\}$ from the purchase of the single item a_i of R. In other words, each term conveys the extent to which purchase of a_i does not lead to the purchase of set $R - \{a_i\}$. The *weakness* of a rule is the average of w-values covering all constituent items. Therefore, when interpreted in terms of *confidence*, *weakness* of an AR reveals the extent (on average) to which, an item is unable to predict the presence of all other items of the rule. The *weakness* value consolidates predictive abilities of the 'n' ARs having the form {*any item of R*} → {*set of all other items of R*}, n being the total number of items in the rule. Thus an AR assuming a low *weakness* value indicates highly confident predictability of the presence of other items given the presence of a single item.

4 A *Weakness*-based Distance between two Association Rules

Consider two association rules R_1 and R_2 having respective *weakness* values w_1 and w_2. R_1 and R_2 may contain items belonging to different domains and hence may originate from transactions sets belonging to different domains. These transaction sets may not have overlapping transactions. This implies that there may not be any common basis to compare R_1 and R_2. However, R_1 and R_2 might exhibit similar behaviour, though they are from different domains. It would then be useful to provide information concerning generality or specificity of ARs with respect to their constituent items. Such information could improve the decision making process.

Consider the rule $\{bread\} \rightarrow \{butter\}$, and assume it has a high w-value (low generality). This low generality indicates that relationships between each of $\{bread, butter\}$ and other items may be revealed in other rules. These relationships may also be worth exploring. Actions taken only on the basis of a single rule having a high w-value (high specificity) could be skewed. This is because an individual rule brings out only a limited aspect of the items' behaviour in the database. Since *weakness* values reflect the presence of relationships among constituents in the database, actions based on rules with equal or near-equal values are likely to yield similar results. Thus the *weakness* of a rule could be used as a heuristic to cluster rules.

In general, distance between two objects may be defined in terms of the various attributes on which they differ. The greater the difference, the greater is the distance between them. Here, we need to define a *Weakness*-based distance function between two ARs. We define distance between R_1 and R_2, as

$$d_w(R_1, R_2) = \frac{|w_1 - w_2|}{w_1 + w_2}, where\ 0 \leq w_1, w_2 \leq 1. \quad (4)$$

Distance between two ARs is the ratio of the difference in their *weakness*-values to their sum. If w_1, w_2, w_3 and w_4 (w_i being the w-value of rule R_i), have values of 0.6, 0.3, 0.7 and 0.2 respectively, then $d_w(R_1, R_2)$=0.333 and $d_w(R_3, R_4)$ =0.555. For a constant sum (of the *weakness* values of two rules), the distance between two rules is directly proportional to the difference of their w-values.

An interesting observation is as follows. Any difference Δw results in a larger distance for low w-values and smaller distance for high w-values.If ($|w_1 - w_2| = |w_3 - w_4|$) and ($w_1 + w_2 \leq w_3 + w_4$), then $d_w(R_1, R_2) > d_w(R_3, R_4)$.

Consider an example. Let w_1=0.4, w_2=0.2, w_3=0.8 and w_4=0.6. Then, $d_w(R_1, R_2)$=0.3333 while $d_w(R_3, R_4)$=0.14285. At a glance, this may seem counter-intuitive. The rationale is as follows. Rules R_1 and R_2 are not able to explain 40% and 20% of the presence of their constituent items respectively. Thus, they are more general than rules R_3 and R_4 whose w-values are 0.8 and 0.6 respectively. R_3 and R_4 have poorer explanatory power than R_1 and R_2, with respect to their constituent items.

In addition, the rationale has an analogical intuitive support. Consider four individuals A, B, C and D. Assume A and B to possess deeper knowledge (of a topic) than C and D. A, B, C and D represent rules R_1, R_2, R_3 and R_4 respectively. Let the absolute difference in the knowledge levels between the individuals in each of $\{A, B\}$ and $\{C, D\}$ be the same. Since A and B are quite knowledgeable, the difference would seem to be larger because it would require more effort to move from A's knowledge level to B's knowledge level. This greater effort may be due to the subtle and conceptually deeper knowledge required. However, it may be relatively easier to bridge the gap between C and D. Fewer facts and straightforward knowledge acquisition may suffice. Similarly, R_1 and R_2 may have good explanatory power and hence the distance between them may seem larger than the distance between R_3 and R_4 - rules that are more specific thus having less explanatory power.

Another analogy could be with respect to the amount of investment needed to increase sales under differing market conditions. Consider a market (say M_1) that is near saturation. This essentially means that the potential for increasing sales in M_1 is lower than that for an unpenetrated market (say M_2). Hence in a saturated market (M_1), a firm may need to invest a large amount (I_1) in sales promotion, advertising and other activities in order to increase its sales by say ΔS. However in the case of M_2, a smaller investment ($I_2 << I_1$) could achieve the same increase (ΔS). Hence, the same investment is likely to result in a larger increase (of sales) in M_2. In terms of the investment required, the distance between the two sales figures could be considered to be larger in a saturated market, than the distance between the same two sales figures in an unsaturated market.

It is easy to establish the metric properties of distance function $d_w(R_i, R_j)$. The proof is given in Annexe, as theorem 2. The intuitive justification of $d_w(R_i, R_j)$ along with its being a metric enables its usage as a distance function for clustering association rules.

5 Clustering Association Rules using $d_w(R_i, R_j)$

ARs with high *weakness* values explain only a small fraction of their constituent items' presence in a database. Hence, decisions made on the basis of ARs having high *weakness* values may not be applicable to many of the future transactions involving the rule's items. This is despite the rule having a high *confidence* value. On the other hand, ARs with low *weakness* values cover a major portion of their constituent items. Such rules are likely to bring out more general characteristics of their constituent items. The generality/specificity brought out by *weakness* does not depend on the actual *support* value of the rule. This is because the *weakness* value of a rule is mainly concerned with the *support* of the rule expressed as a fraction $\frac{S_R}{S_{a_i}}$ of the *support* of each of its constituent items.

Rules having different *weakness* values may require different strategies with respect to the marketing of constituent items. *Weakness* values help retail outlet managers in getting a prior idea about the possible effectiveness of their retail decisions. Two ARs having identical or near equal values of *weakness* indicate that they are 'similar.' This similarity is due to the fact that they explain their constituent items' presence to about the same extent. Hence, retail decisions based on the two rules may turn out to be equally effective. This might be independent of the domains of membership of the items in the rules. If ARs are segregated into groups based on their *weakness* values, then there may not be any need to examine all the rules. Only those groups that have low *weakness* values may need to be examined for making more effective decisions. In addition, a retail manager can have a similar strategy with respect to retail decisions for all items described by the rules in the same cluster. This is because all rules in a cluster are similar with respect to their explanatory power about their items' presence in the database.

Weakness may also be used as an interestingness measure. Rules can then be ranked according to their *weakness* values. Rules that rank low may be selected for further action. However, in this process we may lose information about natural groups present in the rule set. In addition, we may impose artificial thresholds and cut-offs. This could reduce the intuitiveness of the whole scheme. Hence, during analysis and decision making, a retail manager may find *weakness*-based clustering of ARs a more effective strategy than its usage as an interestingness measure.

Distance measures are measures of dissimilarity with large values denoting low similarity. From Equation 4, it may be observed that two rules are close to each other if: (1) absolute value of the difference between their individual *weakness* values is small; and/or (2) sum of their individual *weakness* values is large.

We now demonstrate the efficacy of $d_w(R_i, R_j)$ in clustering ARs extracted from an artificial dataset. Table 1 represents an artificial transaction dataset consisting of 100 transactions. We have consolidated identical transactions and indicated their frequencies in the adjacent column. Each market-basket is the set of items purchased by a customer during a single buying-instance. Here, each transaction is a subset of the set of nine items $\{Bread, Butter, Jam, Milk, Chocolate, Biscuit, Pen, Pencil, Eraser\}$. Although the transaction set is small, it is sufficient to bring out the efficacy and usefulness of the proposed clustering scheme. Table 2 displays the support values of the various items present in the transactions given in Table 1. It may be seen from Table 1 that the set of 100 transactions contains fifteen unique market baskets. Set $\{Pen, Pencil, Eraser\}$ is the most frequent market-basket with 13 occurrences while each of $\{Butter, Milk\}$, $\{Chocolate, Pencil, Eraser\}$ and $\{Pen, Eraser\}$ occurs as three transactions. Market-basket in a transaction is different from an item-set. Market-basket $\{Pen, Eraser\}$ is purchased separately in three transactions. On the other hand, item-set $\{Pen, Eraser\}$ occurs in a total of 16 transactions. This is because $\{Pen, Eraser\}$ forms a part of market-basket $\{Pen, Pencil, Eraser\}$. There are 13 transactions containing only the items *Pen*, *Pencil* and *Eraser*. Fourteen ARs were mined from the transaction set using a *support* and *confidence* thresholds of 0.1 and 0.5 respectively. These ARs, along with values of other parameters are listed in Table 3.

Table 1. An artificial transaction dataset

Transaction	Nos.	Transaction	Nos.
$\{Bread, Butter\}$	6	$\{Bread, Jam\}$	5
$\{Bread, Milk\}$	4	$\{Bread, Butter, Milk\}$	10
$\{Milk, Chocolate\}$	6	$\{Chocolate, Biscuit\}$	8
$\{Milk, Chocolate, Biscuit\}$	11	$\{Butter, Milk\}$	3
$\{Pen, Pencil, Eraser\}$	13	$\{Pencil, Eraser\}$	7
$\{Chocolate, Pencil, Eraser\}$	3	$\{Pen, Eraser\}$	3
$\{Chocolate, Biscuit, Pencil\}$	5	$\{Bread, Butter, Milk, Jam\}$	4
$\{Bread, Jam, Milk\}$	12	—	—

Table 2. Support values of items present in transactions of Table 1

Item	Support	Item	Support	Item	Support
Bread	0.41	Chocolate	0.33	Butter	0.23
Biscuit	0.24	Jam	0.21	Pen	0.16
Milk	0.50	Pencil	0.28	Eraser	0.26

Some interesting observations may be made from the *weakness* values of various ARs. The *weakness* values range from 0.13634 (R_6) to 0.662778 (R_7). R_6 and R_7 have two common items namely, *Chocolate* and *Biscuit*. The higher w-value of R_7 may be due to two reasons. The *support* of R_7 (0.11) is much lower than that of R_6 (0.24). Hence R_7 is not able to account for the presence of *Chocolate* and *Biscuit* as much as R_6. Secondly, the presence of *Milk* in R_7 further increases its *weakness* value. This is because R_7 is able to explain the presence of *Milk* in only 11 of the 50 transactions (22.0 %) that contain *Milk*. The inability to substantially cover its constituent *Milk* increases its *weakness*. A high *support* value does not necessarily guarantee a low *weakness* value. This is demonstrated by the weakness value of R_3. It has a *support* of 0.30 but a w-value of 0.334146. This is because the *support* of R_3, though high, is not sufficient to cover the presence of *Bread* and *Milk*. *Bread* and *Milk* are present in 41 and 50 transactions respectively.

Table 3. Association Rules extracted from transaction set of Table 1

No.	Rule	Support	Confidence	Weakness
R_1	{*Butter*} → {*Bread*}	0.20	0.86957	0.321315
R_2	{*Jam*} → {*Bread*}	0.21	1.00	0.243902
R_3	{*Bread*} → {*Milk*}	0.30	0.7317	0.334146
R_4	{*Butter*} → {*Milk*}	0.17	0.73913	0.460435
R_5	{*Butter, Milk* } → {*Bread*}	0.14	0.82353	0.589947
R_6	{*Chocolate*} → {*Biscuit*}	0.24	0.72727	0.136364
R_7	{*Milk, Biscuit* } → {*Chocolate*}	0.11	1.00	0.662778
R_8	{*Pen*} → {*Pencil, Eraser*}	0.13	0.8125	0.407738
R_9	{*Pen* } → {*Pencil*}	0.13	0.8125	0.361607
R_{10}	{*Pencil*} → {*Eraser*}	0.23	0.82143	0.146978
R_{11}	{*Pen*} → {*Eraser*}	0.16	1.00	0.192308
R_{12}	{*Jam, Milk* } → {*Bread*}	0.16	1.00	0.509284
R_{13}	{*Jam* } → {*Milk*}	0.16	0.76190	0.459048
R_{14}	{*Chocolate* } → {*Milk*}	0.17	0.51515	0.572424

As the cardinality of the set of items in a rule increases, the *support* of the rule decreases. However, this does not necessarily imply that *weakness* should also increase. The *weakness* of a rule is computed as the average *weakness* over all of its constituents. *Support* of a rule as a fraction of the *support* of its constituents may vary from item to item. Hence, in some cases it may happen that the *weakness* of a rule may actually be lower than the *weakness* value of its sub-rule. A typical situation is as follows. Consider a rule $AB \rightarrow C$ and its sub-rule $A \rightarrow B$. Let $sup(AB \rightarrow C) = sup(A \rightarrow B) = 0.2$. Let the *support* values of the various items be related as follows: $sup(A)$, $sup(B) > sup(C)$. For example, $sup(A)$=0.5, $sup(B)$=0.4 and $sup(C)$=0.3. Then, $w(AB \rightarrow C)$ (0.4778) $< w(A \rightarrow B)$ (0.55). This is because rule $AB \rightarrow C$ is able to explain its constituent C to a large extent. This leads to a reduction in the *weakness* of rule $AB \rightarrow C$. However, in general, since the *support* of a rule is always less than or equal to the *support* of its sub-rule, the *weakness* value of a sub-rule is generally lower than that of its corresponding rule. This may be observed in rules R_{12} $\{Jam, Milk\} \rightarrow \{Bread\}$ and R_{13} $\{Jam\} \rightarrow \{Milk\}$. *Weakness* value of R_{12} is 0.509284 while that of R_{13} is 0.49048. Note that R_{12} and R_{13} have the same *support* value (0.16). The only difference is that R_{12} has an additional item *Bread* as one of its constituents. R_{12} covers *Bread* to a lower extent than its coverage of the other constituent *Jam*. Hence, the explanatory power of R_{12} with respect to set $\{Jam, Milk, Bread\}$ gets reduced. This increases the *weakness* of the rule. An examination of Table 3 does not reveal any direct relationship between *confidence* and *weakness* values of a rule. This is because *weakness* as a parameter is an algebraic combination of the confidence values of many related rules.

Table 4. Formation of Clusters (Average linkage Method with $d_w(R_i, R_j)$)

Note: Clusters are identified by one of their member rules

Step	C_x	C_y	**Merging Distance** $d_w(C_x, C_y)$	**No. of Members in new Cluster**
1	R_{13}	R_4	0.002	2
2	R_{14}	R_5	0.015	2
3	R_3	R_1	0.020	2
4	R_{10}	R_6	0.037	2
5	R_3	R_9	0.049	3
6	R_{13}	R_{12}	0.051	3
7	R_{14}	R_7	0.066	3
8	R_8	R_{13}	0.077	4
9	R_{11}	R_2	0.118	2
10	R_8	R_{14}	0.140	7
11	R_3	R_8	0.207	10
12	R_{10}	R_{11}	0.209	4
13	R_{10}	R_3	0.435	14

The fourteen rules listed in Table 3 were clustered using the agglomerative hierarchical clustering (average-linkage) method with *weakness*-based distance $d_w(R_i, R_j)$ as the similarity measure. Statistical package **SYSTAT11** was used. Hierarchical procedures for clustering may be agglomerative or divisive. In agglomerative hierarchical procedures, each object or observation (an AR in this case) is initially in a separate cluster. At each step, the two clusters that are most similar (nearest to each other) are combined to obtain a new aggregate cluster. The process is repeated until all objects are in a single cluster. Hence, this procedure produces $(N-1)$ cluster solutions where N is the number of objects or observations. In the average linkage method, similarity between two clusters is represented as the average distance from all objects in one cluster to all objects in another cluster. This approach tends to combine elements with low variance. The pair-wise distance matrix with the distance between two rules given by $d_w(R_i, R_j)$ formed the input to the SYSTAT11 package. Table 4 details the formation of rule clusters while Figure 1 gives the corresponding dendogram. The first cluster is formed by combining rules R_{13} and R_4 whose distance is 0.002.

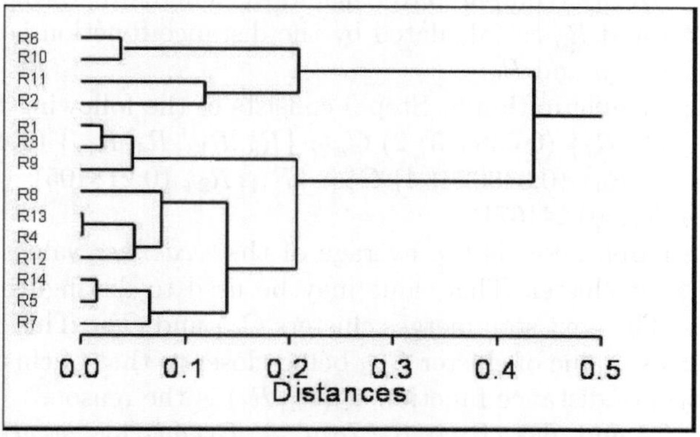

Fig. 1. Dendogram obtained using average linkage method and $d_w(R_i, R_j)$

Table 5. Cluster configuration at each step of the clustering process depicted in Table 4

Note: Singleton clusters are not depicted.

No.	Clusters
1	$\{R_{13}, R_4\}$
2	$\{R_{14}, R_5\}$, $\{R_{13}, R_4\}$
3	$\{R_3, R_1\}$, $\{R_{14}, R_5\}$, $\{R_{13}, R_4\}$
4	$\{R_{10}, R_6\}$, $\{R_3, R_1\}$, $\{R_{14}, R_5\}$, $\{R_{13}, R_4\}$
5	$\{R_3, R_1, R_9\}$, $\{R_{10}, R_6\}$, $\{R_{14}, R_5\}$, $\{R_{13}, R_4\}$
6	$\{R_{13}, R_4, R_{12}\}$, $\{R_3, R_1, R_9\}$, $\{R_{10}, R_6\}$, $\{R_{14}, R_5\}$
7	$\{R_{14}, R_5, R_7\}$, $\{R_{13}, R_4, R_{12}\}$, $\{R_3, R_1, R_9\}$, $\{R_{10}, R_6\}$
8	$\{R_8, R_{13}, R_4, R_{12}\}$, $\{R_{14}, R_5, R_7\}$, $\{R_3, R_1, R_9\}$, $\{R_{10}, R_6\}$
9	$\{R_{11}, R_2\}$, $\{R_8, R_{13}, R_4, R_{12}\}$, $\{R_{14}, R_5, R_7\}$, $\{R_3, R_1, R_9\}$, $\{R_{10}, R_6\}$
10	$\{R_8, R_{13}, R_4, R_{12}, R_{14}, R_5, R_7\}$, $\{R_{11}, R_2\}$, $\{R_3, R_1, R_9\}$, $\{R_{10}, R_6\}$
11	$\{R_8, R_{13}, R_4, R_{12}, R_{14}, R_5, R_7, R_3, R_1, R_9\}$, $\{R_{11}, R_2\}$, $\{R_{10}, R_6\}$
12	$\{R_{11}, R_2, R_{10}, R_6\}$, $\{R_8, R_{13}, R_4, R_{12}, R_{14}, R_5, R_7, R_3, R_1, R_9\}$
13	$\{R_{11}, R_2, R_{10}, R_6, R_8, R_{13}, R_4, R_{12}, R_{14}, R_5, R_7, R_3, R_1, R_9\}$

At each stage of the clustering process, we obtain a cluster configuration consisting of clusters and their constituents. The various cluster configurations occurring at each stage of the clustering process are listed in Table 5 with the newly formed cluster listed as the first cluster in each row. Table 5 does not list singleton clusters. Examination of Table 5 throws some light on the clustering process and the resulting cluster configurations. Rules R_{14} and R_5 merge to form a two-rule cluster in Step 2 prior to the merging of rules R_{10} and R_6. The difference in the *weakness* values between R_{14} and R_5 is 0.017523. This is greater than the difference between R_{10} and R_6, 0.010614. However R_{14} and R_5 are *weaker* rules as compared to R_{10} and R_6. Therefore, the actual distance separating R_{14} and R_5 as calculated by the distance function is smaller than that separating R_{10} and R_6.

The cluster configuration at Step 9 consists of the following five clusters:
1) C_{w1}: $\{R_{14}, R_5, R_7\}$ (0.608383) 2) C_{w2}: $\{R_8, R_{13}, R_4, R_{12}\}$ (0.459126)
3) C_{w3}: $\{R_3, R_1, R_9\}$ (0.339023) 4) C_{w4}: $\{R_{11}, R_2\}$ (0.218105)
5) C_{w5}: $\{R_{10}, R_6\}$ (0.141671)
The value in parentheses is the average of the *weakness* values of the rules constituting that cluster. This value may be used to distinguish one cluster from another. The next step merges clusters C_{w1} and C_{w2}. This is despite the average *weakness* value of cluster C_{w3} being closer to that of cluster C_{w2}. The functional form of distance function $d_w(R_i, R_j)$ is the reason.

Consider R_2 and R_{12}. Rule $R_2(Jam \rightarrow Bread)$ has *weakness*-value of 0.243902 while that of $R_{12}(\{Jam, Milk\} \rightarrow Bread)$ is 0.509284. Both rules score very high (a value of 1.00) on *confidence*. However, the difference in their *weakness* brings out another aspect of ARs. The relatively high *weakness* value of R_{12} indicates that every instance of the purchase of any one of its constituent items may not translate into purchases of the other items. On the other hand, it is more likely that purchase of *Jam* or *Bread* leads to purchase of set $\{Jam, Bread\}$. Hence, a scheme for set $\{Jam, Bread\}$ rather than for $\{Jam, Milk, Bread\}$ is likely to enhance sales to a larger extent. We cannot draw such a conclusion from an examination of *confidence* values alone. Grouping rules with comparable *weakness* values may help a manager use equipotent rules as a basis for decision-making.

The presence of a rule and its sub-rules in different clusters may be due to two reasons. One reason might be the difference in *support* between a rule and its sub-rules. A sub-rule's *support* value might be greater than a rule's *support* value. This may result in different *weakness* values for a rule and its sub-rules. Consequently, they may get assigned to different clusters. Another reason could be the existence of differences in the *support* values of items constituting the rules. If the *support* values of items constituting a rule have a wide variation, then different sub-rules may explain their constituents' presence to different extents. This difference in their *weakness* values may result in the rules being placed in different clusters. Consider rule R_7: $\{Milk, Biscuit\} \rightarrow \{Chocolate\}$. It is a member of cluster C_{w1} after Step 9. Its sub-rules R_{14}: $\{Chocolate\} \rightarrow \{Milk\}$ and R_6: $\{Chocolate\} \rightarrow \{Biscuit\}$ are members of clusters C_{w1} and C_{w5} respectively. While C_{w1} is a high-*weakness* cluster, with its elements having an average *weakness* of 0.608383, cluster C_{w5} is a low-*weakness* cluster with its members exhibiting an average *weakness* of 0.141671. The *support* values of *Milk*, *Chocolate* and *Biscuit* also show some variation the respective values being 0.50, 0.33 and 0.24. Low *support* coupled with high variation in the *support* values of its constituents might result in a *weak* rule.

An interesting observation pertains to item *Milk*. *Milk* that is purchased in 50 transactions is the most frequently occurring item in the database. Seven of the fourteen rules in Table 3 have *Milk* as one of their constituents. However, it is surprising to note that rules describing *Milk* belong to one of the high-*weakness* clusters. This may be due to the fact that none of the rules that contains *Milk* covers its presence to a substantial extent. It may be observed from Table 3 that most of the rules containing *Milk* (with the exception of R_3) have low *support*. These rules cover the presence of *Milk* to a partial extent. The high support of *Milk* increases the *weakness* of low *support* rules that contain it. This is because these rules are unable to explain a large portion of its presence. Thus, we may state that a frequently occurring (high *support*) item may be present in many rules having high *weakness* if the item is purchased in many non-overlapping low *support* market baskets. A similar observation can be made with respect to *Bread* (being supported by 41 transactions).

Another observation is with respect to rules in clusters that have relatively lower average *weakness* values. Consider rules $R_{10}(\{Pencil\} \rightarrow \{Eraser\})$ and $R6(\{Chocolate\} \rightarrow \{Biscuit\})$. Note that the *support* of R_{10} (0.23) is quite close to the *support* of its items *Pencil* and *Eraser*, namely 0.28 and 0.26 respectively. This is an example of low *weakness* clusters not containing high *support* rules. The high explanatory power of such a rule is derived from its *support* value being close to the *support* values of its constituent items.

Cluster C_{w4} consisting of R_{11} and R_2 exhibits another noteworthy characteristic. *Support* of $R_2(\{Jam\} \rightarrow \{Bread\})$ and that of its item *Jam* are both 0.21. Thus, R_2 explains the presence of *Jam* completely. This contributes to the low *weakness* of R_2. However R_2 covers the presence of *Bread* only to the extent of 51.21%. Similarly, R_{11} derives its low *weakness* value mainly from its ability to explain its constituent item *Pen* completely. The other items purchased with jam are revealed by R_{12} and R_{13}.

6 Comparative Analysis

The proposed AR clustering scheme is compared with another scheme from recent literature. Sahar [19] proposed a new non-metric distance measure, d_{SC}. d_{SC} defines distance between two rules on the basis of overlap in the set of transactions that each rule covers. In addition, d_{SC} also uses the differences in item-sets that constitute the two rules. In particular, we have "the dissimilarity between two rules is the weighted sum of the dissimilarities between the assumptions (antecedents), consequents, and attribute sets that make up the two rules, where each component is a weighted measure of the dissimilarities of the support ($diff_{sup}$) and the attribute set ($xor\ ratio$)." [19] Distance measure d_{SC} is given by:

$$d_{SC}(A \rightarrow B, C \rightarrow D) = [1 + diff_{sup}(A, C)]\frac{\|A \oplus C\|}{\|A \cup C\|}\gamma_1$$
$$+[1+diff_{sup}(B, D)]\frac{\|B \oplus D\|}{\|B \cup D\|}\gamma_2 + [1+diff_{sup}(A \cup B, C \cup D)]\frac{\|(A \cup B) \oplus (C \cup D)\|}{\|A \cup B \cup C \cup D\|}\gamma_3$$

where, $diff_{sup}$(difference-in-support) is defined as:
$diff_{sup}(A, B) = support(A) + support(B) - 2 \times support(A \cup B)$.
1 is added to $diff_{sup}$ for cases where both $diff_{sup}(A, B) = 0$ and $(A \oplus B) \neq \phi$ hold at the same time. Symbol $\oplus$ denotes the xor operation: $X \oplus Y = (X \backslash Y) \cup (Y \backslash X)$. $\|X\|$ is the cardinality of elements in set X. Choices of values for γ-weights reflect preferences for set inclusion. Sahar [19] gives further details about d_{SC}.

Computation of d_{SC} and that of d_w differ in two respects. 'd_{SC}' directly computes dissimilarity between any two rules R_1 and R_2 by using xor-ratio of sets in the rules, and differences in their transaction coverage (obtained by the $diff_{sup}$ measure). d_w computation is a two step process.The *weakness* values of the two rules get computed in the first step. The second step consists of calculating the d_w value on the basis of the computed *weakness* values. d_{SC} compares the itemsets given by the antecedents, consequents and combinations thereof while computing distance. On the other hand, d_w considers each item of a rule separately while computing *weakness* values.

Table 6. Formation of rule clusters during average linkage clustering method involving d_{SC} distance measure

Note: Clusters are identified by one of their member rules

Step	C_x	C_y	**Merging Distance** $d_{SC}(C_x, C_y)$	**No. of Members in new Cluster**
1	R_9	R_8	0.429	2
2	R_{12}	R_2	0.437	2
3	R_5	R_1	0.442	2
4	R_{11}	R_9	1.098	3
5	R_4	R_5	1.892	3
6	R_{12}	R_{13}	1.958	3
7	R_{10}	R_{11}	2.244	4
8	R_{14}	R_6	2.313	2
9	R_3	R_{12}	2.734	4
10	R_4	R_3	2.773	7
11	R_{14}	R_7	2.875	3
12	R_{14}	R_4	3.980	10
13	R_{10}	R_{14}	4.437	14

The d_{SC}-based distance matrix was computed (for the rule set presented in Table 3) and given as input to the SYSTAT11 package. The 14 ARs were clustered using average-linkage agglomerative hierarchical procedure - the same procedure that was used with respect to d_w. In addition, the study presented in [19] also used agglomerative hierarchical methods for clustering ARs, thus giving us a common basis for comparing the results of the two methods.

Table 6 displays the process of cluster formation while Table 7 displays the cluster configurations at each step of the process. The corresponding dendogram is presented in Figure 2. Some of the differences between the two clustering schemes get revealed during the process of cluster formation. Consider the first step of d_{SC} and d_w-based clustering schemes. In the case of d_w-based clustering scheme, singleton clusters $\{R_{13}\}$ and $\{R_4\}$ are the first to merge. It may be observed from Tables 2 and 3 that R_{13} and R_4 have very close *support* values, 0.16 and 0.17 respectively. This leads to near equal *weakness* values and hence to the proximity of the rules in terms of d_w distance. However, R_{13} and R_4 have $\{Jam\}$ and $\{Butter\}$ as the respective antecedents. Items *Jam* and *Butter* occur together only in four transactions. The same transactions also bring together item set $\{Jam, Butter, Milk\}$. Although the consequents of R_{13} and R_4 are identical, the low *support* for other transaction overlaps, namely antecedent and total itemset, separates R_{13} and R_4 by a higher value for $d_{SC}(R_{13}, R_4)$. Thus the clusters containing R_{13} and R_4 merge only in the 10th step of the clustering process.

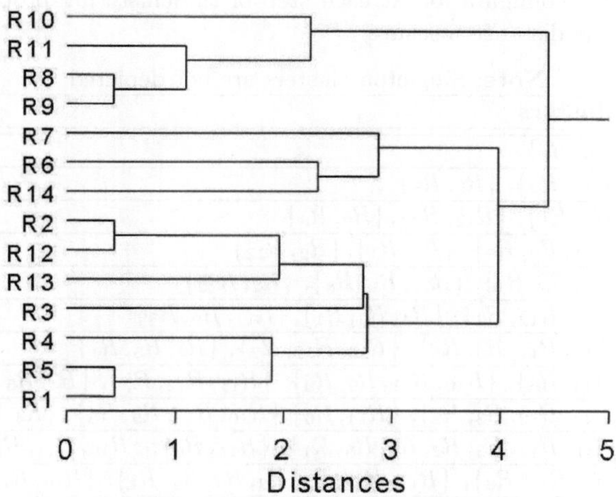

Fig. 2. Dendogram depicting the cluster formation during Average linkage clustering method with d_{SC} distance

The first step of d_{SC}-based clustering scheme brings singleton clusters $\{R_9\}$ and $\{R_8\}$ to form a 2-rule cluster. It is interesting to note that

$R_9(\{Pen\} \rightarrow \{Pencil\})$ is a sub-rule of $R_8(\{Pen\} \rightarrow \{Pencil, Eraser\})$. In addition R_8 and R_9 have the same *support* value of 0.13. This means R_8 and R_9 cover the same set of transactions. Further, there is complete match between the antecedent sets of R_8 and R_9. Hence the contribution due to antecedent dissimilarity towards $d_{SC}(R_8, R_9)$ is 0. Moreover, the consequent $\{Pencil\}$ of R_9 is a subset of the consequent $\{Pencil, Eraser\}$ of R_8. Essentially this means all transactions covered by R_8 are also covered by R_9 thus increasing the similarity between R_8 and R_9 and hence leading to a low $d_{SC}(R_8, R_9)$ value of 0.429167. In the case of d_w-based clustering scheme (Table 5), clusters containing R_8 and R_9 merge at Step 11. This brings out an important difference between the two clustering schemes. d_w-based clustering scheme tends to group rules having equal or near equal weakness values. This is irrespective of the originating transaction sets. Hence, rules that are equally effective in explaining the presence of their constituents are brought together irrespective of the domain. Thus, a retail manager can explore the common strands of purchasing behaviour exhibited by customers in different domains. This might lead to the implementation of effective domain-independent retail strategies. In contrast, d_{SC}-based clustering scheme groups those ARs that describe the same set of transactions. This is due to the usage of itemset matching procedure. Here each cluster consists of rules whose items are members of the same or closely related domains. Exploration of these clusters might lead to the formulation of domain-dependent retail strategies.

Table 7. Cluster configuration at each step of the clustering process depicted in Table 6 using d_{SC} distance measure

Note: Singleton clusters are not depicted.

No.	Clusters
1	$\{R_9, R_8\}$
2	$\{R_{12}, R_2\}$, $\{R_9, R_8\}$
3	$\{R_5, R_1\}$, $\{R_{12}, R_2\}$, $\{R_9, R_8\}$
4	$\{R_{11}, R_9, R_8\}$, $\{R_5, R_1\}$, $\{R_2, R_{12}\}$
5	$\{R_4, R_5, R_1\}$, $\{R_{11}, R_9, R_8\}$, $\{R_2, R_{12}\}$
6	$\{R_{13}, R_{12}, R_2\}$, $\{R_4, R_5, R_1\}$, $\{R_{11}, R_9, R_8\}$
7	$\{R_{10}, R_{11}, R_9, R_8\}$, $\{R_{13}, R_{12}, R_2\}$, $\{R_4, R_5, R_1\}$
8	$\{R_{14}, R_6\}$, $\{R_{10}, R_{11}, R_9, R_8\}$, $\{R_{13}, R_{12}, R_2\}$, $\{R_4, R_5, R_1\}$
9	$\{R_{13}, R_{12}, R_3, R_2\}$, $\{R_{14}, R_6\}$, $\{R_{10}, R_{11}, R_9, R_8\}$, $\{R_4, R_5, R_1\}$
10	$\{R_{13}, R_{12}, R_3, R_2, R_4, R_5, R_1\}$, $\{R_{14}, R_6\}$, $\{R_{10}, R_{11}, R_9, R_8\}$
11	$\{R_7, R_{14}, R_6\}$, $\{R_{13}, R_{12}, R_3, R_2, R_4, R_5, R_1\}$, $\{R_{10}, R_{11}, R_9, R_8\}$
12	$\{R_7, R_{14}, R_6, R_{13}, R_{12}, R_3, R_2, R_4, R_5, R_1\}$, $\{R_{10}, R_{11}, R_9, R_8\}$
13	$\{R_7, R_{14}, R_6, R_{13}, R_{12}, R_3, R_2, R_4, R_5, R_1, R_{10}, R_{11}, R_9, R_8\}$

An observation may be made with respect to the second step. $R_{13}(\{Jam\} \rightarrow \{Milk\})$ and $R_2(\{Jam\} \rightarrow \{Bread\})$ are sub-rules of $R_{12}(\{Jam, Milk\} \rightarrow \{Bread\})$. d_{SC}-based scheme assigns a higher value to pair $\{R_{13}, R_{12}\}$ as compared to pair $\{R_{12}, R_2\}$. This is because antecedent sets of R_{13} and R_{12} match only partially. However, consequent sets of R_{12} and R_2 match completely while their antecedent sets match only partially. Hence, the contribution to antecedent dissimilarity is low while contribution to dissimilarity from their consequents is nil. This results in $d_{SC}(R_{12}, R_2)$ being assigned a lower value as compared to $d_{SC}(R_{13}, R_{12})$. Note that transactions covered by R_{12} are covered by R_2.

d_{SC} tends to bring together rules and their sub-rules in the same cluster. The order of merging of sub-rules with its rule may be predicted on the basis of overlap between transaction sets covered by various components of a rule and corresponding components of its various sub-rules. This is because dissimilarity between two rules is the weighted sum of dissimilarities among the antecedent, consequent and attribute sets that make up the two rules. Comparison of respective antecedents and respective consequents makes d_{SC}-based clusters somewhat intuitive. Step 12 demonstrates this. d_{SC}-based cluster configuration at Step 11 is: C_{1-11}: $\{R_7, R_{14}, R_6\}$, C_{2-11}: $\{R_{13}, R_{12}, R_3, R_2, R_4, R_5, R_1\}$ and C_{3-11}: $\{R_{10}, R_{11}, R_9, R_8\}$. At Step 12, C_{1-11}: $\{R_7, R_{14}, R_6\}$ merges with C_{2-11}: $\{R_{13}, R_{12}, R_3, R_2, R_4, R_5, R_1\}$ at a distance of 3.98. It may be observed that item *Milk* is common across the rules in clusters C_{1-11} and C_{2-11}. On the other hand, rules in C_{3-11}: $\{R_{10}, R_{11}, R_9, R_8\}$ have items from set $\{Pen, Pencil, Eraser\}$. None of the items from set $\{Pen, Pencil, Eraser\}$ occurs in rules belonging to clusters C_{1-11} and C_{2-11}.

Clusters generated by both the schemes reveal different features of the rules. d_{SC}-based clustering scheme cluster rules that show some degree of overlap with respect to the corresponding transactions sets. Hence, from a practical decision-making point of view, a user may not be interested in all the rules in a d_{SC}-generated cluster. d_{SC} is concerned only with those transactions in which all items of the two rules occur. Although this measures one aspect of co-occurrence, other equally important characteristics that deal with individual items present in a rule are not covered.

The motivation for employing clustering techniques lies in reducing the rule quantity problem in AR mining. Cluster configurations obtained by the application of d_w and d_{SC} may be viewed differently. Clustering using d_w tends to group rules with '*similar*' values of *weakness*. Since low-*weakness* clusters may be interesting for certain applications, rules in these clusters may be examined further. Cluster characteristics like average *weakness* may be used to define low-*weakness* clusters. This may help in the selection of appropriate clusters for further examination. Rules in other clusters of the configuration need not be examined.

Unlike d_w, d_{SC} distance measure is not metric. This is because set intersection and union operations on which d_{SC} is based, do not define an equivalence relationship. More particularly, they are not transitive. Hence the triangular inequality may not hold. d_{SC}-based clustering is useful in clustering rules originating from the same portion of a database. Here rules, their sub-rules, and rules showing overlap in their transaction sets get grouped together. Sometimes this results in bringing together rules "involving different items but serving equal purposes" [14]. The clusters may thus bring together rules that may belong to the same or closely related domains. However, a rule and its sub-rules may vary a great deal on parameters, like explanatory power, etc. Hence, a user may have to examine different clusters to find rules having the same specificity or generality. d_{SC}-based clustering may not facilitate certain kinds of decision-making. This is especially true when grouping has to be based on parameters across domains. d_{SC}-based clustering may not be able to distinguish between rules that reflect routine customer purchasing behaviour and those rules that reflect non-routine behaviour. On such occasions, distance measures like d_w may be helpful.

7 Conclusion

The rule immensity problem in AR mining hinders comprehensibility and hence effective decision-making. Here, we explored the utility of *clustering* as a possible approach towards mitigating this problem. In contrast to the predominantly syntactic and transaction coverage overlap-based approaches towards clustering ARs in literature [3, 14, 1, 4, 10, 19], we introduced the notion of *weakness* - an inherent property of an AR.

Weakness reveals the extent to which an AR is unable to explain the presence of its constituent items in a transaction database. *Weakness* values give an indication of a rule's specificity/generality. Highly specific rules that pertain to a limited aspect of an item's behaviour are characterised by a large value of *weakness*. Further, *weakness* values may be useful in identifying non-routine patterns such as seasonal purchases. A *weakness*-based mechanism for clustering ARs was developed and explored. Some properties of *weakness* were then examined. Following this, a distance function (d_w) that used *weakness* as its basis was developed and its intuitiveness was discussed with illustrations. Its usage in clustering ARs was demonstrated with an example.

The proposed AR clustering scheme was compared with d_{SC} -based AR clustering scheme [19]. Since d_{SC} makes use of overlap of transaction sets covered by two rules in defining distance between rules, it tends to cluster an AR with its sub-rules. Such clusters may summarize a domain, but may not aid in certain decision-making processes. This is especially true when the decision-making involves aspects of commonality across domains. On the other hand, the proposed d_w-based clustering scheme views ARs in a different light. The scheme exploits one of the inherent properties, namely *weakness*, instead of overlap between corresponding transaction sets. This facilitates categories of decision-making where the goal is to identify rules having similar explanatory power. This, being domain-independent, facilitates certain marketing decisions that may be applied across all domains. Consequently, examination may be limited to only those clusters that have some relevance to a specific decision. This in turn mitigates the rule immensity problem to some extent.

Application of clustering techniques is a promising and useful approach towards mitigating the effects of rule immensity in AR mining. Our future research agenda will involve exploration of alternative clustering methods and distance measures for AR clustering.

References

1. Jorge A. Hierarchical clustering for thematic browsing and summarization of large sets of association rules. In *Proceedings of the 2004 SIAM International Conference on Data Mining*, 2004.
2. Kosters W. A., Marchiori E., and Oerlemans A. J. Mining clusters with association rules. In *Proceedings of Third Symposium on Intelligent Data Analysis (IDA 99)*, volume 1642 of *LNCS*, pages 39–50. Springer-Verlag, 1999.
3. Baesens B., Viaene S., and Vanthienen J. Post-processing of association rules. In *Proceedings of the Workshop on Post-Processing in Machine Learning and Data Mining: Interpretation, Visualization, Integration, and Related Topics within The Sixth ACM SIGKDD International Conference on Knowledge Discovery and Data Mining*, Boston, MA, USA, 2000. ACM SIGKDD.
4. Lent B., Swami A., and Widom J. Clustering association rules. In *Proceedings of the Thirteenth International Conference on Data Engineering*, pages 220–231, April 1997.
5. Liu B., Hu M., and Hsu W. Multi-level organization and summarization of the discovered rules. In *Proceedings of the ACM SIGKDD International Conference on Knowledge Discovery and Data Mining (KDD 2000)*, pages 208–217, Boston, USA, 2000. ACM SIGKDD, ACM Press.
6. Padmanabhan B. and Tuzhilin A. Unexpectedness as a measure of interestingness in knowledge discovery. *Decision Support Systems*, 27(3):303–318, 1999.
7. Aggarwal C. C., Procopiuc C., and Yu P. S. Finding localized associations in market basket data. *IEEE Transactions on Knowledge and Data Engineering*, 14(1):51–62, 2002.
8. Adomavicius G. and Tuzhilin A. Expert-driven validation of rule-based user models in personalization applications. *Data Mining and Knowledge Discovery*, 5(1-2):33–58, 2001.
9. Dong G. and Li J. Interestingness of discovered association rules in terms of neighborhood-based unexpectedness. In *Proceedings of the Second Pacific-Asia Conference on Knowledge Discovery and Data Mining*, pages 72–86. Springer-Verlag, 1998.
10. Toivonen H., Klemettinen M., Ronkainen P., Hatonen K., and Mannila H. Pruning and grouping discovered association rules. In *Proceedings of the MLnet Workshop on Statistics, Machine Learning and Knowledge Discovery in Databases*, Herakhion, Crete, Greece, 1995.
11. Bayardo Jr. R. J., Agrawal R., and Gunopulos D. Constraint-based rule mining in large, dense databases. *Data Mining and Knowledge Discovery*, 4(2-3):217–240, 2000.
12. Grabmeier J. and Rudolph A. Techniques of cluster algorithms in data mining. *Data Mining and Knowledge Discovery*, 6:303–360, 2002.

13. Miller R. J. and Yang Y. Association rules over interval data. In *Proceedings of the ACM SIGMOD Conference on Management of Data*, pages 452–461. ACM SIGMOD, ACM Press, 1997.
14. Gupta G. K., Strehl A., and Ghosh J. Distance based clustering of association rules. In *Proceedings of Intelligent Engineering Systems through Artificial Neural Networks: ANNIE (1999)*, volume 9, pages 759–764, 1999.
15. Jain A. K., Murty M. N., and Flynn P. J. Data clustering: A review. *ACM Computing Surveys*, 31(3):264–323, 1999.
16. Wang K., Tay S. H. W., and Liu B. Interestingness-based interval merger for numeric association rules. In *Proceedings of the International Conference on Data Mining and Knowledge Discovery*, pages 121–128, New York City, August 1998. AAAI.
17. Kaufman L. and Rousseeuw P. J. *Finding Groups in Data: An Introduction to Cluster Analysis.* Wiley, 1990.
18. Anderberg M. R. *Cluster Analysis for Applications.* Academic Press, 1973.
19. Sahar S. Exploring interestingness through clustering: A framework. In *Proceedings of the IEEE International Conference on Data Mining (ICDM 2002)*, pages 677–680. IEEE, IEEE Computer Society Press, 2002.

Annexes

Theorem 1. *The range of w_R is $[0, (\frac{n-1}{n})(1 - minsup)]$*

Proof:
The minimum value of w_R is 0. This value is obtained when the support of the rule is the same as the support of each constituent item. Each individual term in the summation of Equation 2 can attain a maximum value of $(1-minsup)$. This happens when S_R is restricted to its minimum value of minsup, while S_{a_i} attains its maximum value of 1. Let us assume that all terms of the summation contribute a value of $(1 - minsup)$ to w_R. Then, the summation yields,

$$w_R = \frac{1}{n}\sum_{a_i}(1 - \frac{S_R}{S_{a_i}}) = \frac{1}{n}\sum_{a_i}(1 - minsup) = \frac{n}{n}(1 - minsup) = 1 - minsup$$

This means $S_{a_i} = 1 \forall a_i \in \{a_1, a_2, ..., a_n\}$. However, if all items of R have a support value of 1 in the database, then the support of R cannot be minsup. The support becomes 1. This is because all transactions in the database contain each and every element present in R. S_R=1 forces its w-value to be 0.

Hence, if R is to have a support-value of minsup, then there should be at least one item, whose support is minsup. Then, the contribution due to this item say a_k, to w_R is 0, whereas each of the other $(n-1)$ items may contribute a maximum of $(1 - minsup)$ to w. Thus,
$max(w_R) = \frac{1}{n}\sum_{a_i}(1 - \frac{S_R}{S_{a_i}}) = \frac{1}{n}[\sum_{a_i(\forall i/i \neq k))}(1 - minsup) + (1 - \frac{minsup}{minsup})] = \frac{n-1}{n}(1 - minsup)$
Thus, the maximum value that w_R can attain is $(\frac{n-1}{n})(1 - minsup)$. □

To summarize, w_R can assume a value in the range $[0, \frac{(n-1)}{n}(1-minsup)]$, where n is the number of items in the rule. The minimum value is when the support for each item equals the support of the rule itself. It attains the maximum value when $S_R = minsup$ and $S_{a_i} = 1 \forall a_i \in \{a_1, a_2, ..., a_n\}, i \neq k$ and $S_{a_k} = minsup$, for some $k \in \{1, ..., n\}$.

Theorem 2. *Distance measure $d_w(R_i, R_j)$ is a metric.*

Proof:

Property 1: $d_w(R_i, R_j) \geq 0 \forall i, j$

Note that $d_w(R_i, R_j) = \frac{|w_i - w_j|}{w_i + w_j}; 0 \leq w_i, w_j \leq 1$.
$w_i = w_j \Rightarrow |w_i - w_j| = 0 \Rightarrow d_w(R_2, R_1) = 0$.
$w_i \neq w_j$
$\Rightarrow |w_i - w_j| > 0 \wedge (w_i + w_j) > 0$ *(Since,* $0 \leq w_i, w_j \leq 1$*)*
$\Rightarrow d_w(R_i, R_j) = \frac{|w_i - w_j|}{w_i + w_j} > 0$
Hence property 1.

Property 2: $d_w(R_i, R_i) = 0$

This property follows from the definition of $d_w(R_i, R_j)$.
$\Rightarrow d_w(R_i, R_i) = \frac{|w_i - w_i|}{w_i + w_i} = 0$ *(Since* $|w_i - w_i| = 0$*)*

Property 3: $d_w(R_i, R_j) = d_w(R_j, R_i)$

$d_w(R_i, R_j) = \frac{|w_i - w_j|}{w_i + w_j}$
Now, $|w_i - w_j| = |w_j - w_i| \forall i, j$ *(Property of the mod function).*
Therefore, $d_w(R_i, R_j) = d_w(R_j, R_i)$

Property 4: $d_w(R_i, R_k) \leq d_w(R_i, R_j) + d_w(R_j, R_k)$

We know that,
$d_w(R_i, R_j) = \frac{|w_i - w_j|}{w_i + w_j}$, $d_w(R_i, R_k) = \frac{|w_i - w_k|}{w_i + w_k}$ *and* $d_w(R_j, R_k) = \frac{|w_j - w_k|}{w_j + w_k}$
We also know that $w_i, w_j, w_k \geq 0$.*For ease of notation we assume:* $w_i = a$, $w_j = b$ *and* $w_k = c$. *We need to prove that*
$\frac{|a-b|}{a+b} + \frac{|b-c|}{b+c} \geq \frac{|a-c|}{a+c}$, *(I)*

Without any loss of generality, we may assume that $a \geq b \geq c$. *This translates to four cases that we consider below.*

Case 1: a=b=c
In this case Expression I is trivially true. The three weakness values are equal and represent the same point.

Case 2: $a = b > c$
We consider the L.H.S. of expression I.
$L.H.S. = \frac{|a-b|}{a+b} + \frac{|b-c|}{b+c} = \frac{|b-c|}{b+c}$ *(Since* $a = b$*)*
$= \frac{b-c}{b+c}$ *(Since,* $(b > c) \wedge (b, c) > 0 \Rightarrow |b - c| = (b - c)$*)*
$= \frac{a-c}{a+c}$ *(Since,* $a = b$*)*
$= \frac{|a-c|}{a+c}$ *(Since,* $a > c$*)*
$= R.H.S.$

Case 3: $a > b = c$
Considering the L.H.S. of expression I, we have,
$L.H.S. = \frac{|a-b|}{a+b} + \frac{|b-c|}{b+c}$
$= \frac{|a-b|}{a+b}$ *(Since,* $b = c$*)*
$= \frac{a-b}{a+b}$ *(Since,* $(a > b) \wedge (a, b) \geq 0 \Rightarrow |a - b| = (a - b)$*)*
$= \frac{a-c}{a+c}$ *(Since,* $b = c$*)*
$= \frac{|a-c|}{a+c}$ *(Since,* $a > c$*)* $= R.H.S.$
Case 2 and Case 3 deal with the condition when two of the three rules have equal weakness values. Therefore, the distance from one rule to any one of the other two rules is the same.

Case 4: $a > b > c$
Evidently inequality I translates to: $\frac{a-b}{a+b} + \frac{b-c}{b+c} \geq \frac{a-c}{a+c}$
$\Leftrightarrow \frac{(a-b)(b+c)+(b-c)(a+b)}{(a+b)(b+c)} \geq \frac{a-c}{a+c}$
$\Leftrightarrow \frac{ab-b^2+ac-bc+ab-ac+b^2-bc}{(a+b)(b+c)} \geq \frac{a-c}{a+c}$
$\Leftrightarrow \frac{2(ab-bc)}{(a+b)(b+c)} \geq \frac{a-c}{a+c}$
$\Leftrightarrow \frac{2b(a-c)}{(a+b)(b+c)} \geq \frac{a-c}{a+c}$
$\Leftrightarrow \frac{2b}{(a+b)(b+c)} \geq \frac{1}{a+c}$ *(Since,* $a > c \Rightarrow (a - c) > 0$*)*
$\Leftrightarrow \frac{2b}{(a+b)(b+c)} - \frac{1}{a+c} \geq 0$
$\Leftrightarrow 2b(a + c) - (a + b)(b + c) \geq 0$ *(Since,* $(a + b)(b + c)(c + a) > 0$*)*
$\Leftrightarrow 2ab + 2bc - ab - b^2 - ac - bc \geq 0$
$\Leftrightarrow ab - b^2 + bc - ac \geq 0$
$\Leftrightarrow b(a - b) + c(b - a) \geq 0$
$\Leftrightarrow b(a - b) \geq c(a - b)$ *(II)*
$b > c$ *[Assumption of Case 4]*
$\Rightarrow (a - b)b > (a - b)c$ *(Since,* $(a > b) \Rightarrow (a - b) > 0$*)*
This proves II establishing the validity of I with respect to case 4.
Therefore, $b(a - b) > c(a - b)$ *[Hence proved.]*
This means $\frac{|a-b|}{a+b} + \frac{|b-c|}{b+c} \geq \frac{|a-c|}{a+c}$ *holds for all the four cases thus proving the triangular inequality. Hence the distance function* $d_w(R_i, R_j)$ *is a metric.*

Part III

Rule quality and validation

A New Probabilistic Measure of Interestingness for Association Rules, Based on the Likelihood of the Link

Israël-César Lerman[1] and Jérôme Azé[2]

[1] Irisa-Université de Rennes 1, Campus de Beaulieu, 35042 Rennes Cédex
lerman@irisa.fr

[2] Laboratoire de Recherche en Informatique, Université Paris-Sud, 91405 Orsay Cédex aze@lri.fr

Summary. The interestingness measures for pattern associations proposed in the data mining literature depend only on the observation of relative frequencies obtained from 2×2 contingency tables. They can be called "absolute measures". The underlying scale of such a measure makes statistical decisions difficult. In this paper we present the foundations and the construction of a probabilistic interestingness measure that we call likelihood of the link index. This enables to capture surprising association rules. Indeed, its underlying principle can be related to that of information theory philosophy; but at a relational level. Two facets are developed for this index: symmetrical and asymmetrical. Two stages are needed to build this index. The first is "local" and associated with the two single boolean attributes to be compared. The second corresponds to a discriminant extension of the obtained probabilistic index for measuring an association rule in the context of a relevant set of association rules. Our construction is situated in the framework of the proposed indices in the data mining literature. Thus, new measures have been derived. Finally, we designed experiments to estimate the relevance of our statistical approach, this being theoretically validated, previously.

Key words: Probabilistic Intestingness Measure, Association Rules, Independence Random Model, Contingency Tables.

1 Introduction

Seeking for a relevant interestingness measure in the context of a given data base is a fundamental task at the heart of data mining problems. We assume that the data are given by a a set of objects described by a set of boolean attributes. Let us denote them by $\mathcal{O}$ and $\mathcal{A}$, respectively. The crossing between $\mathcal{O}$ and $\mathcal{A}$ leads to an incidence table. This indicates for each object the subset of attributes (properties) associated with its definition. Let us denote by n and p the cardinalities of $\mathcal{O}$ and $\mathcal{A}$.

Israël-César Lerman and Jérôme Azé: *A New Probabilistic Measure of Interestingness for Association Rules, Based on the Likelihood of the Link*, Studies in Computational Intelligence (SCI) **43**, 207–236 (2007)
www.springerlink.com

Let α_i^j specify the value of the attribute a^j for the object o_i: $\alpha_i^j = a^j(o_i), 1 \leq i \leq n, 1 \leq j \leq p$. The two possible values for a given α_i^j are "true" and "false". Generally the "true" and the "false" values are coded by the numbers 1 and 0, respectively. Without loss of generality we may suppose a "true" value as more significant than a "false" one. This is generally expressed in terms of statistical frequencies: the number of objects where a given attribute is "true" is lower than that where it is "false". We will represent a boolean attribute a by its extension $\mathcal{O}(a)$ which represents the subset of objects where a is true. Thus, $\mathcal{A}$ is represented by a set of parts of the object set $\mathcal{O}$.

The set $\mathcal{O}$ is generally obtained from a training set provided from a universe of objects $\mathcal{U}$. On the other hand, the boolean attribute set $\mathcal{A}$ can be obtained from conjunctions of more elementary attributes, called itemsets. Determining "significant" itemsets is a crucial problem of "Data Mining" [1, 6]. However, this problem will not be addressed here.

Let us now introduce some notations. With the attribute a of $\mathcal{A}$ we associate the negated attribute $\neg a$ that we represent by the complementary subset of $\mathcal{O}(a)$ in $\mathcal{O}$. For a given pair (a, b) of $\mathcal{A} \times \mathcal{A}$, we introduce the conjunctions $a \wedge b$, $a \wedge \neg b$, $\neg a \wedge b$ and $\neg a \wedge \neg b$ that are respectively represented by $\mathcal{O}(a) \cap \mathcal{O}(b)$, $\mathcal{O}(a) \cap \mathcal{O}(\neg b)$, $\mathcal{O}(\neg a) \cap \mathcal{O}(b)$ and $\mathcal{O}(\neg a) \cap \mathcal{O}(\neg b)$. Cardinalities of these sets are respectively denoted by $n(a \wedge b)$, $n(a \wedge \neg b)$, $n(\neg a \wedge b)$ and $n(\neg a \wedge \neg b)$. Finally, $n(a)$ and $n(b)$ will designate the set cardinalities of $\mathcal{O}(a)$ and $\mathcal{O}(b)$. These cardinalities appear in the contingency table crossing the two binary attributes $\{a, \neg a\}$ and $\{b, \neg b\}$. Consider also the ratios of these cardinalities over the number of objects n. These define the following relative frequencies or proportions: $p(a \wedge b)$, $p(a \wedge \neg b)$, $p(\neg a \wedge b)$ and $p(\neg a \wedge \neg b)$.

Relative to the entire set of attribute pairs $\mathcal{A} \times \mathcal{A}$ the objective consists in setting up a reduced subset of pairs (a^j, a^k), $1 \leq j < k \leq p$, such that a true value for the attribute a^j has a real tendency to imply a true value for that a^k. In order to measure numerically such a tendency that we denote by $(a \rightarrow b)$ for the ordered pair of boolean attributes (a, b), many association coefficients have been proposed. Inspired by different metric principles, they have not necessarily comparable behaviours for pattern association in a given application domain. Methodological comparisons between these measures are provided in the most recent research works [14, 16, 27]. Logical, statistical and semantical facets of a collection of 15 interestingness measures are analyzed in [14]. Comparison behaviour study of 20 indices is considered in [16]. Pairwise indices are compared according to the similarity of the rankings that they determine on a set of rules. Moreover, in this contribution 8 formal criteria are considered to characterize in a global manner the properties of a measure. The desired properties proposed in [27] are substantially different from the latter ones. Relative to an ordered pair of boolean attributes (a, b) belonging to $\mathcal{A} \times \mathcal{A}$, these properties are more local and directly associated with a transformation of the respective entries of the contingency table crossing $\{a, \neg a\}$ with $\{b, \neg b\}$ that we have introduced above. In the invariance properties considered in [27], the studied interestingness measures are taken one by one. However, some investigation about the transformation of one type of measure to another, is required. This aspect is considered in section 4, where we focus more particularily on the Confidence, Loevinger and Gras's entropic indices.

Most of the interestingness measures take into account each attribute pair independently. The formulation of such a given measure M for an ordered pair (a, b) depends only on the above mentioned proportions. M is considered as an *absolute measure*. For a couple $\{(a, b), (a', b')\}$ of attribute pairs, the underlying numerical scale enables to answer the question of determining the most stressed association between $(a \rightarrow b)$ and $(a' \rightarrow b')$. However, without loss of generality, by assuming

$$M(a \rightarrow b) > M(a' \rightarrow b') \tag{1}$$

we cannot evaluate how much the intensity of $(a \rightarrow b)$ is significantly greater than that of $(a' \rightarrow b')$. Moreover, for a given ordered pair (a, b) occurring in two different data bases, it is difficult to situate comparitively by means of an absolute measure M the intensity evaluation in each of both data bases. In particular, the size value n does not intervene in the mathematical expression of M.

Our method consists of evaluating in a relative way the degree of implication $a \rightarrow b$ by using an original notion of probabilistic index measuring how much unlikely in terms of probability the pattern is strengthened. A random model of no relation or independence is introduced associating with the observed incidence table a random one. Let us denote by $\mathcal{N}$ this random model. Then the general idea consists of substituting the initial scale given by M for that corresponding to the following equation:

$$P^{\mathcal{N}}(a \rightarrow b) = Pr\left\{M(a^* \rightarrow b^*) \leq M(a \rightarrow b) \mid \mathcal{N}\right\} \tag{2}$$

where (a^*, b^*) is an ordered pair of independent random attributes associated with (a, b) according to the model $\mathcal{N}$. There are two forms of this random model. The first one that we call "context free" model is local and does only concern the observed ordered pair (a, b) of boolean attributes (see section 2 and subsection 4.2). The second form qualified by "in the context" takes into account mutual comparison between all the attribute pairs or a relevant part of them (see section 3 and subsection 4.3). The second version of the random model $\mathcal{N}$ has a conditional meaning. For this model and more precisely, the above measure M is replaced by a standardized version M_s with respect to a relevant subset of $\mathcal{A}\times\mathcal{A}$ (see section 3 and subsection 4.3). In these conditions, the likelihood of the link probabilistic index can be written:

$$P_s^{\mathcal{N}}(a \rightarrow b) = Pr\left\{M_s(a^* \rightarrow b^*) \leq M_s(a \rightarrow b) \mid \mathcal{N}\right\} \tag{3}$$

Such a probabilistic interestingness measure provides clearer answers to the above evaluation questions. Moreover, in the context of a given data base a threshold value filtering the strongest rules is more easily controlled.

Thus our approach refers to the philosophy of the information theory, but at the level of the observed mutual relations. We need to quantify interesting implicative events by means of an index associated with a probability scale. The valuation of a given event, defining an association rule, takes into account all its possible positions relatively to an interesting potential subset of association rules associated in the context of $\mathcal{A} \times \mathcal{A}$.

In fact, for a given form of the independence relation $\mathcal{N}$, there is some invariance property in the probabilistic evaluation given by (3) with respect to the initial choice of the measure M. More precisely, a set of interestingness measures can be divided into stable classes associated with different probability scales, respectively. For example, relative to the collection of 21 measures given in [27] and for the no relation random model denoted below by $\mathcal{N}_1$ (see subsection 2.1), the following interestingness measures lead to exactly the same likelihood of the link probabilistic measure: Support, Confidence, Interest, Cosine, Piatetsky-Shapiro, Jaccard, Kappa and ϕ coefficient.

Consequently our starting point in building the likelihood of the link probabilistic measure will be one of the respective entries of the 2×2 contingency table considered above: $n(a \wedge b)$, $n(a \wedge \neg b)$, $n(\neg a \wedge b)$ or $n(\neg a \wedge \neg b)$. With a clear intuitive sense it suffices to consider $n(a \wedge b)$ for the symmetrical association and $n(a \wedge \neg b)$ for the asymmetrical case. Remind that for the latter we have to set up $(a \rightarrow b)$ more strongly than $(b \rightarrow a)$.

Indeed, in the statistical literature, the symmetrical equivalence case ($a \rightarrow b$ and $b \rightarrow a$) has preceded the asymmetrical implicative one $(a \rightarrow b)$. And in fact, several proposed indices in the data mining literature have a perfect symmetrical nature, that is to say that their expressions are invariant by substituting the ordered boolean pair (a, b) for (b, a). For example, relative to the above mentioned measures one may cite: Support, Interest, Cosine, Jaccard, Piatetsky-Shapiro, Kappa and ϕ.

The local context free form of the likelihood of the link probabilistic indices have been established first. The symmetrical version [19, 20, 18] has preceded the asymmetrical one [9, 23, 11]. The associated probabilistic scale was able to reveal fine structural relations on the attribute set $\mathcal{A}$, only when the number n of objects is lower than 10^3. But nowadays it is often necessary to work with large data (for example n greater than 10^6). And, in such situation, the latter scale becomes not enough discriminant in order to distinguish between different high values of the computed indices.

The "in the context" random model integrates in its construction the previous local model. But the probability scale becomes finely discriminant for any magnitude of n. This global model has been extensively validated theoretically [17, 7] and experimentally [20] in the framework of our hierarchical classification *LLA* (Likelihood of the Link Analysis) method (*Classification Ascendante Hiérarchique par Analyse de la Vraisemblance des Liens*) [20, 18, 21]. This validation aspect represents one important contribution of the analysis presented in this paper.

This analysis leads us to set up new absolute measures expressed in terms of the above relative frequencies $p(a \wedge b)$, $p(a \wedge \neg b)$, $p(\neg a \wedge b)$ and $p(\neg a \wedge \neg b)$ (see subsections 2.2 and 4.1). These measures appear as components of the χ^2 statistic.

Therefore, we begin in the second section by describing the probabilistic construction of the likelihood of the link index in the symmetrical comparison case and for the "context free" random model. The "in the context" random model will be expressed in section 3. Section 4 is devoted to implicative similarity which directly reflects the asymmetric nature of an association rule notion. Analysis of classical indices is performed in this section. Otherwise, local ("context free") and global ("in the context") interpretations are developed. In this section we will present the Probabilistic Normalized Index which enables to discriminate comparisons between association rules. Precisely, section 5 is devoted to experimental results validating the behaviour of this new index with respect to that locally built. We end in section 6 with a general conclusion giving the benefits and the prospects of this work.

2 "Context free" comparison between two boolean attributes

2.1 Building no relation (independence) hypothesis

Let (a, b) be a pair of boolean attributes provided from $\mathcal{A} \times \mathcal{A}$. We have introduced (see above) the set theoretic representation for this pair. We have also defined the cardinal parameters $n(a \wedge b)$, $n(a \wedge \neg b)$, $n(\neg a \wedge b)$ and $n(\neg a \wedge \neg b)$. Without loss of generality assume the inequality $n(a) < n(b)$.

Two distinctive but related problems have to be considered in comparing two boolean attributes a and b. The first consists of evaluating the degree of symmetrical equivalence relation ($a \leftrightarrow b$). The second concerns asymmetrical implicative relation ($a \rightarrow b$) called "association rule". Statistical literature has mainly focused the symmetrical association case. Many coefficients have been proposed for pairwise comparison of a set $\mathcal{A}$ of boolean attributes. All of them can be expressed as functions of the parameters $(n(a \wedge b), n(a), n(b), n)$. Most of them can be reduced to functions of the relative frequencies $(p(a \wedge b), p(a), p(b))$ associated with the absolute frequencies $n(a \wedge b)$, $n(a)$ and $n(b)$, relative to n. Thus, the parameter $n(a \wedge b)$ representing the number of objects where the conjunction $a \wedge b$ is true, appears as a fundamental basis of an association coefficient construction. We call this index a "raw" association coefficient. As a matter of fact, each of the association coefficients proposed in the literarture corresponds to a type of normalization of this raw index, with respect to the sizes $n(a)$ and $n(b)$. Indeed, tendency to a high or low values of $n(a \wedge b)$ is associated with high or low values of both parameters, respectively.

In these conditions, the first step of the normalization process we have adopted consists in introducing a probabilistic model of independence (or no relation) defined as a correspondence:

$$(\mathcal{O}(a), \mathcal{O}(b), \mathcal{O}) \rightarrow (\mathcal{X}, \mathcal{Y}, \Omega) \tag{4}$$

Three versions of this general model that we designate by $\mathcal{N}$, are considered. They lead to three distinct analytical forms of an association coefficient. Ω is associated with the object set $\mathcal{O}$, exactly ($\Omega = \mathcal{O}$) or randomly defined. For a given Ω, $\mathcal{X}$ and $\mathcal{Y}$ are defined by two independent random subsets of Ω, associated with $\mathcal{O}(a)$ and $\mathcal{O}(b)$, repectively. More precisely, the random model is built in such a way that Ω, $\mathcal{X}$ and $\mathcal{Y}$ respect exactly or on average the cardinalities n, $n(a)$ and $n(b)$, respectively. The random subsets $\mathcal{X}$ and $\mathcal{Y}$ can also be denoted by $\mathcal{O}(a^*)$ and $\mathcal{O}(b^*)$ where a^* and b^* are two independent random attributes associated with a and b, respectively.

Denoting by

$$s = n(a \wedge b) = card\,(\mathcal{O}(a) \cap \mathcal{O}(b)) \tag{5}$$

the above-mentioned raw index, the random raw index is defined by:

$$\mathcal{S} = n(a^* \wedge b^*) = card\,(\mathcal{X} \cap \mathcal{Y}) \tag{6}$$

The first form of normalization is obtained by standardizing s with respect to the probability distribution of S:

$$q(a, b) = \frac{s - \mathcal{E}(\mathcal{S})}{\sqrt{var(\mathcal{S})}} \tag{7}$$

where $\mathcal{E}(\mathcal{S})$ and $var(\mathcal{S})$ designate the mathematical expectation and the variance of $\mathcal{S}$.

By using normal distribution for the probability law of $\mathcal{S}$, this coefficient leads to the probabilistic index:

$$\mathcal{I}(a,b) = Pr\left\{\mathcal{S} \leq s|\mathcal{N}\right\} = Pr\left\{q(a^*,b^*) \leq q(a,b)|\mathcal{N}\right\} \tag{8}$$

that we call "local" likelihood of the link association index. Remind that "local" refers to the logical independence of its construction relative to the attribute set $\mathcal{A}$ from which a and b are taken.

For this index the similarity between a and b is measured by a probability value stating how much improbable is the bigness of the observed value of the raw index s. This probability is defined and computed under the independence hypothesis $\mathcal{N}$. Clearly, the index (8) is nothing but the complement to 1 of a $P-value$ in the sense of statistical hypotheses. However, its meaning does not refer to a conditional test [4] but to a conditional probabilistic evaluation, when the sizes $n(a)$ and $n(b)$ are given.

As mentioned above three fundamental forms of the no relation (independence) hypothesis $\mathcal{N}$ have been set up [23, 20]. Let us denote them $\mathcal{N}_1$, $\mathcal{N}_2$ and $\mathcal{N}_3$. These are distinguished in their ways of associating with a given subset $\mathcal{O}(c)$ of $\mathcal{O}$, a random subset $\mathcal{L}$ of an Ω set corresponding to $\mathcal{O}$. Let us designate by $P(\Omega)$ the set of all subsets of Ω organized into levels by the set inclusion relation. A given level is composed by all subsets having the same cardinality.

Let us now make clearer the hypotheses $\mathcal{N}_1$, $\mathcal{N}_2$ and $\mathcal{N}_3$.

For $\mathcal{N}_1$, $\Omega = \mathcal{O}$ and $\mathcal{L}$ is a random element from the $n(c)$ level of $P(\Omega)$, provided by an uniform probability distribution. Then $\mathcal{L}$ is a random subset of $\mathcal{O}$ of size $n(c)$.

For $\mathcal{N}_2$, $\Omega = \mathcal{O}$. But the random model includes two steps. The first consists of randomly choosing a level of $P(\mathcal{O})$. Then $\mathcal{L}$ is defined as a random element of the concerned level, provided by an uniform distribution. More precisely, the level choice follows the binomial distribution with n and $p(c) = \frac{n(c)}{n}$ as parameters. Under these conditions, the probability of the kth level, $1 \leq k \leq n$ is given by $C_n^k p(c)^k p(\neg c)^{n-k}$, where $p(\neg c) = \frac{n(\neg c)}{n}$.

$\mathcal{N}_3$ is defined by a random model with three steps. The first consists of associating with the object set $\mathcal{O}$ a random object set Ω. The only requirement for Ω concerns its cardinality N which is supposed following a Poisson probability law, its parameter being $n = card(\mathcal{O})$. The two following steps are similar to those of the random model $\mathcal{N}_2$. More precisely, for $N = m$ and an object set sized by m, $\mathcal{L}$ is a random part of Ω_0. $\mathcal{L}$ is defined only for $m \geq n(c)$ and in this case we define $\gamma = \frac{n(c)}{m}$. In these conditions, the probability to choose the level k of $\mathcal{P}(\Omega_0)$ is defined by the binomial probability $C_m^k \gamma^k (1-\gamma)^{n-k}$. And for a given level, the random choice of $\mathcal{L}$ is done uniformly on this level.

We established [23, 20] that the distribution of the random raw index $\mathcal{S}$ is:

- hypergeometric of parameters $(n, n(a), n(b))$, under the model $\mathcal{N}_1$;
- binomial of parameters $(n, p(a) * p(b))$, under the model $\mathcal{N}_2$;
- of Poisson of parameters $(n, n * p(a) * p(b))$, under the model $\mathcal{N}_3$.

2.2 The different versions of a statistically standardized index

The normalized form of the raw index s, according to equation (8) for the random models $\mathcal{N}_1, \mathcal{N}_2$ and $\mathcal{N}_3$, respectively, leads to the following indices [23]:

$$q_1(a,b) = \sqrt{n} * \frac{p(a \wedge b) - p(a) * p(b)}{\sqrt{p(a) * p(b) * p(\neg a) * p(\neg b)}} \tag{9}$$

$$q_2(a,b) = \sqrt{n} * \frac{p(a \wedge b) - p(a) * p(b)}{\sqrt{p(a) * p(b) * [1 - p(a) * p(b)]}} \tag{10}$$

and

$$q_3(a,b) = \sqrt{n} * \frac{p(a \wedge b) - p(a) * p(b)}{\sqrt{p(a) * p(b)}} \tag{11}$$

Notice the perfect symmetry of $q_1(a,b)$ according to the following meaning:

$$q_1(a,b) = q_1(\neg a, \neg b) \tag{12}$$

As mentioned above (see introduction) we assume the positive form of the boolean attributes established in such a way that the proportional frequency of the "true" value is less than 0.5. Under this condition we have the following inequalities:

$$q_2(a,b) > q_2(\neg a, \neg b) \tag{13}$$

and

$$q_3(a,b) > q_3(\neg a, \neg b) \tag{14}$$

The last inequality is clearly more differentiated than the previous one. Therefore, we only consider the two most differentiated indices $q_1(a,b)$ and $q_3(a,b)$. One more reason for distinguishing these two indices concerns both formal and statistical aspects. Indeed, by considering the χ^2 statistic associated with the 2×2 contingency table crossing $(a, \neg a)$ and $(b, \neg b)$, we obtain:

$$\begin{aligned}\chi^2\{(a,\neg a),(b,\neg b)\} &= [q_1(a,b)]^2 \\ &= [q_3(a,b)]^2 + [q_3(a,\neg b)]^2 + \\ &\quad [q_3(\neg a,b)]^2 + [q_3(\neg a,\neg b)]^2\end{aligned} \tag{15}$$

Thus, $q_3(a,b)$ defines the direct contribution of the entry (a,b) to the χ^2 statistic.

Dividing by $\sqrt{n}$ the indices q_1 and q_3, one obtains the respective associated indices γ_1 and γ_3. Correlative interpretation of these can be provided. Both are comprised between -1 and $+1$. But depending on $p(a)$ and $p(b)$, γ_3 is included into a more narrow interval than that of γ_1:

$$\gamma_1(a,b) = \frac{p(a \wedge b) - p(a) * p(b)}{\sqrt{p(a) * p(b) * p(\neg a) * p(\neg b)}} \tag{16}$$

and

$$\gamma_3(a,b) = \frac{p(a \wedge b) - p(a) * p(b)}{\sqrt{p(a) * p(b)}} \tag{17}$$

Now, let us designate by d_{ab} the density of the joint empirical probability with respect to the product of marginal probabilities, namely:

$$d_{ab} = \frac{p(a \wedge b)}{p(a) * p(b)} \tag{18}$$

The latter index corresponds to the interest measure [6, 27]. It is directly related to that t^{ab} introduced in [4] by means of the equation:

$$t^{ab} = d_{ab} - 1 \tag{19}$$

Then, $\gamma_1(a,b)$ and $\gamma_3(a,b)$ can be expressed in terms of d_{ab} and the marginal proportional frequencies. More precisely, we have:

$$\gamma_1(a,b) = \sqrt{\frac{p(a) * p(b)}{p(\neg a) * p(\neg b)}} * (d_{ab} - 1) \tag{20}$$

$$\gamma_3(a,b) = \sqrt{p(a) * p(b)} * (d_{ab} - 1) \tag{21}$$

2.3 Behaviour of the likelihood of the link local probabilistic index

The general expression of this index is given in (8) where the no relation (independence) hypothesis $\mathcal{N}$ is not yet specified. By substituting $\mathcal{N}$ with $\mathcal{N}_1$, $\mathcal{N}_2$ or $\mathcal{N}_3$, one has to replace q by q_1, q_2 or q_3, respectively. Even for n relatively high, an exact evaluation of the probability $Pr(\mathcal{S} \leq s|\mathcal{N}_i)(i = 1, 2$ or $3)$ can be obtained by means of a computer program. Cumulative distribution functions of hypergeometric, binomial or of Poisson laws are refered to the subscipt value i, $i = 1, 2$ ou 3, respectively. Nevertheless, for n large enough (for example greater than 100) and $p(a) * p(b)$ not too small, the probability law of $\mathcal{S}$ can be very accurately approximated by the normal distribution:

$$\mathcal{I}_i(a,b) = Pr\left\{\mathcal{S} \leq s|\mathcal{N}_i\right\} = \Phi\left[q_i(a,b)\right] = \Phi\left[\sqrt{n} \times \gamma_i(a,b)\right] \tag{22}$$

where Φ denotes the standardized cumulative normal distribution function and where $i = 1, 2$ or 3.

Consequently, if d_{ab} is clearly greater than unity and for n large enough, the local probabilistic index becomes very close to 1. On the contrary, if d_{ab} is lower than 1 and for n enough large, this index tends to 0. Thus, for n large enough and whatever the computing accuracy reached, this local probabilistic index is only able to discriminate two classes of attribute couples: the positively and the negatively related.

Now, as it is usual in statistical inference, let us imagine the object set $\mathcal{O}$ obtained by means of a random sampling in a universe $\mathcal{U}$ of objects. Then, let us denote at the level of $\mathcal{U}$, by $\pi(a)$, $\pi(b)$ and $\pi(a \wedge b)$ the object proportional frequencies where the boolean attributes a, b and $a \wedge b$ have the "true" value. Note that $\pi(a)$, $\pi(b)$ and $\pi(a \wedge b)$ can be interpreted as probabilities associated with a "true" value for a random object taken with equiprobability distribution from $\mathcal{U}$. In these conditions let us designate by $\rho_i(a,b)$ the mathematical expression corresponding to $\gamma_i(a,b)$ and defined at the level of $\mathcal{U}$, $i = 1, 2$ or 3, respectively. In this fashion, $\gamma_i(a,b)$ defines an estimation of $\rho_i(a,b)$, whose accuracy is an increasing function of n [17].

Now, let us indicate by $\delta(a,b)$ the expression associated with (18), but at the level of $\mathcal{U}$. Clearly, we have the following properties for n enough large:

- $\delta(a,b) < 1$, $\mathcal{I}_i(a,b)$ [*cf.* (22)] tends to 0;
- $\delta(a,b) > 1$, $\mathcal{I}_i(a,b)$ tends to 1;

but, for $\delta(a,b) = 1$, $\mathcal{I}_i(a,b)$ can be considered as an observed value of an uniformly distributed random variable on the $[0,1]$ interval.

However, our statistical framework is restricted to the object set $\mathcal{O}$ and it is in this context that we have to achieve a probabilistic discriminant index. As shown above, this index cannot be obtained if we only have to compare in an absolute manner two boolean attributes a and b independently of the attribute set $\mathcal{A}$ from which they come. And indeed, if our universe is limited to this single couple of attributes $\{a,b\}$, the above proposed index $\mathcal{I}_i(a,b), \forall i = 1, 2$ or 3, is sufficient. In fact, the objective consists of mutual comparison of many attribute pairs and generally of the whole set $P_2(\mathcal{A})$ of attribute pairs.

3 "In the context" comparison between two attributes

The context is determined by the set of attribute pairs of a set $\mathcal{A}$ of p boolean attributes:

$$\mathcal{A} = \{a^j | 1 \leq j \leq p\} \tag{23}$$

In this context, a probabilistic similarity between two boolean attributes will be proposed. This similarity will have a relative meaning with respect to the context. Retain from $\mathcal{A} \times \mathcal{A}$ the following cardinal structure:

$$\left\{ \left(n(a^j \wedge a^k), n(a^j \wedge \neg a^k), n(\neg a^j \wedge a^k), n(\neg a^j \wedge \neg a^k) \right) | 1 \leq j < k \leq p \right\} \tag{24}$$

Also denote

$$\{n(a^j) | 1 \leq j \leq p\} \tag{25}$$

the sequence of the cardinalities of the subsets $\mathcal{O}(a^j), 1 \leq j \leq p$. At the same time, introduce the associated sequence of the proportional frequencies, relative to the total number of objects n. Thus the mathematical tables (24) and (25) give:

$$\left\{ \left(p(a^j \wedge a^k), p(a^j \wedge \neg a^k), p(\neg a^j \wedge a^k), p(\neg a^j \wedge \neg a^k) \right) | 1 \leq j < k \leq p \right\} \tag{26}$$

and

$$\{p(a^j)|1 \leq j \leq p\} \tag{27}$$

Now, reconsider the index $q_3(a,b)$ [*cf.* (11)] locally built by a centring and reducing process with respect to the no relation hypothesis $\mathcal{N}_3$. We have emphasized above the interesting asymmetrical property of this index where the similarity between rare attributes is clearly stressed [*cf.* (13)]. Precisely, we assume that the set of the p boolean attributes is established in such a way that:

$$n(a^j) \leq n(\neg a^j), 1 \leq j \leq p \tag{28}$$

As expressed in the introduction, a system of values such as (26) cannot be evaluated in the same way relatively to the induced equivalence or implicative relations, for any magnitude of the number n of objects. And that matters from two points of view: statistical and semantical. Moreover, the associations rules should be situated in a relative way.

To answer these two requirements let us first reconsider $q_3(a,b)$ [*cf.* (11)]. Now, in order to compare in a mutual and relative manner the set of attribute pairs, introduce the empirical variance of the index q_3 on the set $P_2(\mathcal{A})$ of two element parts of $\mathcal{A}$. This variance can be written:

$$var_e(q_3) = \frac{2}{p*(p-1)} \sum \left\{ \left[q_3(a^j, a^k) - moy_e(q_3)\right]^2 |1 \leq j < k \leq p \right\} \tag{29}$$

where

$$moy_e(q_3) = \frac{2}{p*(p-1)} \sum \{q_3(a^j, a^k)|1 \leq j < k \leq p\} \tag{30}$$

defines the mean of the index q_3 on $P_2(\mathcal{A})$.

For relative comparison between two attributes belonging to $\mathcal{A}$, we introduce the globally normalized index q_3. For example, consider the comparison between two given attributes a^1 and a^2. The new index $q_3^g(a^1,a^2)$ is defined as follows:

$$q_3^g(a^1,a^2) = \frac{q_3(a^1,a^2) - moy_e(q_3)}{\sqrt{var_e(q_3)}} \tag{31}$$

This index corresponds to the relative and directed contribution of $q_3(a^1,a^2)$ to the empirical variance $var_e(q_3)$.

Under these conditions, the likelihood of the link probabilistic index is conceived with respect to a global independence hypothesis where we associate with the attribute set $\mathcal{A}$ [*cf.* (23)] a random attribute set:

$$\mathcal{A}^* = \{a^{j*} | 1 \leq j \leq p\} \tag{32}$$

where the different attributes are mutually independent with respect to the no relation hypothesis $\mathcal{N}_3$. This index can be written:

$$P_g(a^1, a^2) = Pr\left\{q_3^g(a^{1*}, a^{2*}) \leq q_3^g(a^1, a^2) | \mathcal{N}_3\right\} \tag{33}$$

where

$$q_3^g(a^{1*}, a^{2*}) = \frac{q_3(a^{1*}, a^{2*}) - moy_e(q_3^*)}{\sqrt{var_e(q_3^*)}} \tag{34}$$

Theoretical and experimental proofs [17, 7] show that the probabilistic index can be computed by means of the following equation:

$$P_g(a^1, a^2) = \Phi\left[q_3^g(a^1, a^2)\right] \tag{35}$$

where Φ denotes the standardized normal cumulative distribution.

Let us point out that the table of probabilistic similarity indices

$$\left\{P_g(a^j, a^k) | 1 \leq j < k \leq p\right\} \tag{36}$$

is that taken into account in the LLA ascendant hierarchical classification method [18, 21]. Otherwise and for any data analysis method working with dissimilarities, our approach can provide the following table of dissimilarity indices:

$$\{\mathcal{D}(j, k) = -Log_2[P_g(a^j, a^k)] | 1 \leq j < k \leq p\} \tag{37}$$

that we call "informational dissimilarity table".

4 Implicative similarity index

4.1 Indices independent of n and of the context

So far we have been interested in the evaluation of the symmetrical equivalence relation degree between two boolean attributes a and b belonging to a set $\mathcal{A}$ of boolean attributes [*cf.* (23)]. For this purpose we refered to independence statistical hypothesis as a basis to establish an adequate measure. The built indices are clearly situated with respect to this hypothesis. Now, we have to evaluate the asymmetrical implicative relation of the form $a \rightarrow b$. Such a relation is completely satisfied at the level of the object set

$\mathcal{O}$ if the subset $\mathcal{O}(a)$ characterized by a true value of a is included in the subset $\mathcal{O}(b)$ characterized by a true value of b. In practice this event is very rare. When, without having strict inclusion, there are clear experimental situations defined at the level of the object set $\mathcal{O}$, where partial inclusion is more or less strong. And in such situations, we have to evaluate the tendency of b knowing a.

Let us reconsider briefly the complete inclusion situation observed at the level of the object set $\mathcal{O}$. It can be interesting to study it by introducing parametrization with $(n, p(a), p(b))$ where $p(a \wedge \neg b) = 0$. One can also [26] be interested in some aspects of the probability law of the conditional relative frequency $p(b^*|a^*)$ under a very specific model. For the latter it is assumed that both cardinalities are known: the size $N(a)$ of the $\mathcal{U}$ subset where a is true and the size $n(a)$ of the $\mathcal{O}$ subset where a is true.

Now, let us consider the most realistic and the most frequent case where the number of objects $n(a \wedge \neg b)$ is small without being null. In these conditions, we are interested in evaluating the relative smallness of the number of objects $n(a \wedge \neg b)$ where a is true and where b is false. T hese objects contradict the implicative relation $a \rightarrow b$. In order to evaluate this smallness, most of the proposed indices try to neutralize the size influence of $n(a)$ and $n(b)$. However notice that for $n(a)$ and $n(b)$ fixed, $n(a \wedge \neg b)$ "small", $n(a \wedge b)$ "large", $n(\neg a \wedge b)$ "small" and $n(\neg a, \neg b)$ "large" correspond to concordant phenomenons. Thus, some proposed indices for the asymmetrical implicative case have a perfect symmetrical nature with respect to the ordered pair (a, b). Most if not all of them can be situated with respect to the empirical independence hypothesis defined by $p(a \wedge b) = p(a) * p(b)$. According to [26] the easiest one is $(p(a \wedge b) - p(a) * p(b))$. The latter corresponds to the common numerator of $q_1(a, b)$, $q_2(a, b)$ and $q_3(a, b)$ [*cf.* (9,10,11)]. This index and $\gamma_1(a, b)$ are reported in [26] but at the level of the object universe $\mathcal{U}$. γ_1 is nothing else than the K. Pearson coefficient [25]. A symmetrical index called "interest measure" is also proposed in [6]. This index has been denoted by d_{ab} [*cf.* (18)]. It is directly related to the contribution of the entry (a, b) to the χ^2 statistic. This contribution can also be expressed by $[q_3(a, b)]^2$.

Nevertheless, many of the proposed indices for evaluating the strength of an implication $a \rightarrow b$ are asymmetrical. Clearly, they stress the smallness of $n(a \wedge \neg b)$ with respect to the bigness of $n(a \wedge b)$. Let us describe some of them. For coherent reasons in the following development but without explicit intervention, suppose the inequality $n(a) \leq n(b)$ which makes possible the total inclusion of $\mathcal{O}(a)$ in $\mathcal{O}(b)$.

The easiest and the most direct index that we have to mention is that called "the confidence index" [1]. It is defined by the conditional proportion $p(b|a) = \frac{p(a \wedge b)}{p(a)}$. It varies from 0 to 1. The 0 value is associated with disjunction between $\mathcal{O}(a)$ and $\mathcal{O}(b)$, when the 1 value is reached in case of inclusion of $\mathcal{O}(a)$ into $\mathcal{O}(b)$. An interesting comparison analysis of this basic index with different indices proposed in the literature is developed in [14]. The Loevinger index [24] is also a very classical and very known one. Respective to the introduced notations [*cf.* (18)] it can be defined by the following equation:

$$\mathcal{H}(a,b) = 1 - d(a,\neg b) \tag{38}$$

Let us now suppose the two "natural" inequalities $p(a) \leq p(b)$ and $p(a) \leq p(\neg b)$. Under these conditions, the index value varies from 1 in case of complete inclusion $\mathcal{O}(a) \subset \mathcal{O}(b)$, goes through the 0 value for the statistical independence and reachs the negative value $-\left[\frac{p(b)}{p(\neg b)}\right]$ in case where $\mathcal{O}(a) \subset \mathcal{O}(\neg b)$; that is to say, where the opposite implication $a \rightarrow \neg b$ holds.

The Loevinger index can also be written:

$$\mathcal{H}(a,b) = \frac{p(a \wedge b) - p(a) * p(b)}{p(a) * p(\neg b)} \tag{39}$$

Then, it corresponds to an asymmetrical reduction with respect to (a,b) of the first index proposed by G. Piatetsky-Shapiro [26].

One can also clearly situate the $\mathcal{H}(a,b)$ index with respect to $\gamma_3(a,\neg b)$. $\mathcal{H}(a,b)$ is obtained by reducing the centred index $[p(a \wedge \neg b) - p(a) * p(\neg b)]$ by means of $p(a) * p(\neg b)$ when the reduction is performed with $\sqrt{p(a) * p(\neg b)}$ in the $-\gamma_3(a,\neg b)$ index. Consequently, $-\gamma_3(a,\neg b)$ defines a new implicative index only depending on $[p(a \wedge b), p(a), p(b)]$. Obviously the 0 value characterizes statistical independence. On the other hand, the logical implication $a \rightarrow b$ is obtained for the value $\sqrt{p(a) * p(\neg b)}$. Otherwise, the index value can decrease till the value $-p(b) * \sqrt{\frac{p(a)}{p(\neg b)}}$.

Therefore and clearly, one can propose the following discriminant "free context" index:

$$-\gamma_3(a,\neg b) = \frac{-p(a \wedge \neg b) + p(a) * p(\neg b)}{\sqrt{p(a) * p(\neg b)}} \tag{40}$$

This index is exactly the opposite of the direct contribution of the entry $(a,\neg b)$ to the χ^2/n coefficient. The index $-\gamma_3(a,\neg b)$ appears coherent with the "in the context" construction process of the likelihood of the link probabilistic index (see the above section 3). It can be seen that the two extreme value limits are included in the interval $[-1,+1]$ when the lowest negative boundary of $\mathcal{H}(a,b)$ can potentially reach any negative value. More precisely and under the inequalities ($p(a) \leq p(b)$ and $p(a) \leq p(\neg b)$), we establish that the two boundaries are comprised in the interval $[-p(b), p(\neg b)]$.

Now, by considering the coherence condition $p(a) \leq p(b)$ which enables the complete inclusion $\mathcal{O}(a) \subset \mathcal{O}(b)$ and, as said in the introduction, by assuming the significant conditions $p(a) \leq p(\neg a)$ and $p(b) \leq p(\neg b)$ one can establish that the minimal and maximal values of $\gamma_3(a, \neg b)$ are -0.5 and $+0.5$, respectively. The above limit $p(\neg b)$ is then reduced. Consequently and under the mentioned coherence conditions, we can obtain an index whose value ranges from 0 to 1 by setting:

$$\eta_3(a,b) = 0.5 - \gamma_3(a, \neg b) \tag{41}$$

Otherwise and under the above coherent conditions, the minimal boundary for the index $\mathcal{H}(a,b)$ is greater or equal to -1. Then an index denoted by $\mathcal{K}(a,b)$, deduced from $\mathcal{H}(a,b)$ and comprised between 0 and 1 can be put in the following form:

$$\mathcal{K}(a,b) = \frac{1}{2}\left(\mathcal{H}(a,b) + 1\right) \tag{42}$$

Notice that the two new indices have the common value $\frac{1}{2}$ in case of statistical independence.

It has been shown that the presented above indices can be expressed in terms of the different components of the χ^2 statistic associated with the 2×2 contingency table defined by the crossing $\{a, \neg a\} \times \{b, \neg b\}$. We are now going to present two indices which employ the mutual information statistic associated with this contingency table. Three formal versions can be considered for this statistic:

$$\begin{aligned}
\mathcal{E} &= p(a \wedge b)\log_2(d(a,b)) + p(a \wedge \neg b)\log_2(d(a,\neg b)) \\
&\quad + p(\neg a \wedge b)\log_2(d(\neg a,b)) + p(\neg a \wedge \neg b)\log_2(d(\neg a,\neg b)) && (43) \\
&= E(a) - p(b)E(a|b) - p(\neg b)E(a|\neg b) && (44) \\
&= E(b) - p(a)E(b|a) - p(\neg a)E(b|\neg a) && (45)
\end{aligned}$$

where $E(x)$ denotes the entropy of the binary distribution $(p(x), p(\neg x))$ and where $E(x|y)$ denotes that of the conditional distribution $(p(x|y), p(\neg x|y))$, x and y, indicating two boolean attributes.

Precisely, the Goodman & Smith [8] J-measure corresponds to the sum of the first two terms of the previous first equation (43). In these, the boolean attribute a is taken positively, when the negated attribute $\neg a$ is considered in the sum of the last two terms of (43). A second index proposed by R. Gras [10] has fundamentally an entropic conception. It is called "inclusion" index and it takes the following form:

$$\tau(a,b) = \sqrt{G(b|a) * G(\neg a|\neg b)} \tag{46}$$

where $G(x|y)$ is defined by the square root of

$$G^2(x|y) = \begin{cases} 1 - E^2(x|y) \text{ if } p(\neg x \wedge y) \leq \frac{1}{2} * p(y) \\ 0 \text{ if not} \end{cases} \tag{47}$$

This index employs the conditional entropies $E(b|a)$ and $E(\neg a|\neg b)$ which are associated with the binary distributions $(p(b|a),\, p(\neg b|a))$ and $(p(\neg a|\neg b),\, p(a|\neg b))$, respectively. The former entropy can be obtained as a constitutive component of the equation (45) expressing the mutual information $\mathcal{E}$, when the latter entropy is defined as a component element of the preceding equation (44).

Notice that a high value of the index $1 - E^2(b|a)$ expresses two opposite inclusion tendencies. The first one consists of $\mathcal{O}(a) \subset \mathcal{O}(b)$ and the other logical opposite $\mathcal{O}(a) \subset \mathcal{O}(\neg b)$ $(E(b|a) = E(\neg b|a))$. Similarly, a high value of $1 - E^2(\neg a|\neg b)$, reflects two opposite inclusion tendencies corresponding to $\mathcal{O}(\neg b) \subset \mathcal{O}(\neg a)$ and $\mathcal{O}(\neg b) \subset \mathcal{O}(a)$ $(E(\neg a|\neg b) = E(a|\neg b))$. However, taking into account the condition included in equation (47), a strictly positive value of $\tau(a,b)$ is constrained by $p(b|a) > p(\neg b|a)$. Consequently, a strictly positive value of the inclusion index cannot occur if one of both conditional probabilities is lower than 0.5. Now, there may be situations where $p(b|a)$ is high enough (clearly greater than 0.5) and where $p(\neg a|\neg b)$ is low enough (notably lower than 0.5). And in these, there is no reason to reject *a priori* an implicative tendency value for $a \rightarrow b$. This weakness of the inclusion index is somewhat balanced by its quality consisting of taking into account both implicative evaluations: $a \rightarrow b$ and $\neg b \rightarrow \neg a$.

4.2 Comparing local implicative of the link likelihood and entropic intensity indices

Conceptually, for associating symmetrically two boolean attributes a and b, the likelihood of the link probabilistic approach evaluates under a random model of no relation how much unlikely is in probability terms the relative bigness of $n(a \wedge b)$. The respective influences of the $n(a)$ and $n(b)$ sizes are neutralized in this model. This idea has been extensively developed in the framework of the LLA ascendant hierarchical classification of descriptive attributes [20]. It has been adapted by R. Gras [9], [23] in the asymmetrical implicative case. In the latter, one has to evaluate how much is unlikely the smallness of $n(a \wedge \neg b)$ with respect to a random no relation model $\mathcal{N}$, neutralizing the respective influences of $n(a)$ and $n(b)$. The index can be written:

$$\begin{aligned}\mathcal{I}(a,b) &= 1 - Pr\left\{n(a^* \wedge \neg b^*) < n(a \wedge \neg b)|\mathcal{N}\right\} \\ &= Pr\left\{n(a^* \wedge \neg b^*) \geq n(a \wedge \neg b)|\mathcal{N}\right\}\end{aligned} \tag{48}$$

where (a^*, b^*) denotes the random ordered attribute pair associated with (a, b) under a random model $\mathcal{N}$ defining an independence hypothesis.

Let us designate by u the index $n(a \wedge \neg b)$ and by u^*, the random associated index $n(a^* \wedge \neg b^*)$ under the hypothesis of no relation $\mathcal{N}$. Then, the standardized index (by centring and reducing u) takes the following form:

$$q(a, \neg b) = \frac{u - \mathcal{E}(u^*)}{\sqrt{var(u^*)}} \tag{49}$$

where $\mathcal{E}(u^*)$ and $var(u^*)$ denote the mathematical expectation and the variance of u^*, respectively.

As expressed above three versions of the random model $\mathcal{N}$ have been set up: $\mathcal{N}_1$, $\mathcal{N}_2$ and $\mathcal{N}_3$. These lead for the random index u^*, to hypergeometric, binomial and of Poisson probability laws, respectively. $\mathcal{N}_1$ and $\mathcal{N}_3$ are the most differentiated models [23]. By designating $q_i(a, \neg b)$, the index u standardized with respect to $\mathcal{N}_i$, we have:

$$q_1(a, \neg b) = q_1(\neg a, b) = -q_1(a, b) = -q_1(\neg a, \neg b) \tag{50}$$

Remark that for the random model $\mathcal{N}_1$, the implicative form of the index is exactly equivalent to the symmetrical case. For $\mathcal{N}_3$, we have with the condition $p(a) < p(b)$:

$$|q_3(a, \neg b)| > |q_3(b, \neg a)| \tag{51}$$

A natural condition in order to consider the evaluation of the association rule $a \rightarrow b$ can be defined by a negative value of $n(a \wedge \neg b) - \left(\frac{n(a)*n(\neg b)}{n}\right)$. This expression represents the numerator of $q_3(a, \neg b)$. It is identical to that of $n(b \wedge \neg a) - \left(\frac{n(b)*n(\neg a)}{n}\right)$ associated with the index $q_3(b, \neg a)$ corresponding to the opposite implication $b \rightarrow a$. However, since $n(b) > n(a)$, the latter is more difficult to accept. The inequality (51) consists of a coherent statement since, for the local likelihood of the link index associated with $\mathcal{N}_3$, we have [*cf.* (48)]:

$$\mathcal{J}_3(a, b) > \mathcal{J}_3(b, a) \tag{52}$$

To be convinced of this property, consider the excellent normal approximation for n enough large, of the probability Poisson law of $n(a^* \wedge \neg b^*)$ and $n(b^* \wedge \neg a^*)$, respectively, under the $\mathcal{N}_3$ model:

$$\mathcal{J}_3(a,b) = 1 - \Phi(q_3(a,\neg b)) \tag{53}$$

and

$$\mathcal{J}_3(b,a) = 1 - \Phi(q_3(b,\neg a)) \tag{54}$$

where Φ is the cumulative distribution function of the standardized normal law.

Moreover, the necessary and sufficient condition to have (52) is $n(a) < n(b)$. Mostly we have $p(a) < p(b) < \frac{1}{2}$. With this condition one obtains the following inequalities which comprise that (52):

$$\mathcal{I}_3(\neg a,\neg b) < \mathcal{J}_3(b,a) < \mathcal{J}_3(a,b) < \mathcal{I}_3(a,b) \tag{55}$$

where $\mathcal{I}_3$ indicates the local likelihood of the link probabilistic index defined in (22).

Now, consider two situations that we denote by I and II where $\mathcal{O}(a)$ and $\mathcal{O}(b)$ have the same relative position. We mean that the proportional frequencies induced by $n(a \wedge b)$, $n(a \wedge \neg b)$ and $n(\neg a \wedge b)$ remain constant. But we suppose variation for the relative frequency induced by $n(\neg a \wedge \neg b)$ between I and II. Notice that every similarity index symmetrical or asymmetrical based on relative proportions defined into $\mathcal{O}(a \vee b)$ cannot distinguish I and II, see for example the famous Jaccard index [13] or the confidence index $(a \rightarrow b) = \frac{n(a \wedge b)}{n(a)}$. This statement becomes false in case of an association rule whose conception depends logically on $\mathcal{O}(\neg a \wedge \neg b)$. In order to illustrate this point and to explain the behaviour of the local probabilistic index which is directly related to $q_3(a,\neg b)$, consider for $n = 4000$ the two following situations I and II. The situation I is described by $n(a \wedge b) = 200$, $n(a \wedge \neg b) = 400$ and $n(\neg a \wedge b) = 600$; when, the situation II relative to the attribute ordered pair that we denote by (a', b'), is characterized by $n(a' \wedge b') = 400$, $n(a' \wedge \neg b') = 800$ and $n(\neg a' \wedge b') = 1200$. Thus, $n(a' \wedge b')$, $n(a' \wedge \neg b')$ and $n(\neg a' \wedge \neg b')$ are obtained as twice $n(a \wedge b)$, $n(a \wedge \neg b)$ and $n(\neg a \wedge \neg b)$, respectively. In these conditions, it is not surprising to observe that the situation I corresponds to a strong implication $a \rightarrow b$ according to the q_3 value, $q_3(a,\neg b) = -3.65$; when, for the situation II, the implication $a' \rightarrow b'$ vanishes ($q_3(a',\neg b') = 2.98$). In fact, by comparing the respective sizes of $\mathcal{O}(a)$ and $\mathcal{O}(b)$ for one hand and, $\mathcal{O}(a')$ and $\mathcal{O}(b')$ for the other hand, we can realize that the inclusion degree of $\mathcal{O}(a)$ into $\mathcal{O}(b)$ is much more exceptional than that of $\mathcal{O}(a')$ into $\mathcal{O}(b')$. This latter should reach the value 578 for $n(a' \wedge b')$ to have $q_3(a',\neg b') = q_3(a,\neg b) = -3.65$. This phenomenon is amplified when we multiply all the cardinalities by a same coefficient greater than 1. Thus, with a multiplicative factor 10 one obtains $q_3(a, \neg b) = -11.55$ et $q_3(a', \neg b') = 9.43$. Moreover, with a multiplicative factor equal to 100 one obtains $q_3(a, \neg b) = -36.51$ and $q_3(a', \neg b') = 29.81$.

Consequently, for n enough large the local probabilistic index $\mathcal{J}_3(a, b)$ looses its discriminant power for pairwise comparing many implication rules. The solution proposed by [10] consists of combining a geometric mean the initial index $\mathcal{J}_3(a, b)$ (denoted by φ) with the inclusion index $\tau(a, b)$ [*cf.* (46)], in order to obtain the index called by R. Gras "entropic intensity":

$$\Psi(a, b) = \sqrt{\varphi(a, b) * \tau(a, b)} \tag{56}$$

But indices $\varphi(a, b)$ and $\tau(a, b)$ mixed into a unique one, are very different in their conceptions though they are related in an implicit way logically and statistically. This relation is difficult to analyze in spite of an established formal link between the χ^2 and the mutual information statistics [3]. In these conditions and if n is not too large in order to allow significant contributions of the two components of the entropic intensity, we cannot control the respective parts of these components in the index value $\Psi(a, b)$. Now, for n large enough and increasing, $\varphi(a, b)$ becomes closer and closer of 0 or 1 and then the index $\Psi(a, b)$ tends quickly to 0 or to $\sqrt{\tau(a, b)}$. Deeper formal and statistical analysis would be interesting to be concretely provided for this combined index in the framework of interesting experimental results.

The next subsection is devoted to our approach in which a pure probabilistic framework is maintained.

4.3 Implicative contextual similarity of the likelihood of the link

To build an implicative probabilistic index which remains discriminant whatever the value of n we proceed to a global reduction of the implicative similarities of the form $q_3(a, \neg b)$, with respect to an interesting set of ordered attribute pairs. This solution has been previously proposed, but with less consistency in [23]. More precisely, this normalization has to be performed in the context of a data base comprising attribute couples of which the corresponding association rules have to be compared mutually and in a relative manner. Under these conditions, we are placed into a dependency statistical structure and that could be more or less strong.

This method consists simply in transposing to the asymmetrical case the normalization adopted for the symmetrical one [*cf.* section 3]. Excellent experimental resuts have been obtained in practicing ascendant hierarchical classification according to the LLA likelihood of the link method [20, 18, 21].

A significant problem concerns choosing the set of attribute ordered pairs as a basis for normalization. In the following experimental scheme this basis is defined by all the distinct attribute couples of $\mathcal{A}$ [*cf.* 23]. We will denote the latter by its graph:

$$\mathcal{G}_0 = \{(j,k)|1 \leq j \neq k \leq p\} \tag{57}$$

In [23] a selective choice have been recommended where the reference graph is:

$$\mathcal{G}_1 = \{(j,k)|(1 \leq j \neq k \leq p) \wedge (n(a^j) < n(a^k))\} \tag{58}$$

so the full absorption of $\mathcal{O}(a^j)$ by $\mathcal{O}(a^k)$ can be achieved.

In the data mining community, an implication as $a \rightarrow b$ is taken into account only if indices such like $support(a \rightarrow b) = \frac{n(a \wedge b)}{n}$ and $confidence(a \rightarrow b) = \frac{n(a \wedge b)}{n(a)}$ are respectively higher than thresholds s_0 and c_0 defined by an expert [1, 12]. In this context, we will focus on the data base defined by the graph:

$$\begin{aligned}\mathcal{G}_2 = \{(j,k)|(1 \leq j \neq k \leq p) \wedge (n(a^j) < n(a^k)) \\ \wedge\ (support(a^j \rightarrow a^k) > s_0) \\ \wedge\ (confidence(a^j \rightarrow a^k) > c_0)\}\end{aligned} \tag{59}$$

Now and with respect to a given graph $\mathcal{G}_i$ ($i = 0, 1$ or 2), global normalization leads us to replace the "local" index $q_3(a^j, a^k)$, by the "global" index $q_3^g(a^j, a^k)$ defined as follows:

$$q_3^g(a^j, a^k) = \frac{q_3(a^j, a^k) - mean_e\{q_3|\mathcal{G}_i\}}{\sqrt{var_e\{q_3|\mathcal{G}_i\}}} \tag{60}$$

where $mean_e\{q_3|\mathcal{G}_i\}$ and $var_e\{q_3|\mathcal{G}_i\}$ represent the empirical mean and variance of q_3 on $\mathcal{G}_i$, respectively.

The no relation or independence random model considered is that $\mathcal{N}_3$ (see above). But here this model has to be interpreted globally by associating with the whole attribute set $\mathcal{A}$ a set $\mathcal{A}^*$ of independent random attributes [*cf.* (32)]. In these conditions, $q_3^g(a^{j*}, a^{k*})$ follows a standardized normal law whose cumulative distribution function is, as above, denoted by Φ. Thus, in order to evaluate an association rule $(a^j \rightarrow a^k)$ taking its place in $\mathcal{G}_i$, the likelihood of the link probabilistic discriminant index is defined by the equation:

$$\mathcal{J}_n(a^j, a^k) = 1 - \Phi\left[q_3^g(a^j, a^k)\right] \tag{61}$$

5 Experimental results

5.1 Experimental scheme

Two kinds of experiments have been performed using the "mushrooms" database [5]. This database contains 8124 individuals described by 22 attributes and additionally an attribute class. The latter has been considered at the same level as the others attributes and all of them have been transformed into boolean attributes. By this way, the database is made of 125 attributes describing 8124 individuals. Our choice of this data base is motivated by studying the behaviour of the tested indices on real data and not on artificial ones.

The experimental process can be described as follows:

1. Compute all the couples (a, b) such that $n(a) < n(b)$
2. For each couple, compute $q_3(a, \neg b)$ and $\mathcal{J}(a, b)$ [*cf.* (48)]
3. Then compute q_3 mean and standard deviation for all the couples (a, b) and normalize the q_3 index before using the normal law

In the first experiments, selection of couples (a, b) is only controlled by the following condition: $n(a) \neq 0$, $n(b) \neq 0$ and $n(a \wedge b) \neq 0$.
In the second experiments, some support and confidence constraints are used to control the selected couples. Remind that the support and the confidence of a couple (a, b) are equal to $Pr(a \wedge b)$ and $Pr(b|a) = \frac{Pr(a \wedge b)}{Pr(a)}$, respectively. All the selected couples in the second experiments, are such that $support(a, b) > s_0$ and $confidence(a, b) > c_0$.

These second experiments allow us to analyze the behaviour of several indices. The main goal of all these experiments is to observe the behaviour of the built indices when the size of $\mathcal{O}$ increases without any modification of the cardinalities of $\mathcal{O}(a)$, $\mathcal{O}(b)$ and $\mathcal{O}(a) \cup \mathcal{O}(b)$. In order to do that, a couple (a, b) is selected according to the constraints of the experimental scheme. Then, the value of n is increased without modifying the values of $n(a)$, $n(b)$ and $n(a \wedge b)$. This technique can be compared to adding occurences for which all the concerned attributes have a false value.

We used the following algorithm to realize our experiments:

(1) Select all the couples (a, b) verifying the experimental scheme constraints
(2) For each couple, compute $q_3(a, \neg b)$
(3) Compute mean and standard-deviation of all the q_3 values
(4) For each couple, compute the local likelihood of the link probabilistic index and the globally normalized one

(5) Increase n: $n \leftarrow n + constant$
(6) Repeat (2-5) until $n < threshold$

5.2 Detailed results

The first set of experiments (no constraints on the couples (a, b)) allow us to show many relevant properties for our normalized index.

- this index is discriminant whatever is the n value
- moreover, results show that the behaviour of our index is more ignificant for rules $a \to b$ when $n(a) < n(b)$

Figures 1 and 2 show that the normalized index has a discriminant behaviour unlike the local index (always equal to 1), for attributes a and b considered.

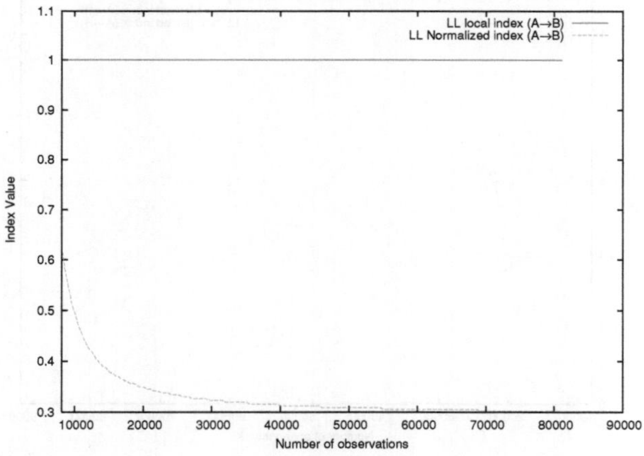

Fig. 1. $n(a) = 192$, $n(b) = 1202$, $n(a \wedge b) = 96$, $s_0 = 0$, $c_0 = 0$

We can also show, on Figure 3, that the normalized index reachs low values, even though the local index tends to unity and this, when $n(a \wedge b)$ is small compared to $n(a)$ (see Figure 4 and 3).

5.3 Robust version of the normalized probabilistic index on a selected attribute couples

In the second series of experiments we focus on couples (a, b) whose support and confidence values are higher than thresholds defined by the user. For all of our experiments, we have used the thresholds $s_0 = 0.1$ and $c_0 = 0.9$. Then, the couples (a, b) under study will be such that $support(a \to b) > 0.1$ and $confidence(a \to b) > 0.9$. In fact these, thresholds are usually considered in

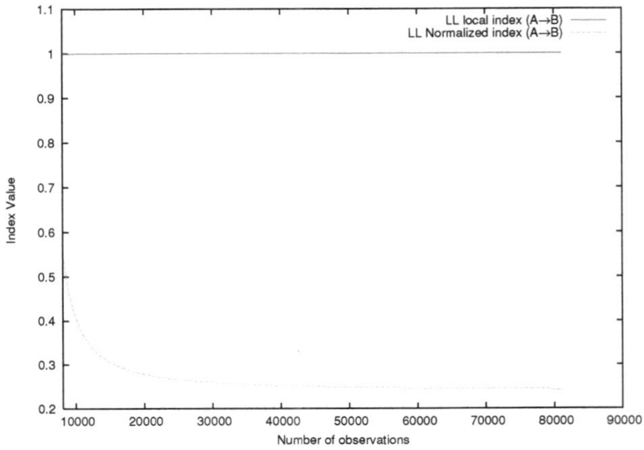

Fig. 2. $n(a) = 192$, $n(b) = 828$, $n(a \wedge b) = 64$, $s_0 = 0$, $c_0 = 0$

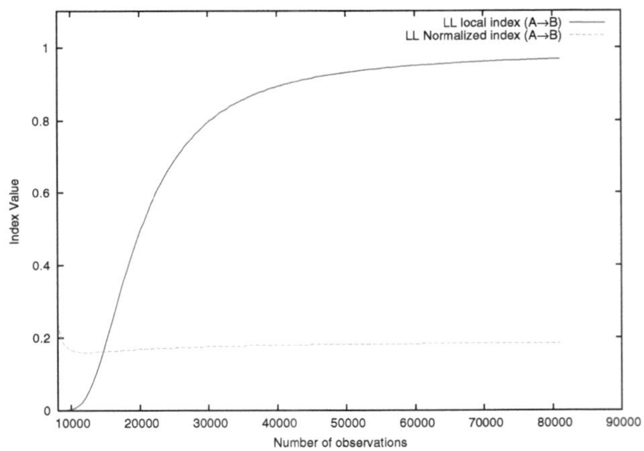

Fig. 3. $n(a) = 452$, $n(b) = 2320$, $n(a \wedge b) = 52$, $s_0 = 0$, $c_0 = 0$

rule extraction context [15, 2]. Furthermore, similar values have been used with the mushrooms database.

Achieved results show that, for this reduced set of attribute couples (a, b), our normalized index is always more discriminant than the local one. Moreover, as we can see on Figure 7, the normalized index is more discriminant in this experimental conditions than in the previous ones (see Figure 6). The shown curves are associated with the configuration presented in Figure 5.

In the second series of experiments, strong relationships characterize all the studied couples. In this case, with n increasing the relationship presented in Figure 5 becomes less and less significant relative to all the

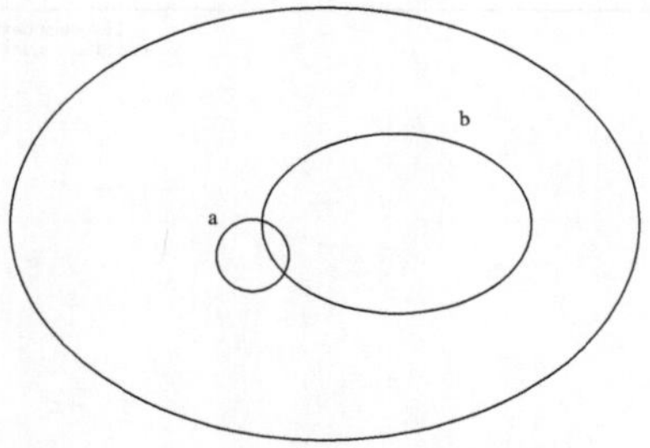

Fig. 4. $n(a) = 452$, $n(b) = 2320$, $n(a \wedge b) = 52$

other relations with the same conditions. Therefore, the filtering step based on thresholds s_0 and c_0 allows us to only retain a set of relations having high values for all the considered indices.

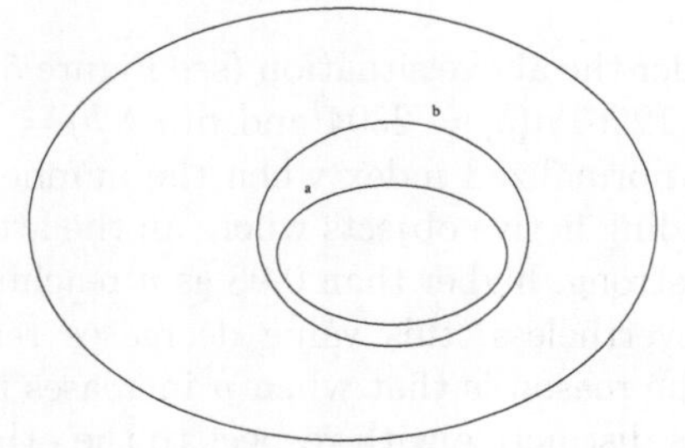

Fig. 5. $n(a) = 1296$, $n(b) = 2304$, $n(a \wedge b) = 1296$

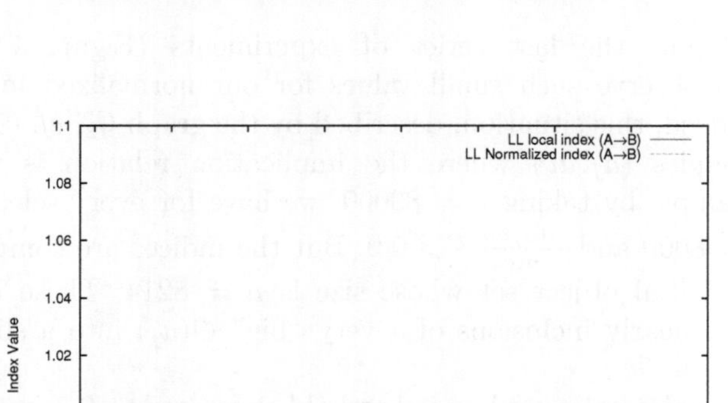

Fig. 6. $n(a) = 1296$, $n(b) = 2304$, $n(a \wedge b) = 1296$, $s_0 = 0$, $c_0 = 0$

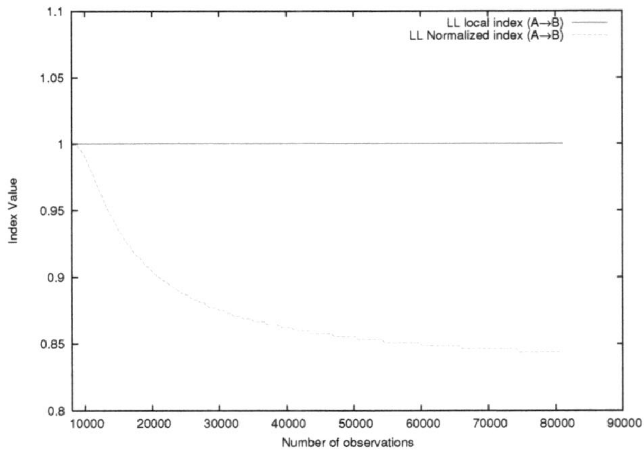

Fig. 7. $n(a) = 1296$, $n(b) = 2304$, $n(a \wedge b) = 1296$, $s_0 = 0.1$, $c_0 = 0.9$

Let us now consider the above situation (see Figure 5) where $\mathcal{O}(a) \subset \mathcal{O}(b)$. In this case, $n(a) = 1296$, $n(b) = 2304$ and $n(a \wedge b) = 1296$. Figure 6 shows the evolution of the normalized index when the number of objects increases from $n = 8124$ by adding fictive objects where all the attributes are false. The index value remains strong, higher than 0.98 as n reaching 80000 and starting from $n = 8124$. Nevertheless, this value decreases reaching a stable level when n increases. The reason is that when n increases the above implication becomes less and less distinctive with respect to the other implications. Such situation must concern cases where $n(a)$ and $n(b)$ are high and near each other.

Consider now the last series of experiments (Figure 7), it is not surprising to observe such small values for our normalized index when n increases. Indeed, this situation, described by the graph $\mathcal{G}_2$ [*cf.* (59)] concerns attribute couples (a_j, a_k) where the implication relation is very strong. Thus for example by taking $n = 80000$, we have for every selected (a_j, a_k), $n(aj \wedge a_k) > 8000$ and $\frac{n(a_j \wedge a_k)}{n(a_j)} > 0.9$. But the indices are computed on the basis of the initial object set whose size is $n = 8214$. These indices must correspond to nearly inclusions of a very "big" $\mathcal{O}(a_j)$ into a slightly bigger $\mathcal{O}(a_k)$.

When the minimal confidence threshold decreases to 0.5, we observe, on Figure 8, that the relation considered becomes more significant with respect to the other relations. This behaviour is not surprising and can be explained by the fact that the set of studied relations includes relations with lower degree of confidence than in the case presented in Figure 7. As a consequence of all our analysis one can say that the new probabilistic normalized index by global reduction is discriminant and reflects in some way the statistical surprise of a rule in the context of other rules.

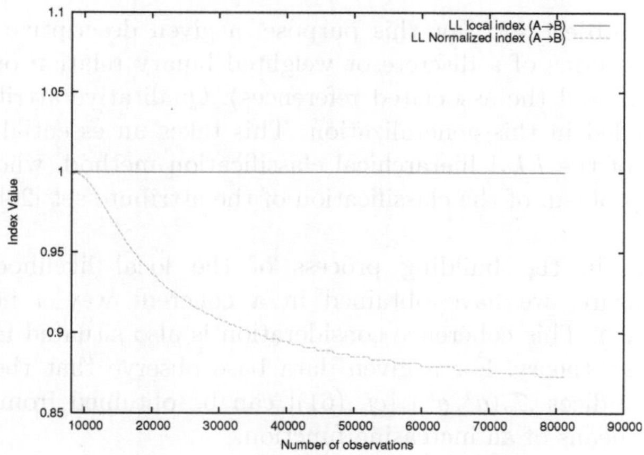

Fig. 8. $n(a) = 1296$, $n(b) = 2304$, $n(a \wedge b) = 1296$, $s_0 = 0.1$, $c_0 = 0.5$

6 Conclusion

In order to evaluate an association rule $a \to b$, many indices have been proposed in data mining literature. Most of them have an absolute meaning and only depend on the concerned ordered pair (a, b) of attributes. This dependence is expressed in terms of proportional frequencies defined by the crossing between the two attributes on a learning set, $p(a)$, $p(b)$ and $p(a \wedge b)$. The formal aspects of these indices are analyzed and mutually compared. In their construction, the importance of empirical statistical independence is set up. Mainly, these indices are compared with our approach in defining probabilistic similarity measure associated with a notion of likelihood of the link with respect to independence hypothesis. Indeed, this notion is expected to capture that of "interestingness measure" setting up surprising rules. This similarity can be symmetrical translating equivalence relation degree or asymmetrical, reflecting asymmetrical implicative relation.

The first conception of the likelihood of the link similarity measure is local, i.e. only depends on the attribute pair to be compared. Unfortunately, this local version looses its discriminant power when the data size becomes large enough. And one of the major objectives of this paper consists in building a likelihood of the link probabilistic index associated in a specific relative manner with the preceding one and finely discriminant for large data bases. It is globally built by normalization with respect to an interesting set of association rules. Then the new index is contextual. The resulting increase of the computing complexity remains linear with respect to the size of the latter rule set.

This conceptual construction has been extremely extended in order to compare mutually in a symmetric way a set of complex attributes observed on a training object set. For this purpose, a given descriptive attribute is interpreted in terms of a discrete or weighted binary relation on the object set $\mathcal{O}$ (see [22] and the associated references). Qualitative attributes of any sort are included in this generalization. This takes an essential part in the development of the *LLA* hierarchical classification method, when the latter adresses the problem of the classification of the attribute set [20].

Otherwise, in the building process of the local likelihood link implicative measure, we have obtained in a coherent way a new absolute measure $\eta_3(a, b)$. This coherence consideration is also situated in the framework of the χ^2 theory. For a given data base observe that the contextual probabilistic indices $\mathcal{J}_n(a^j, a^k)$ [*cf.* (61)] can be obtained from the indices $\eta_3(a^i, a^k)$ by means of an increasing function.

However the specificity of the numerical values of $\mathcal{J}_n$ with respect to those of η_3 enables precisely to better distinguish between different association rules according to the unlikelihood principle. Note that this principle constitutes a basis of the information theory philosophy. Moreover, $\mathcal{J}_n$ enables easily to compare (relatively) a given association rule $a \rightarrow b$ in the contexts of two different sets of rules. Finally and mathematically, several indices conceived locally lead to the unique probabilistic index $\mathcal{J}_n$ by the global likelihood of the link construction (cf 4.1).

The experimental validation includes different interesting aspects. Our main goal was to compare the respective behaviours of the local and the global probabilistic indices. In this, we have clearly shown the discriminant ability of the global normalized index. For this purpose the classical "mushrooms" data base has been employed.

Similar but more global experiments have been performed on several data bases available at `http://www.ics.uci.edu/~mlearn/`: car, monks-1, monks-2, monks-3, nursery, tic-tac-toe and votes. More precisely, for a given number of instances n (see above), the mean and the variance of the local and normalized indices have been computed on a selected subset of association rules. The empirical distributions of these statistical parameters (mean and variance) over n enable to realize that the normalized index is clearly more discriminant than the local one, for all of the mentioned data bases. These experimental results will be reported in a future work.

More importantly, it is interesting in the future to continue the experimental analysis by studying with the same experimental scheme the behaviour of other indices such that those listed in [27], $\psi(a, b)$, called "entropic intensity" or $\eta_3(a, b)$, mentionned above.

Finally, when using $\mathcal{J}_n$ in order to evaluate the interest of a given rule in the context of a data base, boundaries could be defined by the expert knowledge. For this purpose one can for example consider inclusion situations such as the one provided in Figure 5.

References

1. R. Agrawal, T. Imielinski, and A. N. Swami. Mining Association Rules between Sets of Items in Large Databases. In *Proceedings of the 1993 ACM SIGMOD International Conference on Management of Data*, pages 207–216, 1993.
2. Yves Bastide, Rafik Taouil, Nicolas Pasquier, Gerd Stumme, and Lofti Lakhal. Pascal : un algorithme d'extraction des motifs fréquents. *Techniques et Science Informatiques*, 21(21):65–95, 2002.
3. J. P. Bénzecri. Théorie de l'information et classification d'après un tableau de contingence. In *L'Analyse des Donnnées, Tome 1: La Taxinomie*, pages 207–236. Dunod, 1973.
4. J-M. Bernard and C. Charron. L'analyse implicative bayésienne, une méthode pour l'étude des dépendances orientées : Données binaires. *Revue Mathématique Informatique et Sciences Humaines*, 134:5–38, 1996.
5. C.L. Blake and C.J. Merz. UCI Repository of machine learning databases, 1998.
6. S. Brin, R. Motwani, and C. Silverstein. Beyond market baskets: generalizing association rules to correlations. In *Proceedings of ACM SIGMOD'97*, pages 265–276, 1997.
7. F. Daudé. *Analyse et justification de la notion de ressemblance entre variables qualitatives dans l'optique de la classification hiérarchique par AVL.* PhD thesis, Université de Rennes 1, juin 1992.
8. R. M. Goodman and P. Smyth. Information-theoretic rule induction. In *ECAI 1988*, pages 357–362, 1988.
9. R. Gras. *Contribution à l'étude expérimentale et à l'analyse de certaines acquisitions cognitives et de certains objectifs didactiques en mathématiques.* PhD thesis, Doctorat d' État, Université de Rennes 1, octobre 1979.
10. R. Gras, P. Kuntz, and H. Briand. Les fondements de l'analyse statistique implicative et quelques prolongements pour la fouille des données. *Revue Mathématique et Sciences Humaines*, 154-155:9–29, 2001.
11. R. Gras and A. Larher. L'implication statistique, une nouvelle méthode d'analyse de données. *Mathématique Informatique et Sciences Humaines*, 18(120):5–31, 1992.
12. S. Guillaume. *Traitement des données volumineuses. Mesures et algorithmes d'extraction de règles d'association et règles ordinales.* PhD thesis, Université de Nantes, décembre 2000.
13. P. Jaccard. Nouvelles recherches sur la distribution florale. *Bulletin de la Société Vaudoise en Sciences Naturelles*, 44:223–270, 1908.
14. S. Lallich and O. Teytaud. Évaluation et validation de l'intérêt des règles d'association. In *Mesures de Qualité pour la Fouille des Données*, pages 193–218. Cépaduès, 2004.
15. R. Lehn. *Un système interactif de visualisation et de fouille de règles pour l'extraction de connaissances dans les bases de données.* PhD thesis, Institut de Recherche en Informatique de Nantes, Décembre 2000.

16. P. Lenca, P. Meyer, B. Vaillant, B. Picouet, and S. Lallich. Évaluation et analyse multicritère des mesures de qualité des règles d'association. In *Mesures de Qualité pour la Fouille des Données*, pages 219–245. Cépaduès, 2004.
17. I. C. Lerman. Justification et validité statistique d'une échelle [0,1] de fréquence mathématique pour une structure de proximité sur un ensemble de variables observées. *Publications de l'Institut de Statistique des Universités de Paris*, 29:27–57, 1984.
18. I. C. Lerman. Foundations of the Likelihood Linkage Analysis (LLA) classification method. *Applied Stochastic Models and Data Analysis*, 7:63–76, 1991.
19. I.C. Lerman. Sur l'analyse des données préalable à une classification automatique; proposition d'une nouvelle mesure de similarité. *Mathématiques et Sciences Humaines*, 8:5–15, 1970.
20. I.C. Lerman. *Classification et analyse ordinale des données*. Dunod, 1981.
21. I.C. Lerman. Likelihood Linkage Analysis (LLA) classification method (Around an example treated by hand). *Biochimie*, 75:379–397, 1993.
22. I.C. Lerman. Coefficient numérique général de discrimination de classes d'objets par des variables de types quelconques. Application des données génotypiques. *Revue de Statistique Appliquée*, in press, 2006.
23. I.C. Lerman, R. Gras, and H. Rostam. Élaboration et évaluation d'un indice d'implication pour des données binaires I et II. *Revue Mathématique et Sciences Humaines*, 74 and 75:5–35, 5–47, 1981.
24. J. Loevinger. A systematic approach to the construction and evaluation of tests of ability. *Psychological Monographs*, 61:1–49, 1947.
25. K. Pearson. On a criterion that a given system of deviations from the probable in the case of a correlated system of variables is such that it can reasonably be supposed to have arisen from random sampling. *Philosophical Magazine*, 50(5):157–175, 1900.
26. G. Piatetsky-Shapiro. Discovery, analysis, and presentation of strong rules. In *Knowledge Discovery in Databases*, pages 229–248. MIT Press, 1991.
27. P-N. Tan, V. Kumar, and J. Srivastava. Selecting the Right Interestingness Measure for Association Patterns. In *Proceedings of the 8th ACM SIGKDD Conference on Knowledge Discovery and Data Mining*, 2002.

Towards a Unifying Probabilistic Implicative Normalized Quality Measure for Association Rules

Jean Diatta[1], Henri Ralambondrainy[1], and André Totohasina[2]

[1] IREMIA, Université de La Réunion
15, Avenue René Cassin - B.P. 7151
97715, Saint-Denis. Messag Cedex 9, France
{jdiatta,ralambon}@univ-reunion.fr

[2] Département de Mathématiques et Informatique
ENSET, Université d'Antsiranana -B.P. 0
Antsiranana 201 Madagascar
totohasina@yahoo.fr

Summary. We define the so-called normalized probabilistic quality measures (PQM) for association rules; that is, PQMs whose values lay between minus one and plus one, and which take into account reference situations such as incompatibility, repulsion, independence, attraction, and logical implication, between the antecedent and the consequent of association rules. Moreover, we characterize the PQMs that can be normalized and propose a way to normalize them. On the other hand, we consider a normalized and implicative PQM called M_{GK}. It appears that M_{GK} is the normalized PQM associated to most of the PQMs proposed in the literature. Furthermore, it satisfies additional properties, including Piatetsky-Shapiro's principles and Freitas's.

Key words: Negative association rule, Probabilistic quality measure, Normalization, Data mining, Logical implication.

1 Introduction

The association rule mining problem is among the most popular data mining techniques [1, 16, 17, 6, 4, 22, 14]. Association rules are useful for discovering relationships among data in huge databases. The standard example consists of a large dataset of supermarket sales transactions. Each transaction reveals the items purchased by a particular client at the supermarket. The challenge is then to find relevant ordered itemset pairs (U, V) revealing an interesting relationship between the antecedent U and the consequent V of the association rule [2, 21, 12]. Several rule interestingness measures, also called rule quality

Jean Diatta et al.: *Towards a Unifying Probabilistic Implicative Normalized Quality Measure for Association Rules*, Studies in Computational Intelligence (SCI) **43**, 237–250 (2007)
www.springerlink.com

measures, have been proposed in the literature, each aiming to capture the rules that meet some given properties [11, 13].

This chapter is concerned with probabilistic quality measures (PQM). We define a class of PQMs said to be normalized in the following sense:

- their values lay between minus one and plus one and
- they take into account incompatibility, repulsion, independence, attraction, and logical implication between the antecedent and the consequent of association rules.

Moreover, we characterize the PQMs that can be normalized and propose a way to construct its so-called associated normalized PQM, for each of them.

On the other hand, we consider the normalized and implicative PQM M_{GK} earlier introduced in [10] and whose mathematical properties have been studied in [19]. It appears that M_{GK} is the normalized PQM associated to most of the PQMs proposed in the literature. Furthermore, M_{GK} satisfies additional properties such as Piatetsky-Shapiro's principles [15] and Freitas's principle [8], and takes into account deviation from equilibrium [5]. The chapter is organized as follows.

Section 2 briefly recalls the definition of association rules in the framework of binary contexts. In Section 3 we consider probabilistic quality measures and review some of their expected properties. The normalization of PQMs is addressed in Section 4 and the PQM M_{GK} introduced in Section 5. A short conclusion is included at the end of the chapter.

2 Association rules

2.1 Binary contexts

A *binary variable/entity context* (or simply *binary context*) is a pair $(\mathcal{E}, \mathcal{V})$, where $\mathcal{E}$ is an entity set and $\mathcal{V}$ a set of variables on $\mathcal{E}$ taking their values in $\{0,1\}$. A binary context $(\mathcal{E}, \mathcal{V})$ is said to be *finite* if both of $\mathcal{E}$ and $\mathcal{V}$ are finite. To any variable subset V is associated its dual V', consisting of entities for which each variable belonging to V takes the value 1; that is

$$V' = \{e \in \mathcal{E} : V(e) = 1\} := \{e \in \mathcal{E} : \forall v \in V\, [v(e) = 1]\}.$$

As binary variables take their values in $\{0,1\}$, they can be thought of as attributes that entities may or may not possess. Thus, they are sometimes called *items* and a set of binary variables is accordingly called an *itemset*.

In the sequel, $\mathcal{E}$ will denote a finite entity set, $\mathcal{V}$ a finite set of binary variables, and $\mathbb{K}$ the binary context $(\mathcal{E}, \mathcal{V})$. Moreover, we will consider the finite discrete probabilized space $(\mathcal{E}, \mathcal{P}(\mathcal{E}), P)$, where P is the intuitive probability such that for an event E in $\mathcal{P}(\mathcal{E})$, $P(E) = \frac{|E|}{|\mathcal{E}|}$. Each variable v will be considered as a Bernoulli random variable defined on the sample space $\mathcal{E}$, such that $P(v = 1) = P(v^{-1}(1)) = \frac{|v^{-1}(1)|}{|\mathcal{E}|}$.

Table 1 presents a binary (entity/variable) context $\mathbb{K}_1 := (\mathcal{E}_1, \mathcal{V}_1)$, where $\mathcal{E}_1 = \{e_1, e_2, e_3, e_4, e_5\}$ and $\mathcal{V}_1 = \{v_1, v_2, v_3, v_4, v_5\}$. If we let $V = \{v_2, v_3\}$, then $V' = \{e_2, e_3, e_5\}$ so that $P(V') = 3/5$.

Table 1. A finite binary context $\mathbb{K}_1$

	v_1	v_2	v_3	v_4	v_5
e_1	1	0	1	1	0
e_2	0	1	1	0	1
e_3	1	1	1	0	1
e_4	0	1	0	0	1
e_5	1	1	1	0	1

2.2 Association rules

Definition 1. *An* association rule *in the context* $\mathbb{K}$ *is an ordered pair* $r := (U, V)$ *of itemsets, denoted* $U \to V$*, where* V *is not empty. The itemsets* U *and* V *are called the* antecedent *and the* consequent *of* r*, respectively.*

Typically, association rules express relationships among data in a given context. Thus, to catch interesting relationships, we need to generate only interesting association rules, *i.e.*, those meeting some given properties. The interestingness of an association rule is usually assessed by means of a so-called interestingness measure or quality measure. A *quality measure* is a real-valued function μ defined on $\mathcal{P}(\mathcal{V}) \times \mathcal{P}(\mathcal{V})$. We often write $\mu(U \to V)$ instead of $\mu(U, V)$. The most used quality measures are certainly the so-called "Support" and "Confidence" [1], respectively defined by:

$$\mathrm{Supp}(U \to V) = P(U' \cap V')$$

and

$$\mathrm{Conf}(U \to V) = P(V' | U').$$

Table 2 presents some association rules in the context $\mathbb{K}_1$ defined in Table 1, along with their support and confidence. For simplicity, itemsets are denoted as finite words; for instance, $v_2 v_5$ denotes the pair $\{v_2, v_5\}$.

Table 2. Association rules in $\mathbb{K}_1$ along with their support and confidence

rule	support	confidence
$\emptyset \to v_2 v_5$	0.8	4/5
$\emptyset \to v_3$	0.8	4/5
$v_1 v_3 \to v_2 v_5$	0.4	2/3

3 Probabilistic quality measures

Before defining probabilistic quality measures, let us first set some notations. Let us denote by the following for two itemsets U and V:

- n_U the number $|U'|$ of realizations of U;
- n_{UV} the number $|U' \cap V'|$ of examples of the association rule $U \rightarrow V$;
- $n_{U\overline{V}}$ the number $|U' \cap \overline{V'}|$ of counter-examples of the association rule $U \rightarrow V$;
- $\overline{U}$ the logical negation of U, *i.e.*, for an entity e, $\overline{U}(e) = 1$ if and only if $U(e) = 0$;
- $(\overline{U})' = \mathcal{E} \setminus U'$.

Definition 2. *A quality measure μ will be said to be probabilistic if for an association rule $U \rightarrow V$, $\mu(U, V)$ can be entirely expressed in terms of the probabilities $P(U')$, $P(V')$ and $P(U' \cap V')$.*

Remark 1. In a binary context, a probabilistic quality measure depends exclusively on the four parameters n, n_U, n_V and n_{UV}. Indeed, consider the contingency table K_{UV} (see Table 3). When the marginal values n, n_U and n_V are fixed, the knowledge of one value of the table, for instance n_{UV}, determines the other values. Now the probabilities $P(U')$, $P(V')$ and $P(U' \cap V')$ are computed using the entries of this contingency table.

Table 3. The contingency table K_{UV}

$U \diagdown V$	V	$\overline{V}$	
U	n_{UV}	$n_{U\overline{V}}$	n_U
$\overline{U}$	$n_{\overline{U}V}$	$n_{\overline{U}\overline{V}}$	$n_{\overline{U}}$
	n_V	$n_{\overline{V}}$	n

In the sequel, all the quality measures will be supposed to be probabilistic and continuous.

3.1 Motivations

As mentioned above, Support and Confidence are the most used association rule quality measures. Indeed, given a data mining context, one usually looks for support-confidence valid association rules; that is, association rules whose support and confidence are respectively at least equal to user-given minimum thresholds. Many authors have pointed out that this can lead to erroneous situations such as those mentioned below:

- Support-confidence valid rules are not always relevant. Indeed, assume that the minimum support and minimum confidence thresholds are set to 50% and 50%, respectively. Let U and V be two independent itemsets with $P(U') = 70\%$ and $P(V') = 90\%$. Then, the rule $U \rightarrow V$ will be considered as valid since its support is equal to $P(U' \cap V') = P(U') \times P(V') = 63\%$ and its confidence is equal to $P(V'|U') = P(V') = 90\%$.
- Support-confidence validity may prevent capture of relevant rules. Indeed, as $\mathrm{Supp}(U \rightarrow V) \leq \mathrm{Conf}(U \rightarrow V)$, a rule having a low support (less than the minimum fixed threshold) may have a high confidence (greater than the minimum fixed threshold).

Several rule interestingness measures have been proposed in the literature [5], each attempting to take into account reference situations such as incompatibility, independence, deviation from equilibrium and so on. In this section, we broadly outline basic requirements for a rule quality measure, derived from comparative studies of PQMs [15, 11].

- *Piatetsky-Shapiro's principles* [15]:
  According to Piatetsky-Shapiro, a rule quality measure μ must satisfy the following conditions:
  1. $\mu(U \rightarrow V)$ must be null when U and V are independent;
  2. $\mu(U \rightarrow V)$ must strictly increase with the number n_{UV} of examples when the parameters n, n_U, n_V are fixed;
  3. $\mu(U \rightarrow V)$ must strictly decrease with the cardinality n_V of V' when the parameters n, n_{UV}, n_U (or n_V) are fixed.
- *Freitas's principle* [8]:
  According to Freitas, a quality measure μ must be a non symmetric function, *i.e.*, for two different itemsets U and V, $\mu(U \rightarrow V)$ must not be equal to $\mu(V \rightarrow U)$.
- *Sensitivity to incompatibility*:
  The antecedent U and the consequent V of a rule are said to be *incompatible* if they cannot be realized simultaneously, *i.e.*, if $P(V'|U') = 0$. The minimal value of a PQM is expected to correspond to incompatibility between the antecedent and the consequent of the rule.
- *Sensitivity to equilibrium*:
  There is an *indetermination* or *equilibrium* situation when the number of examples and that of counter-examples are balanced, *i.e.*, when $P(V'|U') = \frac{1}{2}$. The value of a PQM in a situation of equilibrium is expected to be around $\frac{1}{2}$.
- *Sensitivity to logical implication*:
  There is a *logical implication* between the antecedent U and the consequent V of a rule if V is realized whenever U is, *i.e.*, if $P(V'|U') = 1$. The maximal value of a PQM is expected to correspond to logical implication between the antecedent and the consequent of the rule.
- *Agreement with logical principle*:
  In mathematical logic, and implication $U \rightarrow V$ is equivalent to its

contraposition $\overline{V} \to \overline{U}$. It may be expected that a PQM μ takes this feature into account in the sense that $\mu(U \to V) = \mu(\overline{V} \to \overline{U})$ for any itemsets U, V. Such a PQM will be said to be *implicative.*

4 Normalization of PQMs

4.1 Normalized PQMs

We have seen in the previous section that a PQM μ is expected to satisfy, among others, the following conditions:

- $\mu(U \to V)$ is minimal if U and V are incompatible, *i.e.*, if $P(U' \cap V') = 0$;
- $\mu(U \to V) = 0$ if U and V are independent, *i.e.*, if $P(U' \cap V') = P(U')P(V')$;
- $\mu(U \to V)$ is maximal if U logically implies V, *i.e.*, if $P(V'|U') = 1$.

As the values of a PQM may be signed, to enable easier interpretation of rules, it is preferable that:

- positive values correspond to *positive dependence* between the antecedent U and the consequent V, *i.e.*, when $P(V'|U') > P(V')$;
- negatives values correspond to *negative dependence* between the antecedent U and the consequent V, *i.e.*, when $P(V'|U') < P(V')$.

We will sometimes say that U *favors* V to mean that there is a positive dependence (or *attraction*) between U and V. Similarly, we will sometimes say that U *disfavors* V to mean that there is a negative dependence (or *repulsion*) between U and V. Moreover, the five conditions above lead us to the following definition of a normalized PQM.

Definition 3. *A PQM μ will be said to be* normalized *if it satisfies the following conditions:*

1. $0 < \mu(U \to V) \leq 1$ *if U favors V;*
2. $-1 \leq \mu(U \to V) < 0$ *if U disfavors V;*
3. $\mu(U \to V) = 0$ *if U and V are independent;*
4. $\mu(U \to V) = -1$ *if U and V are incompatible;*
5. $\mu(U \to V) = 1$ *if U logically implies V.*

Roughly speaking, a normalized PQM is a PQM whose values lie in the interval $[-1, +1]$ and reflect reference situations such as incompatibility, repulsion, independence, attraction and logical implication. Thus, the values of a normalized PQM are distributed as shown in Figure 1. The following three questions naturally arise from the definition of a normalized PQM:

(a) Is there any normalized PQM?
(b) In case of positive answer to (a), can any given PQM be normalized?
(c) In case of positive answer to (b), how to normalize a given PQM?

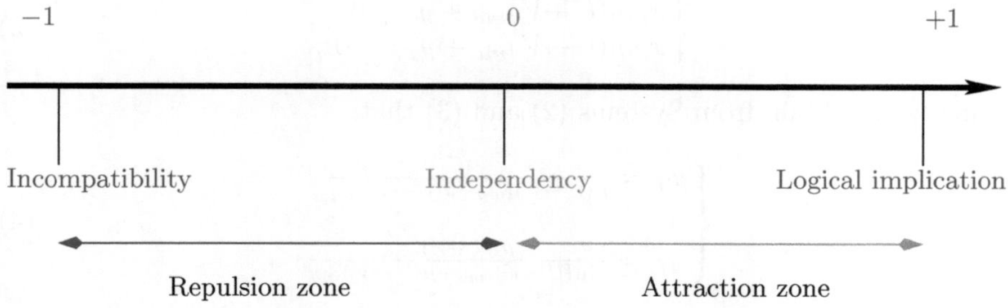

Fig. 1. Distribution of the values of a normalized PQM

The answer to Question (a) is positive (see the PQM M_{GK} in Section 5 below). Moreover, as to Questions (b) and (c), a characterization of the PQMs that can be normalized is given in Section 4.2, as well as a way to normalize them.

4.2 Normalizing a PQM

By analogy to the well-known normalization of random variables, a PQM will be normalized by reducing (if possible) and centering its values. As we distinguish positive dependence from negative one, normalization should bring the PQM values into the union of the two intervals $[-1, 0[$ (negative dependence) and $]0, 1]$ (positive dependence), and the central value 0 should be reached when the antecedent and the consequent are independent.

Let μ be a PQM that can be normalized and let μ_n denote the normalized PQM associate to μ. Then μ_n can be written as:

$$\mu_n(U{\rightarrow}V) = \begin{cases} x_f.\mu(U{\rightarrow}V) + y_f, & \text{if } U \text{ favors } V \text{ or } U \text{ and } V \text{ are independent} \\ x_d\mu(U{\rightarrow}V) + y_d, & \text{if } U \text{ disfavors } V \text{ or } U \text{ and } V \text{ are independent} \end{cases} \quad (1)$$

where x_f, x_d, y_f and y_d are the unknowns. As μ is continuous, it is sufficient to determine the unknowns on reference situations such as incompatibility, independence and logical implication (see Figure 1). Let us denote by the following:

- $\mu(U{\rightarrow}V)_{imp}$ the limit of $\mu(U{\rightarrow}V)$ when $P(V'|U')$ tends towards 1;
- $\mu(U{\rightarrow}V)_{ind}$ the limit of $\mu(U{\rightarrow}V)$ when $P(V'|U')$ tends towards $P(V')$;
- $\mu(U{\rightarrow}V)_{inc}$ the limit of $\mu(U{\rightarrow}V)$ when $P(V'|U')$ tends towards 0.

Then, solutions to Equation (1) are those of both Systems (2) and (3) below:

$$\begin{cases} x_f.\mu(U{\rightarrow}V)_{imp} + y_f = +1 \\ x_f.\mu(U{\rightarrow}V)_{ind} + y_f = 0 \end{cases} \quad (2)$$

and

$$\begin{cases} x_d.\mu(U{\rightarrow}V)_{ind} + y_d = 0 \\ x_d.\mu(U{\rightarrow}V)_{inc} + y_d = -1 \end{cases} \tag{3}$$

Finally, we obtain from Systems (2) and (3) that:

$$\begin{cases} x_f = \frac{1}{\mu(U\rightarrow V)_{imp} - \mu(U\rightarrow V)_{ind}} \\ \\ y_f = -\frac{\mu(U\rightarrow V)_{imd}}{\mu(U\rightarrow V)_{imp} - \mu(U\rightarrow V)_{ind}} \end{cases} \tag{4}$$

and

$$\begin{cases} x_d = \frac{1}{\mu(U\rightarrow V)_{ind} - \mu(U\rightarrow V)_{inc}} \\ \\ y_d = \frac{\mu(U\rightarrow V)_{ind}}{\mu(U\rightarrow V)_{ind} - \mu(U\rightarrow V)_{inc}} \end{cases} \tag{5}$$

We now go on to give a characterization of PQMs that can be normalized. This characterization naturally derives from Systems (2) and (3) above. Indeed, according to the expressions of Solutions (4) and (5), a PQM μ can be normalized only if the following two conditions are satisfied:

- $\mu(U{\rightarrow}V)_{imp} \neq \mu(U{\rightarrow}V)_{ind}$;
- $\mu(U{\rightarrow}V)_{inc} \neq \mu(U{\rightarrow}V)_{ind}$.

Moreover, the following result has been proved in [7].

Proposition 1. *A PQM μ can be normalized if and only if it satisfies the following three conditions for any two itemsets U, V:*

(i) each of the three limits $\mu(U{\rightarrow}V)_{imp}$, $\mu(U{\rightarrow}V)_{ind}$ and $\mu(U{\rightarrow}V)_{inc}$ is finite;
(i) $\mu(U{\rightarrow}V)_{imp} \neq \mu(U{\rightarrow}V)_{ind}$;
(i) $\mu(U{\rightarrow}V)_{inc} \neq \mu(U{\rightarrow}V)_{ind}$.

As a consequence of this result, it can be observed that the PQMs Conviction and Sebag-Schoenauer's measure [20] cannot be normalized.

5 An implicative normalized PQM: M_{GK}

5.1 Definition and properties

In this section, we consider the normalized PQM M_{GK}, earlier introduced in [10]. Properties (i) and (ii) of the next remark may help to understand the definition of the PQM M_{GK}.

Remark 2. Let U and V be two itemsets. Then, the following properties hold.

(i) If U favors V, then $0 < P(V'|U') - P(V') \leq 1 - P(V')$.
(ii) If U disfavors V, then $-P(V') \leq P(V'|U') - P(V') < 0$.

(iii) "U disfavors V" is equivalent to "U favors $\overline{V}$"; indeed
$1 - P(V') < 1 - P(V'|U')$ if and only if $P(\overline{V'}) < P(\overline{V'}|U')$.

Definition 4. *The quality measure M_{GK} is defined by*

$$M_{GK}(U \to V) = \begin{cases} \frac{P(V'|U') - P(V')}{1 - P(V')}, & \text{if } U \text{ favors } V \text{ or } U \text{ and } V \text{ are independent} \\ \frac{P(V'|U') - P(V')}{P(V')}, & \text{if } U \text{ disfavors } V \text{ or } U \text{ and } V \text{ are independent} \end{cases}$$

Remark 3. The expression of M_{GK} can be written in terms of cardinalities as follows:

$$M_{GK}(U \to V) = \begin{cases} \frac{n n_{UV} - n_U n_V}{n_U (n - n_V)}, & \text{if } U \text{ favors } V \text{ or } U \text{ and } V \text{ are independent} \\ \frac{n n_{UV} - n_U n_V}{n_U n_V}, & \text{if } U \text{ disfavors } V \text{ or } U \text{ and } V \text{ are independent} \end{cases}$$

It is easy to check that M_{GK} is a non symmetric PQM. Moreover, M_{GK} satisfies the three Piatetsky-Shapiro principles in addition to the following properties.

Proposition 2. *The PQM M_{GK} satisfies the following properties for any itemsets U, V:*

1. *$M_{GK}(U \to V) = 0$ if and only if U and V are independent*
2. *$M_{GK}(U \to V) > 0$ if and only if U favors V;*
3. *$M_{GK}(U \to V) < 0$ if and only if U disfavors V;*
4. *$M_{GK}(U \to V) = +1$ if and only if U logically implies V;*
5. *$M_{GK}(U \to V) = -1$ if and only if U and V are incompatible;*
6. *If U favors V or U and V are independent, then $M_{GK}(U \to V) = M_{GK}(\overline{V} \to \overline{U})$.*

Proof.
Properties (1) to (5) can be easily checked. Thus, we will prove only Property (6). Let U, V be two itemsets such that U favors V or U and V are independent. Then

$$M_{GK}(\overline{V} \to \overline{U}) = \frac{P(\overline{U'}|\overline{V'}) - P(\overline{U'})}{1 - P(\overline{U'})} = \frac{1 - P(U'|\overline{V'}) - 1 + P(U')}{P(U')}$$
$$= \frac{P(U') - P(U' \cap \overline{V'})/(1 - P(V'))}{P(U')} = \frac{-P(U')P(V') + P(U' \cap V')}{P(U')(1 - P(V'))}$$
$$= \frac{P(V'|U') - P(V')}{1 - P(V')} = M_{GK}(U \to V).$$

□

Recall that with regard to formal logic, a PQM μ is said to be implicative if for any itemsets U, V, $\mu(U \to V) = \mu(\overline{V} \to \overline{U})$. Thus, as we are only interested in rules for which the antecedent favors the consequent and due to Property (6) of Proposition 2, the PQM M_{GK} can be considered as being implicative.

On the other hand, it may be observed that M_{GK} is its own associated normalized PQM. Furthermore, it appears that M_{GK} unifies many of the existing PQMs. Indeed, M_{GK} is the normalized PQM associated to most of the PQMs proposed in the literature, including:

- Support: $\text{Supp}(U \to V) = P(U' \cap V')$.
- Confidence: $\text{Conf}(U \to V) = P(V'|U')$.
- Lovinger's measure: $\text{Lov}(U \to V) = \frac{P(V'|U') - P(V')}{1-P(V')}$.
- Piatetsky-Shapiro's measure: $\text{PiatShap}(U \to V) = nP(U')(P(V'|U') - P(V'))$.
- ϕ-coefficient: $\phi(U \to V) = \frac{P(U' \cap V') - P(U')P(V')}{\sqrt{(P(U')P(V')(1-P(U'))((1-P(V'))}}$.

5.2 Further properties of M_{GK}

In this section, we discuss the ability of M_{GK} to handle negative association rules, as well as the way it takes into account situations such as surprise, deviation from equilibrium and independence.

Independence

For two itemsets U and V, the number

$$\delta(U, V) = \frac{P(V'|U') - P(V')}{P(V')} = \frac{P(V' \cap U') - P(U')P(V')}{P(U')P(V')}$$

measures the deviation ratio of the itemset pair (U, V) from independence, when U favors V. Now, it may be noticed that, when U favors V, $M_{GK}(U \to V) = \delta(U, V)\frac{P(V')}{1-P(V')}$. Thus, $M_{GK}(U \to V) > \delta(U, V)$ if and only if $\frac{P(V')}{1-P(V')} > 1$, *i.e.* $P(V') > 1/2$.

This observation can be interpreted in the following way: for two itemsets U and V such that U favors V, when $P(V') > 1/2$, $M_{GK}(U \to V)$ indicates the degree to which V depends on U: the larger the M_{GK}'s absolute value, the stronger the dependency of V on U.

On the other hand, as can be observed from Remark 1, the definition of M_{GK} implicitly refers to contingency tables. This makes it easy to link M_{GK} with the well-known χ^2, as stated in Proposition 3 below.

Proposition 3. *For two itemsets U and V:*

$$M_{GK}(U \to V) = \begin{cases} \sqrt{\frac{1}{n}\frac{n-n_U}{n_U}\frac{n_V}{n-n_V}\chi^2}, & \text{if } U \text{ favors } V \text{ or } U \text{ and } V \text{ are independent} \\ -\sqrt{\frac{1}{n}\frac{n-n_U}{n_U}\frac{n-n_V}{n_V}\chi^2}, & \text{if } U \text{ disfavors } V \text{ or } U \text{ and } V \text{ are independent} \end{cases}$$

Proposition 3 shows that the dependence significance thresholds of M_{GK} can be easily obtained from those of χ^2 (see [18] or [9, pp 43-47] for Gras's index).

Figure 2 compares the ways the PQM M_{GK} and the χ^2 measure the dependence degree between two itemsets, in five reference situations: positive dependence (Fig. 2 (1)), negative dependence (Fig. 2 (2)), independence (Fig. 2 (3)), incompatibility (Fig. 2 (4)) and logical implication (Fig. 2 (5)). It appears that, in contrast to the χ^2, the PQM M_{GK} specifies the degree of dependence in a scale within the interval $[-1, +1]$, in addition to its orientation.

$U \setminus V$	V	$\overline{V}$	
U	3000	2000	5000
$\overline{U}$	2500	2500	5000
	5500	4500	10000

(1) Positive dependence
$\chi^2 = 101$, $M_{GK} = +0.11$

$U \setminus V$	V	$\overline{V}$	
U	1000	3000	4000
$\overline{U}$	4500	1500	6000
	5500	4500	10000

(2) Negative dependence
$\chi^2 = 2424$, $M_{GK} = -0.54$

$U \setminus V$	V	$\overline{V}$	
U	2200	1800	4000
$\overline{U}$	3300	2700	6000
	5500	4500	10000

(3) Independence
$\chi^2 = 0 = M_{GK}$

$U \setminus V$	V	$\overline{V}$	
U	0	2000	2000
$\overline{U}$	6000	2000	8000
	6000	4000	10000

(4) Incompatibility
$\chi^2 = 3750$, $M_{GK} = -1$

$U \setminus V$	V	$\overline{V}$	
U	3000	0	3000
$\overline{U}$	3000	4000	7000
	6000	4000	10000

(5) Logical implication
$\chi^2 = 2857$, $M_{GK} = 1$

Fig. 2. Comparison of the PQM M_{GK} and the χ^2 in five reference situations

Deviation from equilibrium

Let U and V be two itemsets such that U favors V. If U and V are in situation of equilibrium, *i.e.*, the number n_{UV} of example equals the number $n_{U\overline{V}}$ of counter-example, then:

$$M_{GK}(U \to V) = \frac{1}{2} - \frac{n_V}{2(n - n_V)} = \frac{1}{2} - o(\frac{1}{n}),$$

which tends towards $\frac{1}{2}$ when n becomes sufficiently large. This shows that M_{GK} takes into account deviation from equilibrium in large databases.

Surprise measure

According to [3], the surprise brought by an association rule $U \rightarrow V$ is the quantity defined by

$$\text{Surprise}(U \rightarrow V) = \frac{P(U' \cap V') - P(U' \cap \overline{V'})}{P(V')}.$$

Thus, $\text{Surprise}(U \rightarrow V) = 0$ when U and V are in situation of equilibrium. Moreover, M_{GK} takes surprise background into account, as shown in Proposition 4 below.

Proposition 4. *The following properties hold for any two itemsets U and V.*

(i) $\text{Surprise}(U \rightarrow V) > 0$ *if and only if* $\text{Conf}(U \rightarrow V) > 1/2$.
(ii) If $M_{GK}(U \rightarrow V) > 0$ and $P(V') < 1/2$, then $Surp(U \rightarrow V) > 0$.

Wu et al. [20] payed special attention to so-called *negative rules* of the form $U \rightarrow \overline{V}$, and showed that such rules play an important role in decision making. Actually, to handle negative rules, they used their independently discovered so-called confidence conditional probability increment ratio function, which is nothing else than M_{GK}. Proposition 5 shows how M_{GK} makes it possible to capture both positive and negative association rules.

Proposition 5. *Let U and V be two itemsets and let $\alpha \in]0, 1[$ be a positive real number. Then the following properties hold.*

(i) $M_{GK}(U \rightarrow \overline{V}) = -M_{GK}(U \rightarrow V)$;
(ii) $\alpha < M_{GK}(U \rightarrow V) < 1$ *if and only if* $-1 < M_{GK}(U \rightarrow \overline{V}) < -\alpha$.

6 Conclusion

In this paper, we introduced the concept of a normalized probabilistic quality measure intended to take into account reference situations such as incompatibility, repulsion, independence, attraction, and logical implication, between the antecedent and the consequent of an association rule. Moreover, we characterized the PQMs that can be normalized and proposed a way to normalize them.

On the other hand, we considered the PQM M_{GK}, earlier introduced in [10]. The PQM M_{GK} turns out to be normalized and implicative. Furthermore, it satisfies additional properties such as Piatetsky-Shapiro's principles [15], Freitas's principle [8], and takes into account deviation from equilibrium [5]. Finally, M_{GK} appears to be the normalized PQM associated to most of the PQMs proposed in the literature.

References

1. R. Agrawal, T. Imielinski, and A. Swami. Mining association rules between sets of items in large databases. In P. Buneman and S. Jajodia, editors, *Proc. of the ACM SIGMOD International Conference on Management of Data*, volume 22, pages 207–216, Washington, 1993. ACM press.
2. R. Agrawal and R. Srikant. Fast algorithms for mining association rules. In *Proc. of the 20th VLDB Conference*, pages 487–499, 1994.
3. J. Azé. Une nouvelle mesure de qualité pour l'extraction de pépites de connaissances. *Revue des Sciences et Technologies de l'Information*, 17:171–182, 2003.
4. R. J. Bayardo. Efficiently mining long patterns from databases. In *Proc. of the ACM SIGMOD Conference*, pages 85–93, 1998.
5. J. Blanchard, F. Guillet, H. Brilland, and R. Gras. Assessing rule interestigness with a probabilistic measure of deviation from equilibrium. In *Proc. of Applied stochastic Models and Data Analysis*, pages 334–344, ENST Bretagne, France, 2005.
6. S. Brin, R. Motwani, J. D. Ullman, and S. Tsur. Dynamic itemset counting and implication rules for market basket data. In *Proc. of the ACM SIGMOD Conference*, pages 255–264, 1997.
7. D. Feno, J. Diatta, and A. Totohasina. Normalisée d'une mesure probabiliste de qualité des règles d'association: étude de cas. In *Actes du 2nd Atelier Qualité des Données et des Connaissances*, pages 25–30, Lille, France, 2006.
8. A.A. Freitas. On rule interestingness measure. *Knowledge-Based System*, 12:309–315, 1999.
9. R. Gras. *L'implication statistique. Nouvelle méthode exploratoire de données.* La Penée sauvage, France, 1996.
10. S. Guillaume. *Traitement des données volumineuses. Mesures et algorithmes d'extraction des règles d'association et règles ordinales.* PhD thesis, Université de Nantes, France, 2000.
11. R. J. Hilderman and H. J. Hamilton. Knowledge discovery and interestingness measures: A survey. Technical Report CS 99-04, Department of Computer Science, University of Regina, 1999.
12. J. Hipp, U. Güntzer, and G. Nakhaeizadeh. Algorithms for Association Rule Mining - a General SURVEY and Comparison. *SIGKDD Explorations*, 2:58–64, 2000.
13. M.L. Antonie and O.-R. Zaïane. Mining positive and negative Association Rules: an approach for confined rules. Technical report, Dept of Computing Sciences, university of Alberta, Canada, 2004.
14. Nicolas Pasquier, Yves Bastide, Rafik Taouil, and Lotfi Lakhal. Closed set based discovery of small covers for association rules. In *Proc. 15emes Journees Bases de Donnees Avancees, BDA*, pages 361–381, 1999.
15. G. Piatetsky-Shapiro. Knowledge discovery in real databases. a report on the ijcai-86 workshop. *AI Magazine*, 11(5):68–70, 1991.
16. A Savasere, E. Omiecinski, and S. Navathe. An efficient algorithm for mining association rules in large databases. In *Proc. of the 21th VLDB Conference*, pages 432–444, September 1995.
17. H. Toivonen. Sampling large databases for association rules. In *Proc. of the 22nd VLDB Conference*, pages 134–145, September 1994.

18. A. Totohasina. Notes sur l'implication statistique: dépendance positive orientée, valeurs critiques. Technical report, SCAD, Dept de Maths-Info, Université du Québec à Montréal, 1994.
19. A. Totohasina. Normalization of probabilistic quality measure (in french). In *Proc. French Society of Statistics (SFDS'03), XXVth Days of Statistics*, volume 2, pages 958–988, Lyon 2, France, 2003.
20. X. Wu, C. Zhang, and S. Zhang. Mining both positive and negative rules. *ACM J. Information Systems*, 22(3):381–405, 2004.
21. M. J. Zaki and C.-J. Hsiao. CHARM: An efficient algorithm for closed itemset mining, 1999. Technical Report 99-10, Computer Science, Rensselaer Polytechnic Institute, 1999.
22. M. J. Zaki and M. Ogihara. Theoretical Foundations of Association Rules. In *3rd SIGMOD'98 Workshop on Research Issues in Data Mining and Knowledge Discovery (DMKD)*, pages 1–8, 1998.

Association Rule Interestingness: Measure and Statistical Validation

Stephane Lallich[1], Olivier Teytaud[2], and Elie Prudhomme[1]

[1] Université Lyon 2, Equipe de Recherche en Ingénierie des Connaissances, 5 Avenue Pierre Mendès-France, 69676 BRON Cedex, France
stephane.lallich@univ-lyon2.fr, eprudhomme@eric.univ-lyon2.fr

[2] TAO-Inria, LRI, CNRS-Université Paris-Sud, bat. 490, 91405 Orsay Cedex, France teytaud@lri.fr

Summary. The search for interesting Boolean association rules is an important topic in knowledge discovery in databases. The set of admissible rules for the selected support and confidence thresholds can easily be extracted by algorithms based on support and confidence, such as *Apriori*. However, they may produce a large number of rules, many of them are uninteresting. One has to resolve a two-tier problem: choosing the measures best suited to the problem at hand, then validating the interesting rules against the selected measures. First, the usual measures suggested in the literature will be reviewed and criteria to appreciate the qualities of these measures will be proposed. Statistical validation of the most interesting rules requests performing a large number of tests. Thus, controlling for false discoveries (type I errors) is of prime importance. An original bootstrap-based validation method is proposed which controls, for a given level, the number of false discoveries. The interest of this method for the selection of interesting association rules will be illustrated by several examples.

Key words: Association rules, interestingness measures, measure properties, multiple testing, false discoveries.

1 Introduction

The association between Boolean variables has been studied for a long time, especially in the context of 2×2 cross-tables. As Hajek and Rauch [22] point out, one of the first methods used to look for association rules is the GUHA method, proposed by Hajek *et al.* [21], where the notions of support and confidence appear. Work done by Agrawal *et al.* [1], Agrawal and Srikant [2], Mannila *et al.* [37] on the extraction of association rules from transactional databases has renewed the interest in the association rules.

In such a database, each record is a transaction (or more generally, a case) whereas the fields are the possible items of a transaction. Let n be the number

Stephane Lallich et al.: *Association Rule Interestingness: Measure and Statistical Validation*, Studies in Computational Intelligence (SCI) **43**, 251–275 (2007)
www.springerlink.com

of transactions and p the number of items. A Boolean variable is associated to each item. It takes the value "1" for a given transaction if the considered item is present in this transaction, "0" else. The set of transactions form a $n \times p$ Boolean matrix. To each itemset is associated a Boolean variable which is the conjunction of the Boolean variables associated to each item of the considered itemset.

From the Boolean matrix showing which items are the objects of which transaction, one extracts rules like "if a client buys bread and cheese, he is quite likely to also buy wine". A rule of association is an expression $A \rightarrow B$, where A and B are disjoint itemsets. More generally, this form can be applied to any data matrix, as long as continuous variables are discretized and categorical variables are dichotomized.

As the number of possible association rules grows exponentially with the number of items, selecting the "interesting" rules is paramount. Now, one needs to measure how interesting a rule is, and to validate the truly interesting rules with respect to said measure. Previous work done by the authors on the measure [28, 30] and on the validation of the association rules [45, 29] is synthesized in this chapter. Measuring the interest of a rule requires that the user chooses those best adapted to his data and his goal, targeting of group or prediction. Various criteria are presented. Once a measure has been selected and that rules are assessed using that measure, they still must be validated. One could retain the 50 or 100 rules with the highest scores, but these need not be interesting. Whenever possible, one should set a practical or probabilistic threshold. When the measure exceeds the threshold, either the rule is really interesting (true discovery), or it is merely an artefact of the random choice and the rule is not really interesting (false discovery). Each rule must be tested, which mechanically leads to a multitude of false discoveries (or false positives). The authors propose a bootstrap-based method to select interesting rules while controlling the number of false discoveries.

Criteria that can be used to assess measures appropriate to one's goal are presented in Sect. 2. In Sect. 3, it is shown that the validation of rules identified by the selected measures relies on a multitude of tests and the authors propose a multiple test method that controls the number of false discoveries.

2 Measuring Association Rule Interestingness

In this section, we will first look at the support-confidence approach (Sect. 2.1). Then, the notion of rules, implication and equivalences are examined (Sect. 2.2). A list of measures and several assessment criteria are given in the following subsections (Sects. 2.3 and 2.4). In the last subsection, some common features of these measures are highlighted (Sect. 2.5).

2.1 Appeal and Limitations of the Support-Confidence Approach

Support and Confidence

Let n_a and n_b the respective number of A and B transactions, and let n_{ab} be the number of transactions where A and B items appear simultaneously. The support of the rule $A \to B$ is the proportion of joint A and B transactions:

$SUP\,(A \to B) = p_{ab} = \frac{n_{ab}}{n}$.

whereas the confidence is the proportion of B transactions among the A transactions, that is the conditional frequency of B given A:

$CONF\,(A \to B) = \frac{p_{ab}}{p_a} = \frac{n_{ab}}{n_a} = 1 - \frac{n_{a\bar{b}}}{n_a}$.

"Support-Confidence" Extraction Algorithms

Following *Apriori*, the founding algorithm [2], support-confidence extraction algorithms exhaustively seek the association rules, the support and the confidence of which exceed some user-defined thresholds noted min_{SUP} and min_{CONF}. They look for frequent itemsets among the lattice of itemsets, that is, those itemsets whose support exceeds min_{SUP}, using the principle of antimonotonicity of support on the lattice of itemsets:

- any subset of a frequent itemset is frequent
- any superset of a non-frequent itemset is non-frequent.

Then, for each frequent itemset X, the support-confidence algorithms only keep rules of the type $X \backslash Y \to Y$, with $Y \subset X$, the confidence of which exceeds min_{CONF}.

Pros and Cons of Support-Confidence Approach

The antimonotonicity property of the support makes the support-confidence approach to rule extraction quite appealing. However, its usefulness is questionable, even though the very meaning of support and confidence are translated in easy-to-grasp measures.

First, algorithms of this type generate a very large number of rules, many of them of little interest. Moreover, the support condition, at the core of the extraction process, neglects rules with a small support though some may have a high confidence thus being of genuinely interesting, a common situation in marketing (the so-called nuggets of data mining). If the support threshold is lowered to remedy this inconvenient, even more rules are produced, choking the extraction algorithms.

Finally, the support and confidence conditions alone do not ensure rules with a real interest. Indeed, if the confidence of the rule $A \to B$ is equal to the marginal frequency of B, namely $p_{b/a} = p_b$, which means that A and B are independent, then the rule $A \to B$ adds no information (e.g. $p_a = 0.8$, $p_b = 0.9$, $p_{ab} = 0.72$, $p_{b/a} = 0.9$)!

Hence, measures other than support and confidence must be examined, thus promoting some amount of inductive bias.

Table 1. Notations for the joint distribution of itemsets A and B

$A \setminus B$	0	1	total
0	$p_{\overline{a}\overline{b}}$	$p_{\overline{a}b}$	$p_{\overline{a}}$
1	$p_{a\overline{b}}$	p_{ab}	p_a
total	$p_{\overline{b}}$	p_b	1

2.2 Rule, Implication and Equivalence

Association rules, implication and equivalence must be distinguished; let A and B be two itemsets whose joint distribution is given in Table 1, where 0 means "false" and 1, "true".

First, note that such a table has 3 degrees of freedom when the margins n_a and n_b are not fixed, that is, one can reconstruct the table from knowing only 3 values. The knowledge of 3 not linked values, for example SUP, $CONF$ and $LIFT$ completely determines the joint frequency distribution of A and B (Table 1).

Following proponents of association rules, using a support-confidence approach, attention is focused on examples A, on p_{ab} (support) and on $p_{b/a} = p_{ab}/p_a$ (confidence). Distribution of examples $\overline{A}$ between B and $\overline{B}$ is not taken into consideration.

The various examples and counter-examples of association rules, of implication and of equivalence that can be derived from A and B are displayed in Fig. 1. Rules are on level 1, implications on level 2 and equivalences on level 3. For each of the possible 8 rules, 4 implications and 2 equivalences, a $A \times B$ cross table is derived, where the values of A $(0, 1)$ are the lines and those of B $(0, 1)$ the columns. Each combination is marked as an example $(+)$, a counter-example $(-)$, or not accounted for (o). The rules, implications and equivalences on the left-hand side are positive while those on the right-hand side are negative.

The rule $A \rightarrow B$ has a single counter-example, $A\overline{B}$, and a single example, AB. One can see that a rule and its contrapositive share the same counter-examples but have different examples. The implication $A \Longrightarrow B$ and its contrapositive $\overline{B} \Longrightarrow \overline{A}$ are equivalent to $\overline{A} \vee B$, with $A\overline{B}$ as the only counter-example. Finally, the equivalence $A \Leftrightarrow B$ and its contrapositive $\overline{B} \Leftrightarrow \overline{A}$ correspond to $(AB) \vee (\overline{AB})$; their examples (resp. counter-examples) are the examples (resp. counter-examples) of the 4 covariant rules.

2.3 A List of Measures

Table 2 lists the usual measures of interest for association rules which respect the nature of the association rules, measures that are decreasing with $n_{a\overline{b}}$, margins n_a and n_b being fixed, and distinguish $A \rightarrow B$ from $A \rightarrow \overline{B}$. In the reminder of this chapter, only those measures will be considered. Other measures are given in [23, 44, 20].

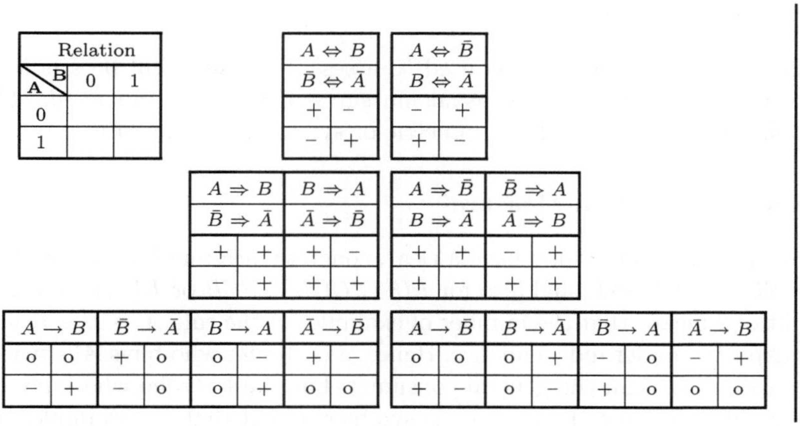

Fig. 1. Examples and counter-examples of rules, implications and equivalencies

Table 2. Usual Measures of interest

Measure	Formula	Acronym	Ref.
Support	p_{ab}	SUP	[1]
Confidence	$p_{b/a}$	$CONF$	[1]
Centered confidence	$p_{b/a} - p_b$	$CENCONF$	
Ganascia	$2p_{b/a} - 1$	GAN	[13]
Piatetsky-Shapiro	$np_a \left(p_{b/a} - p_b\right)$	PS	[39]
Loevinger	$\frac{p_{b/a}-p_b}{p_{\bar{b}}}$	LOE	[36]
Zhang	$\frac{p_{ab}-p_a p_b}{Max\left\{p_{ab}p_{\bar{b}};\, p_b p_{a\bar{b}}\right\}}$	$ZHANG$	[48]
Correlation Coefficient	$\frac{p_{ab}-p_a p_b}{\sqrt{p_a p_{\bar{a}} p_b p_{\bar{b}}}}$	R	
Implication Index	$\sqrt{n}\frac{p_{a\bar{b}}-p_a p_{\bar{b}}}{\sqrt{p_a p_{\bar{b}}}}$	$IMPIND$	[35]
Lift	$\frac{p_{ab}}{p_a p_b}$	$LIFT$	[11]
Least contradiction	$\frac{p_{ab}-p_{a\bar{b}}}{p_b}$	LC	[3]
Conviction	$\frac{p_a p_{\bar{b}}}{p_{a\bar{b}}}$	$CONV$	[10]
Implication Intensity	$P\left[Poisson\left(np_a p_{\bar{b}}\right) \geq np_{a\bar{b}}\right]$	$IMPINT$	[16]
Sebag-Schoenauer	$\frac{p_{ab}}{p_{a\bar{b}}}$	SEB	[42]
Bayes Factor	$\frac{p_{ab}p_{\bar{b}}}{p_{a\bar{b}}p_b}$	BF	[25]

2.4 Assessment Criteria

A number of criteria that can be used to assess a measure will be studied, yielding a critical review of the usual measures of interest. Tan *et al.* [44] undertook a similar exercise for symmetric or symmetrized measures.

The very Meaning of a Measure

Does the measure under study have a clear, concrete meaning for the user? It is so for *SUP* and *CONF*, and also for *LIFT*, *CONV*, *SEB* or *BF*. A measure with a lift of 2 means that the number of examples of the rule $A \rightarrow B$ is twice what is expected under independence. Hence, a customer who buys A is twice as likely to buy B than the general consumer, but similarly, he who buys B is twice as likely to buy A, since lift is symmetric and that the examples of $A \rightarrow B$ are also those of $B \rightarrow A$. $CONV = 2$ means that $n_{a\bar{b}}$ is half the expected number under the independence of A and B. When $SEB = 2$, the odds of "buying B" given "A was bought" is 2, or, he who buys A is twice as likely to buy B than to not buy B, or chances of buying B are 2/3. If $BF = 2$, odds of buying B are doubled if A is bought. Interpreting other measures is not as easy, especially *ZHANG* and *EII*, the entropic form of *IMPINT*.

Measure and Corresponding Rule

A measure must distinguish the various rules associating A and B (Fig. 1).

1. A measure must permit a clear choice between $A \rightarrow B$ and $A \rightarrow \overline{B}$, since the examples of one are the counter-examples of the other. Thus, *Pearl's*, *J-measure* and χ^2 (see [20] for those measures) were eliminated, since they do not account for the positivity or negativity of the rule.
2. Asymmetric measures which respect the nature of transactional rules are preferred: "if those items (A) are in the basket, then quite often those (B) are also". Symmetric measures like *SUP*, *PS*, *LIFT*, or R and its derivatives, give the same assessment of rules $A \rightarrow B$ and $B \rightarrow A$; while these rules have the same examples, they do not have the same counter-examples.
3. Should a measure give the same assessment to $A \rightarrow B$ and $\overline{B} \rightarrow \overline{A}$ [27]? If logical implication requires a strict equality, it is not so in the context of association rules. Indeed, both rules have the same counter-examples but not the same examples. The entropic intensity of implication, or *EII* [18] accounts for the contrapositive and brings the rule and the logical implication closer.

Examples and Counter-Examples

At first glance, one could say that a rule is unexpected whether one pays attention to the exceptionally high number of examples of the rule, n_{ab}, or

Table 3. Behaviour of certain measures in extreme situations

Situation	**Incompatibility**	**Independence**	**Logical rule**
Characterization	$p_{ab} = 0$	$p_{ab} = p_a p_b$	$p_{ab} = p_a$
Support	0	$p_a p_b$	p_a
Confidence	0	p_b	1
Centered confidence	$-p_b$	0	$p_{\bar{b}}$
Ganascia	-1	$2p_b - 1$	1
Piatetsky-Shapiro	$-np_a p_b$	0	$np_a p_{\bar{b}}$
Loevinger	$\frac{-p_b}{p_{\bar{b}}}$	0	1
Zhang	-1	0	1
Correlation Coefficient	$-\sqrt{\frac{p_a p_b}{p_{\bar{a}} p_{\bar{b}}}}$	0	$\sqrt{\frac{p_a p_{\bar{b}}}{p_{\bar{a}} p_b}}$
Implication Index (-)	$-p_b \sqrt{\frac{np_a}{p_{\bar{b}}}}$	0	$\sqrt{np_a p_{\bar{b}}}$
Lift	0	1	$\frac{1}{p_b}$
Least contradiction	$-\frac{p_a}{p_b}$	$2p_a - \frac{p_a}{p_b}$	$\frac{p_a}{p_b}$
Conviction	$p_{\bar{b}}$	1	∞
Implication Intensity	0	0.5	1
Sebag-Schoenauer	0	$\frac{p_b}{1-p_b}$	∞
Bayes Factor	0	1	∞

to the exceptionally low number of counter-examples, $n_{a\bar{b}}$. However, the examples of $A \to B$ are also those of $B \to A$ (Fig. 1), whereas the counter-examples of $A \to B$ are also those of $\overline{B} \to \overline{A}$. This justifies a preference for the counter-examples. To obtain a true difference between those options, one should, following Lerman *et al.* [35], explore the counter-examples and some probabilistic model; it is important that the margin n_a be not fixed, otherwise the number of examples and of counter-examples would be dependant, $n_{ab} + n_{a\bar{b}} = n_a$. *IMPIND*, *IMPINT* and *EII* derived from Lerman's model 3 (see Sect. 2.4) are such measures.

Direction of the Variation in the Measure and Reference Points

We limited our study to the measures that are decreasing with the number of counter-examples, margins n_a and n_b being fixed. Such a measure is maximum when $n_{a\bar{b}} = 0$, that is when $p_{b/a} = 1$, which corresponds to a logical rule. It is minimum when $n_{a\bar{b}} = n_a$, that is $n_{ab} = 0$, and $p_{b/a} = 0$, which means that A and B are incompatible. In fact, a rule is interesting whenever $n_{a\bar{b}} < \frac{n_a n_{\bar{b}}}{n}$, that is when $p_{b/a} > p_b$ ($p_{b/a} = p_b$, when A and B are independent). According to Piatetsky-Shapiro [39], a good measure should be:

a . $= 0$, A and B are independent, $p_{ab} = p_a p_b$
b . > 0, under attraction, $p_{ab} > p_a p_b$
c . < 0, under repulsion, $p_{ab} < p_a p_b$

He proposes *PS*, a symmetric measure whose bounds depend on A and B, $PS(A \to B) = np_a \left[p_{b/a} - p_b\right] = n \left[p_{ab} - p_a p_b\right]$. The conditions b and c above can be replaced by normalizing conditions b' and c' [48], which gives the so-called *ZHANG*:

b'. $= 1$, in case of a logical rule ($p_{b/a} = 1$, i.e. $A \subset B$)

c'. $= -1$, in case of incompatibility ($p_{b/a} = 0$, i.e. $AB = \emptyset$)

The only measures that take fixed reference values in the case of independence and extreme values (Table 3) are *ZHANG* and *BF*. However, the value in case of incompatibility is not very important since the only interesting situations are those where $p_b \leq p_{b/a} \leq 1$.

The lower reference point is thus often the case of independence. In that case, for the measures listed in Table 3, the value is often fixed, most often 0, sometimes 1 (*LIFT*) or 0.5 (*IMPINT*). The only exceptions are *CONF*, *LC*, *SEB* and *GAN*, or again some derived measure like the example and counter-example rate, $ECR = 1 - \frac{1}{SEB}$. As pointed out in Blanchard *et al.* [7], these are measures for which the lower reference point is not independence but rather indetermination ($n_{a\bar{b}} = \frac{n_a}{2}$, that is $p_{b/a} = p_{\bar{b}/a} = 0.5$). Lallich [28] suggested modifying *SEB* so that it be fixed under independence:

$$\frac{p_{\bar{b}}}{p_b} SEB(A \rightarrow B) = \frac{p_{ab} p_{\bar{b}}}{p_{a\bar{b}} p_b} = LIFT(A \rightarrow B) \times CONV(A \rightarrow B) \ .$$

This measure is similar to *Sufficiency* proposed by Kamber and Shingal [26]. It is actually similar to a Bayes factor [25], hence its name and notation *BF*.

On the other hand, the higher reference point is always when no counter-example exists, that is the logical rule. Normalizing to 1 is not always advisable in this case, as all logical rules are given the same interest. *LIFT* would tend to favour that of two rules which has the lower p_b.

Non-Linear Variation

Some authors [18] think it is preferable that the variation of a measure M be slow as the first counter-examples are encountered to account for random noise, then quicker, and then slow again (concave then convex). This is not the case of confidence and of all measures derived through an affine transformation which depends only of the margins n_a and n_b (Table 5). In fact, confidence is an affine function of the number of examples (or counter-examples) which depends only of n_a:

$$CONF\,(A \rightarrow B) = \frac{n_{ab}}{n_a} = 1 - \frac{n_{a\bar{b}}}{n_a} \ .$$

Conversely, to penalize false discoveries, *BF* will be preferred, as it decreases rapidly with the number of counter-examples (convex for values of $n_{a\bar{b}}$ in the neighbourhood of 0).

Impact of the Rarity of the Consequent

Following Piatetsky-Shapiro [39], a measure M must be an increasing function of $1 - p_b$ the rarity of the consequent, for fixed p_a and p_{ab}. Indeed, the rarer the consequent B is, the more "$B \supset A$" becomes interesting. This is especially true when the support condition is not taken into consideration anymore. This is partly what happens when a measure derived from centering confidence on p_b is used. This is also obtained by merely multiplying by p_b or by dividing by $1 - p_b$; thus, the measure $BF = \frac{p_{\bar{b}}}{p_b} SEB$ improves SEB in this respect.

Descriptive vs. Statistical Approaches

Measures can be regarded as descriptive or as statistical [28, 17]. A measure is descriptive if it remains unchanged when all the counts are multiplied by a constant θ, $\theta > 1$. Otherwise, the measure is said to be statistical. It seems logical to prefer statistical measures, as the reliability of its assessment increases with n, the number of transactions. A statistical measure supposes a random model and some hypothesis H_0 concerning the lower reference point, quite often, the independence of A and B [35]. One can consider that the base at hand is a mere sample of a much larger population, or that the distribution of 0's and 1's is random for each item.

We denote by N_x the random variable generating n_x. Under the hypothesis of independence, Lerman *et al.* [35] suggest that a statistical measure can be obtained by standardizing an observed value, say the number of counter-examples $N_{a\bar{b}}$, giving:

$N_{a\bar{b}}^{CR} = \frac{N_{a\bar{b}} - E(N_{a\bar{b}}/H_0)}{\sqrt{Var(N_{a\bar{b}}/H_0)}}$.

This statistical measure is asymptotically standard normal under H_0. A probabilistic measure is given by $1 - X$, where X is the right tail p-value of $N_{a\bar{b}}^{CR}$ for the test of H_0, which is uniformly distributed on $[0, 1]$ under H_0.

Lerman *et al.* [35] propose that H_0 be modelled with up to 3 random distributions (Hyp, Bin, and Poi denoting respectively the hypergeometric, the binomial and the Poisson distributions):

- Mod. 1: n, n_a fixed, $N_{a\bar{b}} \equiv Hyp(n, n_a, p_{\bar{b}})$
- Mod. 2: $N_a \equiv Bin(n, p_a)$; $/N_a = n_a$, $N_{a\bar{b}} \equiv Bin(n_a, p_{\bar{b}})$
- Mod. 3: $N \equiv Poi(n)$; $/N = n$, $N_a \equiv Bin(n, p_a)$; $/N = n, N_a = n_a$, $N_{a\bar{b}} \equiv Bin(n_a, p_{\bar{b}})$

Depending on the model, $N_{a\bar{b}}$ is distributed as a $Hyp(n, n_a, p_{\bar{b}})$, as a $Bin(n, p_a p_{\bar{b}})$ or a $Poi(np_a p_{\bar{b}})$. When standardizing, the expectation is the same, but the variance is model-dependent. Model 1 yields the correlation coefficient R, whereas Model 3 yields *IMPIND* the implication index. The latter has the advantage of being even more asymmetrical. Each statistical measure gives in turn a probabilistic measure; for example, under Model 3, $IMPINT = P(N(0, 1) > IMPIND)$ [16, 19].

Discriminating Power

Statistical measures tend to lose their discriminating power when n is large as small deviations from H_0 become significant. Consider the example of Table 4

Table 4. Displaying dilatation based on one example

		(a),(b),(c)	(a),(b),(c)	(a)	(b)	(c)	(a)	(b)	(c)
p_{ab}	$p_{a\bar{b}}$	$CONF$	R	R^{cr}	R^{cr}	R^{cr}	M	M	M
0.00	0.30	0	-0.65	-2.93	-4.14	-9.26	0.002	0.000	0.000
0.05	0.25	0.17	-0.44	-1.95	-2.76	-6.17	0.025	0.003	0.000
0.10	0.20	0.33	-0.22	-0.98	-1.38	-3.09	0.165	0.084	0.001
0.15	0.15	0.5	0	0	0	0	0.500	0.500	0.500
0.20	0.10	0.67	0.22	0.98	1.38	3.09	0.835	0.916	0.999
0.25	0.05	0.83	0.44	1.95	2.76	6.17	0.975	0.997	1.000
0.30	0.00	1	0.65	2.93	4.14	9.26	0.998	1.000	1.000

where the margins are fixed, $p_a = 0.30$ and $p_b = 0.50$. Examine how the various measures react to changes in $p_{a\bar{b}}$ the proportion of counter-examples. The measures considered here are *CONF*, *R*, R^{CR} (R standardized under independence), and M the p-value of R under independence. The various measures are compared with $n = 20$ (columns (a)), $n = 40$ (columns (b)), and $n = 200$ (columns (c)). Clearly, as n grows, M is less able to distinguish the interesting rules. On the other hand, the ordering remains unchanged. As n is the same for all rules of a given base, one might want to first select the rules that reject independence to the benefit of positive dependence, then considered centered descriptive measures and reason on the ordering induced by those measures.

The contextual approach, developed by Lerman for classification problems, offers a first solution to the loss of discriminating power suffered by statistical measures: consider the probabilistic discriminant index *PDI* [34]. This index is defined as $PDI(A \rightarrow B) = 1 - \Phi\left[IMPINT(A \rightarrow B)^{CR/\mathcal{R}}\right]$, where Φ is the standard Gaussian distribution function and $\mathcal{R}$ is a base of admissible rules. This base can contain all the rules, or only those that meet some conditions, for example conditions on support and confidence, or even the additional condition $n_a < n_b$.

It has been suggested by Gras *et al.* [18] that the statistical measure (*IMPINT)* be weighted by some inclusion index based on the entropy H of B/A and $\overline{A}/\overline{B}$. With $H(X) = -p_x \log_2 p_x - (1-p_x)\log_2(1-p_x)$, these authors also define $H^*(X) = H(X)$, if $p_x > 0.5$, and $H^*(X) = 1$, otherwise. The inclusion index, noted $i(A \subset B)$, is then defined as

$$i(A \subset B) = \left[\left(1 - H^*(B/A)^\alpha\right)\left(1 - H^*(\overline{A}/\overline{B})^\alpha\right)\right]^{\frac{1}{2\alpha}}.$$

In later work [17, 8], the authors recommend using $\alpha = 2$ as it allows a certain tolerance with respect to the first counter-examples, and define the entropic implication index, noted *EII*, as $EII = [IMPINT \times i(A \subset B)]^{\frac{1}{2}}$.

Parameterization of Measures

As shown in [17], the lower reference situation is that of indetermination. This is preferable to independence for predictive rules. More generally, Lallich *et al.* [31] have suggested a parameterized lower situation, adapted to the case of targeting. It can be written as $p_{b/a} = \theta$ or $p_{b/a} = \lambda p_b$. After parameterizing and rewriting the usual measures, it comes that *GAN* and *LOE* are special cases of a single parameterized measure $\frac{p_{b/a}-\theta}{1-\theta}$ for $\theta = 0.5$ and $\theta = p_b$. Moreover, each of the statistical, probabilistic and discriminant measures derived from Lerman *et al.* model-based approach [35] has been parameterized. The null hypothesis can be written as H_0: $\pi_{b/a} = \theta$ (or possibly $\pi_{b/a} = \lambda\pi_b$), with $\pi_{b/a}$ the theoretical confidence of the rule over all possible cases, and π_b the theoretical frequency of B under a right tail alternative. In particular, under Model 3, the parameterized version of *IMPIND* and *IMPINT*, noted $IMPINDG_{|\theta}$ and $IMPINTG_{|\theta}$, are given by:

$$IMPINDG_{|\theta} = \frac{N_{a\bar{b}}-np_a(1-\theta)}{\sqrt{np_a(1-\theta)}};$$

$$IMPINTG_{|\theta} = P(N(0,1) > IMPINDG_{|\theta}) .$$

The parameterized discriminant versions are obtained by transforming the entropy used in constructing *EII* into a penalizing function $\widetilde{H}(X)$, $\widetilde{H}(X) = 1$ for $p_x = \theta$ (instead of 0.5). It is sufficient, in the formula for $H(X)$, to replace p_x by $\widetilde{p}_x$, $\widetilde{p}_x = \frac{p_x}{2\theta}$, if $p_x < \theta$, and $\widetilde{p}_x = \frac{p_x+1-2\theta}{2(1-\theta)}$, if $p_x \geq \theta$. Let $\widetilde{H}^*_{|\theta}(X) = \widetilde{H}_{|\theta}(X)$, if $p_x > 0.5$, and $\widetilde{H}^*_{|\theta}(X) = 1$, otherwise. The generalized inclusion index $i_{|\theta}(A \subset B)$ is given by:

$$i_{|\theta}(A \subset B) = \left[\left(1 - \widetilde{H}^*_{|\theta}(B/A)^{\alpha}\right)\left(1 - \widetilde{H}^*_{|\theta}(\overline{A}/\overline{B})^{\alpha}\right)\right]^{\frac{1}{2\alpha}} .$$

A generalized entropic implication index can then be derived as:

$$GEII_{|\theta} = \left[IMPINT_{|\theta}(A \to B) \times i_{|\theta}(A \subset B)\right]^{\frac{1}{2}} .$$

Establishing a Threshold

It is important that the measures considered allow the establishment of a threshold able to retain only the interesting rules, without resorting to classifying all of them [28]. Classically, the threshold is defined in relation to the cumulative probability of the observed measure under H_0 for a given model. Note that the threshold is not a risk level for the multitude of tests, but merely a control parameter. By definition, it is possible to set such a threshold directly for *PDI* and *IMPINT*. Other measures do not allow such direct a calculation. It is quite complex for *ZHANG* because of the standardization, and for *EII* because of the correction factor.

Ordering Induced by a Measure

Two measures M and M' give the same order to the rules of a transactional base if and only if for all pairs of rules extracted from the base:

$$M(A \to B) > M(A' \to B') \Longleftrightarrow M'(A \to B) > M'(A' \to B')$$

This defines an equivalence relation on the set of possible measures [28]. For example, *SEB* orders like *CONF* because it can be written as a monotonic increasing transformation of *CONF*, $SEB = \frac{CONF}{1-CONF}$. Similarly, one can show that LOE orders like *CONV*, and that *PDI* orders like *IMPINT*. It should be pointed out that if the consequent is given, that is p_b fixed as it is the case for association rules in supervised learning, then *LIFT*, *CONV*, *LOE* and *SEB* order like *CONF*.

2.5 Various Measures of the Interest of a Rule

We have shown that alternatives to *SUP* and *CONF* are necessary to identify interesting rules and we have proposed several selection criteria. Now, let us specify the link between the usual measures and confidence, highlighting those that are affine transformations of confidence.

First, let's stress that the support is the index of association for Boolean variables proposed by Russel and Rao [41]; then, it will be pointed out that the usual indices of proximity defined on logical variables (see [33]) are not useful for the assessment of association rules because of their symmetrical treatment of Boolean attributes.

Affine Transformations of Confidence

Several measures can be written as a standardization of confidence via some affine transformation [28], namely $M = \theta_1 (CONF - \theta_0)$, whose parameters only depend on the relative margins of the $A \times B$ cross table and possibly on n (Table 5). Most often, the change in location indicates a departure from independence, $p_{b/a} - p_b$, and the change in scale depends on the ultimate goal. Conversely, changes in scale inform on what distinguishes two measures centered on p_b. There are two notable exceptions, *LIFT* for which the comparison to p_b is merely a change of scale and LC which centers confidence at 0.5.

Measures Derived from Confidence via Some Affine Transformation

All these measures improve on confidence but, by construction, inherit its principal characteristics. For fixed margins, they are affine functions of the number of counter-examples. Moreover, these measures remain invariant under changes in n when θ_1 and θ_0 parameters do not depend on n, which is the case for all of them except *PS* and *IMPIND*.

The lift is interpreted as the quotient of the observed and expected number of examples, assuming the independence of A and B. As an expression of the number of examples, it is symmetrical, since the rules $A \rightarrow B$ and $B \rightarrow A$ have the same examples (Fig. 1). LC is another transformation of confidence, but centered on 0.5 rather than on p_b, a better predictive that targeting tool.

Pearson's correlation R between two itemsets can be positive (see Fig. 1, examples and counter-examples of $A \Leftrightarrow B$) or negative (see Fig. 1, $A \Leftrightarrow \overline{B}$). R

Table 5. Measures derived from confidence via some affine transformation

Measure	**center (θ_0)**	**scale (θ_1)**
Centered confidence	p_b	1
Ganascia	0.5	2
Piatetsky-Shapiro	p_b	np_a
Loevinger	p_b	$\frac{1}{p_{\overline{b}}}$
Zhang	p_b	$\frac{p_a}{Max}$
Correlation Coefficient	p_b	$\frac{\sqrt{p_a}}{\sqrt{p_{\overline{a}} p_b p_{\overline{b}}}}$
Implication Index	p_b	$\sqrt{n}\sqrt{\frac{p_a}{p_{\overline{b}}}}$
Lift	0	$\frac{1}{p_b}$
Least contradiction	0.5	$2\frac{p_a}{p_b}$

is linked to the χ^2 of independence between A and B used by Brin *et al.* [10], since $\chi^2 = nR$, with $nR^2 \approx N(0,1)^2$, assuming independence. Contrary to χ^2, R distinguishes the cases $A \rightarrow B$ and $A \rightarrow \overline{B}$. The correlation coefficient R can be written as:

$$R = \frac{p_{ab} - p_a p_b}{\sqrt{p_a p_{\overline{a}} p_b p_{\overline{b}}}} = \frac{\sqrt{p_a}}{\sqrt{p_{\overline{a}} p_b p_{\overline{b}}}} [CONF - p_b] \ .$$

This can be simplified as $R = \frac{p_{ab} - p_b^2}{p_b p_{\overline{b}}} = \frac{1}{p_{\overline{b}}} [CONF - p_b] = LOE$, when A and B have the same marginal distribution ($p_a = p_b$), and as $R = 2CONF - 1$ (i.e. GAN), when this distribution is balanced ($p_a = p_b = 0.5$). Thus, R and *CONF* can be seen as equivalent for cross tables with balanced margins. The cross table is then symmetrical, meaning that the 4 covariant rules connecting A and B or their complement have the same confidence, as well as the 4 contravariant rules.

Other Measures

The measures that cannot be reduced to an affine transformation of confidence are *CONV*, *SEB* and *BF*, as well as measures derived from the implication index. CONV can be expressed as a monotonic increasing function of LOE, $CONV = (1 - LOE)^{-1}$. $CONV$ is analogous to $LIFT$ applied to the counter-examples:

$$CONV(A \rightarrow B) = \frac{p_{\overline{b}}}{p_{\overline{b}/a}} = \frac{p_a p_{\overline{b}}}{p_{a\overline{b}}} = LIFT(A \rightarrow \overline{B})^{-1} \ .$$

SEB is a monotonic increasing transformation of confidence, as well as – just like *BF* – an affine transformation of conviction with fixed margins:

$$SEB = \frac{CONF}{1 - CONF} = \frac{1}{p_{\overline{b}}} (CONV - 1) \, ; \, BF = \frac{1}{p_b} (CONV - 1) \ .$$

Statistical measures are based on *IMPIND* which is an affine transformation of *CONF* with fixed margins, namely $IMPINT = P(N(0,1) > IMPIND)$, and its discriminating versions *EII* and *PDI*.

Strategy

The user must choose the measures the most appropriate to his objective and to the characteristics of his data; criteria proposed in this section may be found of help. The user can also opt for some automated decision-making procedure to decide on the most appropriate measure [32].

Because support and confidence are more easily understood, and because support condition is antimonotonic, support-confidence algorithms are often applied first to transactional databases. A set of admissible rules for the selected support and confidence thresholds is then obtained. Such sets comprise a large number (m) of rules, not always interesting. The most interesting rules can be identified with the help of the selected measures. If the support condition is released and if the interesting rules are sought for directly with the selected measures, the number of rules becomes excessively large. Then, one may be restricted to simple rules [3].

3 Validating Interesting Rules

Rules that are truly interesting for the user are those for which the real world value, for the selected measure, exceeds some preset threshold. Most often, as the transactional database is seen as a mere sample of the universe of all possible transactions, one only knows some empirical evaluation of those rules. The problem becomes the selection of the rules whose empirical values significantly exceed the threshold. This means testing each one of the m rules, that is, m tests.

For example, one can seek rules significantly far from the independence of A and B, which leads to selecting rules for which confidence $p_{b/a}$ is significantly larger than the threshold p_b. The hypothesis of independence, noted H_0, is given by $\pi_{b/a} = \pi_b$, where $\pi_{b/a}$ is the theoretical confidence (or confidence over all possible transactions), whereas π_b is the prior theoretical frequency of B. For each rule, H_0 is tested against the right-tail alternative of positive dependence noted H_1 given by $\pi_{b/a} > \pi_b$. If one is seeking predictive rules, one would select rules for which the confidence $p_{b/a}$ is significantly larger than 0.5, that is, testing H_0: $\pi_{b/a} \leq 0.5$ against H_1: $\pi_{b/a} > 0.5$. If the objective is targeting of group, one can also seek rules for which the confidence $p_{b/a}$ is significantly larger than the threshold λp_b, that is, a lift larger than some set value $\lambda > 1$; this is equivalent to testing H_0: $\pi_{b/a} \leq \lambda \pi_b$ against H_1: $\pi_{b/a} > \lambda \pi_b$.

These various situations could be analyzed with a measure other than confidence. If q measures are available, one needs a total of qm tests. Moreover, certain measures have a complicated algebraic expression (e.g. *EII*) which impedes the elaboration of a parametric test. In summary, the validation of interesting rules requires the ability to develop a multitude of tests using some possibly non-parametric device.

This multiplicity of tests inflates the number of false discoveries (rules wrongly selected). Indeed, if m tests are developed, each with a probability of Type I error set at α_0, even if no rule is truly interesting, the procedure creates on average $m\alpha_0$ false positives. Controlling multiple risk is rarely a topic in data mining literature. A noteworthy exception is the work of Meggido and Srikant [38] on the significance of association rules with respect to independence, who simulate the number of false discoveries for a given level of Type I risk. On the other hand, this topic is well covered in biostatistics (see Sect. 3.1). The authors have proposed in earlier work methods to control multiple risk using statistical learning theory and VC-dimension [45], or bootstrap [30]. In practice, because they make no allowance for false discoveries among the m rules, these methods have little power, yet ignoring significant rules. The authors have proposed *BS_FD* [29] to test the significance of rules; this method controls the number of false discoveries and uses an original bootstrap criterion. The general case with any threshold is exposed below.

First, the problem of controlling risk with multiple tests will be reviewed (Sect. 3.1), as well as procedures that control risk using p-values (Sect 3.2). Then, *BS_FD* will be introduced (Sect. 3.3) and will be applied to selecting the most interesting association rules (Sect. 3.4).

3.1 Constructing Multiple Tests

Significance Test for a Rule

Consider a rule $A \to B$ and some measure of interest M, decreasing with $n_{a\bar{b}}$ and fixed margins. Note M_{obs} the observed value of $M(A \to B)$ on the sample of transactions and μ its theoretical value on a very large set of transactions. The rule is said to be significant under M with respect to μ_0 if $M_{obs} = M(A \to B)$ is significantly larger to some preset value μ_0. A test for the null hypothesis H_0: $\mu = \mu_0$ against the unilateral alternative H_1: $\mu > \mu_0$ is needed. H_0 is rejected whenever M_{obs} is too far from H_0 in the direction of H_1, with a Type I error risk set at $\alpha = \alpha_0$. The p-value for M_{obs} is computed as the probability of obtaining a value as large as M_{obs} assuming H_0 is true, and the rule is selected if the p-value for M_{obs} is less than α_0. Obviously, this requires the knowledge of the distribution of $M(A \to B)$ under H_0.

Risk and Type I Error

The identification of the significant rules under M among the m rules extracted from a transactional database requires m tests. This raises the problem of false discoveries, a recurrent problem in data mining. If m uninteresting rules are tested at the level α_0, then, on average, $m\alpha_0$ rules will mechanically be erroneously selected. For example, with $\alpha_0 = 0.05$, and a base of extracted rules comprising $m = 10,000$ rules, even if all were non-significant, about 500 rules would mechanically be selected.

Table 6. Synthesis of the results of m tests

Reality \ Decision	Acceptation	Reject	Total
H_0 true	U	V	m_0
H_1 true	T	S	m_1
Total	W	R	m

To take into account the multiplicity of tests, the fundamental idea of Benjamini and Hochberg [4] is to consider the number of errors over m iterations of the test, rather than the risk of being wrong on one test (see Table 6, where a upper case represents observable random variates and lower case are fixed yet unknown quantities m_0 and m_1). From this table, these authors derive several indicators. Two most common ones are described next, *FWER* (Family wise error rate) and *FDR* (False Discovery Rate).

FWER is the chance of erroneously rejecting H_0 at least once, $FWER = P(V > 0)$. It is a much too strict criterion for a large number of tests, because it does not allow any false discovery.

The authors [29] proposed the *User Adjusted Family Wise Error Rate*, $UAFWER = P(V > V_0)$, an original and more flexible variant which allows V_0 false discoveries. $UAFWER$ can be controlled at the level δ using a bootstrap-based algorithm (Sect. 3.3).

Several quantities using the expectation of V, the number of false discoveries, possibly standardized, have been proposed to remedy the difficulties inherent to *FWER*. The best known is *FDR* [4], the expected proportion of erroneous selections among the selected rules. When $R = 0$, define $\frac{V}{R} = 0$, that is, $FDR = E(Q)$, where $Q = \frac{V}{R}$ if $R > 0$, 0 otherwise. Then:

$FDR = E(\frac{V}{R} \mid R > 0)P(R > 0)$.

Storey [43] proposed the *pFDR*, a variation of *FDR*, using the knowledge that H_0 has been rejected at least once:

$pFDR = E(\frac{V}{R} \mid R > 0)$.

At the cost of a fixed proportion of erroneous selections, these quantities are less severe, thus augmenting the probability of selecting an interesting rule (increased power). One has $FDR \leq FWER$ and $FDR \leq pFDR$, hence $FDR \leq pFDR \leq FWER$ when m is large, because $P(R > 0)$ goes to 1 as m increases. The problem of controlling the Type I risk is resolved in the literature by the use of p-values. *FWER* and *FDR* will be examined in turn.

3.2 Controlling Multiple Risk with p-values

Several solutions have been proposed to control *FWER* or *FDR*, most recently in the context of gene selection. A remarkable summary of this work can be found in [14].

Control of FWER

Bonferroni Correction

Let us denote by P_r the random variable generating the p-value p_r associated to the test statistics $T_r, r = 1, \ldots, m$. One can show that $FWER = 1 - P(\bigcap_{r=1}^{m} (P_r > \frac{\alpha_0}{m}) \mid H_0)$. Assuming that the rules are independent, then $FWER = 1 - (1 - \frac{\alpha_0}{m})^m \approx \alpha_0$. The Bonferroni correction consists in constructing each test at the level $\frac{\alpha_0}{m}$, in order to set the $FWER$ on α_0. This correction is usually applied by adjusting the m p-values. The adjusted p-value $\widetilde{p}_r$ is defined by $\widetilde{p}_r = \min\{mp_r, 1\}$. All rules having an adjusted p-value smaller than the risk α_0 are selected. If independence cannot be assumed, one only has $\frac{\alpha_0}{m} \leq FWER \leq \alpha_0$. The Bonferroni correction is not a good solution for two reasons:

- FWER is actually not controlled, but somewhere between $\frac{\alpha_0}{m}$ and α_0; it is equal to α_0 only when the rules are mutually independent. Now, rules are not independent, as they share items and because items are dependent.

- FWER is conservative, thus increasing the risk of a Type II error, that is not finding an interesting rule.

Holm's Step-down Procedure

Stepwise procedures examine p-values in increasing order of magnitude, adjusting the critical value as the procedure progresses. Holm [24] considers that a selected variable corresponds to H_0 false, and the critical value is adjusted to only account for the variables remaining to be examined. Since the p-values are sorted in increasing order, with $p_{(i)}$ the ith p-value, H_0 is rejected while $p_{(i)} < \frac{\alpha_0}{m-i+1}$. H_0 is accepted for all p-values following the first acceptance. This procedure, easy to implement, gives good results when the number of tests is rather small, as the adjustment to the critical value has some importance. The procedure is ill-adapted to large numbers of tests.

Control of FDR

Benjamini and Liu's Procedure

Benjamini and Liu [5] proposed a sequential method for the control of *FDR* under the assumption of independence. The p-values are examined in increasing order and the null hypothesis is rejected if the p-value at hand $p_{(i)}$ is less than $\frac{i\alpha_0}{m}$. This procedure ensures that $FDR = \frac{m_0}{m}\alpha_0$ under independence. It is compatible with positively dependent data.

pFDR

In order to estimate $pFDR = E(\frac{V}{R} \mid R > 0)$, the proportion of false detections, Storey [43] proposes the approximation:

$p\hat{FDR}(\delta) = \frac{\hat{\pi}_0 . m . \delta}{\#\{p_i \leq \delta, i=1,\ldots,m\}}$, where

- m is the number of rules to be tested; δ defines the rejection area: hypotheses corresponding to p-values less than or equal to δ are rejected;
- p_i is the ith largest p-value;
- $\pi_0 = \frac{m_0}{m}$ is the proportion of null hypotheses; here, π_0 is estimated by $\hat{f}(1)$, where $\hat{f}$ is a cubic spline of $\hat{\pi_0}(\lambda)$ over λ: $\hat{\pi_0}(\lambda) = \frac{\#\{p_i \geq \lambda, i=1,\ldots,m\}}{m(1-\lambda)}$; $0 < \lambda < 0.95$ represents the acceptation area.

The *pFDR* is defined in terms of a preset rejection area. Once the global *pFDR* is computed, variables are controlled by a step-down procedure using the q-values defined for each p-value as $\hat{q}(p_m) = \hat{\pi_0}.p_m$ and:

$\hat{q}(p_i) = \min\left(\frac{\hat{\pi_0}.m.p_i}{i}, \hat{q}(p_{i+1})\right); i = m-1, \ldots, 1$.

The q-value is to the *pFDR* what the p-value is to Type I error, or what the adjusted p-value is to the FWER. Any rule whose p-value has a corresponding q-value less than *pFDR* is selected.

3.3 Controling UAFWER Using the *BS_FD* Algorithm

We have proposed a bootstrap-based non-parametric method to control *UAFWER*. This method does not require p-values, which is advantageous when the distribution of $M(A \to B)$ under H_0 is unknown (e.g. the discriminant versions of the statistical measures, like *EII* [18] or its generalization *GEII* [31]).

Notations

- $\mathcal{T}$: set of transactions, $n = Card(\mathcal{T})$, p: number of items;
- $\mathcal{R}$: base of admissible association rules with respect to some predefined measures, for example, support and confidence, $m = Card(\mathcal{R})$;
- M: measure of interest; $\mu(r)$: theoretical value of M for rule r; $M(r)$: empirical value of M for r on $\mathcal{T}$;
- V: number of false discoveries, δ: risk level of the control procedure, with V_0 the number of false discoveries not to be exceeded given δ, $\mathcal{R}^*$ a subset of $\mathcal{R}$ comprising the significant rules as determined by M and μ_0.

Objective

The objective is to select the rules r of $\mathcal{R}$ that are statistically significant as measured by M, meaning that $M(r)$ is significantly larger than $\mu_0(r)$, the expected value under H_0. We have suggested various algorithms that use the tools of statistical learning so that 100% of the identified rules be significant for a given α, among others the bootstrap-based algorithm *BS* [30]. Experience has shown that this approach might be too prudent, therefore not powerful enough. Allowing a small number of false discoveries, after Benjamini's work (Sect. 3.1), the authors propose *BS_FD*, an adaptation of *BS* that controls the number of false discoveries.

BS_FD selects rules so that $UAFWER = P(V > V_0)$, which ensures that the number of false discoveries does not exceed V_0 at the level δ. The algorithm guarantees that $P(V > V_0)$ converges to δ when the size of the samples of transactions increases.

Algorithm *BS_FD*

Given $\mathcal{T}$, $\mathcal{R}$, and M, $\mu(r) > \mu_0(r)$ is guaranteed by setting $\mu(r) > 0$, without loss of generality simply by shifting $\mu(r)$ to $\mu(r) - \mu_0(r)$. V_0 false discoveries are allowed at risk δ. Finally, $\#E = Card(E)$.

1. *Empirical assessment.* All rules of $\mathcal{R}$ are measured using M on the set of transactions $\mathcal{T}$, creating the $M(r), r \in \mathcal{R}$.
2. *Bootstrap.* The following operations are repeated l times:
   - Sample with replacement and equal probability m transactions from $\mathcal{T}$, thus creating $\mathcal{T}'$, $Card(\mathcal{T}') = Card(\mathcal{T})$. Some transactions of $\mathcal{T}$ will not be in $\mathcal{T}'$ while some others will be there many times. All rules are measured using M, creating the $M(r)$, $r \in \mathcal{R}$.
   - Compute the differences $M'(r) - M(r)$, then compute $\varepsilon(V_0, i)$, the smallest value such that $\#\{M'(r) > M(r) + \varepsilon(V_0, i)\} \leq V_0$. Hence, $\varepsilon(V_0, i)$ is the $(V_0 + 1)^{\text{st}}$ largest element of the $M'(r) - M(r)$, during the ith iteration, $i = 1, 2...l$.
3. *Summary of bootstrap samples.* There are l values $\varepsilon(V_0, i)$. Compute $\varepsilon(\delta)$, $(1-\delta)^{\text{th}}$ quantile of the $\varepsilon(V_0, i)$: that is, $\varepsilon(V_0, i)$ was larger than $\varepsilon(\delta)$ only $l\delta$ times in l.
4. *Decision.* Keep in $\mathcal{R}^*$ all rules r such that $M(r) > \varepsilon(\delta)$.

Rationale

Bootstrap methods [12] approximate the distance between the empirical and true distributions by the distance between the bootstrap and empirical distributions. At the ith bootstrap iteration, there are V_0 rules whose evaluation augments by more than $\varepsilon(V_0, i)$. Given the definition of $\varepsilon(\delta)$, the number of rules whose evaluation augments by more than $\varepsilon(\delta)$ is larger than V_0 in a proportion δ of the l iterations. Consequently, selecting rules for which $M(r)$ exceeds $\varepsilon(\delta)$, one is guaranteed to have at most V_0 false discoveries at the risk level δ.

Moreover, bootstrap-based methods have solid mathematical foundations [15] which require a clearly posed question. Formally, the objective is that the distribution function of the number of rules such that $\mu(r) < 0$ while $M(r) > \epsilon$, be at least $1 - \delta$ for V_0. One gets $\#\{\mu(r) \leq 0 \text{ et } M(r) > \epsilon\} \leq \#\{M(r) \geq \mu(r) + \epsilon\}$. Theorems on bootstrap applied to a family of functions verifying the minimal conditions [47] yield the approximation of this quantity by $\#\{M'(r) \geq M(r) + \epsilon\}$, which serves as a basis for $\varepsilon(V_0, i)$ and $\varepsilon(\delta)$ described in this section.

Extension to Multiple Measures

In practice, more than one measure will be of interest, for example, *SUP*, *CONF* and a measure of the departure from independence. The extension of *BS_FD*, noted *BS_FD_mm*, is achieved by using as a summary measure the minimum of the various measures. Hence, for 3 measures M_1, M_2 and M_3, one considers $M(r) = min\,\{M_1(r), M_2(r), M_3(r)\}$. Using *BS_FD_mm* on M at the level δ will select rules which comply with M_1, M_2 and M_3, at level δ.

Risk of Type II errors can be optimized by working with Hadamard differentiable transformations of the M_i that will make the measures homogenous [47], for example, p-values or reductions, through standardization.

Complexity of *BS_FD*

The complexity of *BS_FD* is proportional to $l \times m \times n$, assuming that the random number generator operates in constant time. In fact, the complexity of the search for the kth largest element of a table is proportional to the size of the table. The value of l must be large enough so that the finiteness of l impede not the global reliability, and be independent of both m and n. The algorithm is globally linear in $m \times n$, to a constant l linked to the bootstrap.

3.4 Application to the Rules Selection and Experimentation

Selecting Significant Rules According to Independence

Description of Data

The filtering methods presented here were applied to five sets of rules available on HERBS [46]. They were extracted using Borgelt and Kruse's implementation [9] of *Apriori* applied on data sets available from the UCI site [6]: Contraceptive Method Choice (*CMC*), Flags (*Flags*), Wisconsin Breast Cancer (*WBC*), Solar Flare I (*SFI*) and Solar Flare II (*SFII*). The authors computed for each method, the reduction rate of each set of rules after removal of non-significant rules.

Parameterization

For the "5% control", Holm and Bonferroni procedures (cf. Sect. 3.2) were applied with a level of 5%. $pFDR$ is calculated with a rejection rate of 0.1%. The number of false discoveries is shown between parentheses. The rejection zone is chosen so that it will be as acceptable as possible. FDR, described in Sect. 3.2 is used with a threshold set by the last q-value selected by $pFDR$, shown in brackets. Indeed, to compare $pFDR$ and FDR, control levels should be close. Control level for *BS_FD(R)* is set at 5%, with V_0 equal to the result of $pFDR$, on the correlation coefficient R tested against 0.

Table 7. Filtering of some sets of rules

Characteristics	CMC	Flags	WBC	SF I	SF II
# cases	1,473	194	699	323	1066
# rules	2,878	3,329	3,095	5,402	3,596
Covering rate	100%	100%	96.2%	100%	100%
Re-covering rate	259	1,848	646	1,828.6	2,277
Support threshold	5%	50%	10%	20%	20%
Confidence threshold	60%	90%	70%	85%	85%
Results	CMC	Flags	WBC	SF I	SF II
level 5%	1,401	2,181	3,094	2,544	2,558
pFDR	916 (3)	1,200 (3)	3,095 (0)	900 (5)	1,625 (4)
FDR (Benj.)	913 (0.003)	1,198 (0.0027)	/	899 (0.006)	1,626 (0.0022)
BS_FD(R)	794	1,074	3,093	604	738
Holm	742	564	3,094	432	1,020
Bonferroni	731	539	3,042	427	1,006

Results

Table 7 requires some explanations. No filter is efficient on *WBC.* This is because only one rule of the starting set has a p-value above 0.05 (viz. 0.184), an other one is at 0.023, the remaining p-values are less than 0.01, and 3,036 of them are less than 0.00005!

For the 4 other set of rules, merely repeating independence tests shrinks the sets by 51%, 34%, 53%, and 39%. However, not all false discoveries are eliminated. Using control on multiple risk reduces the number of rules selected.

Among those, Bonferroni correction is the most stringent. It produces reductions of 75%, 81%, 65% and 60%. Though stringent, it lacks power, avoiding false positives but creating false negatives. Holm's procedure gives similar results; it is inefficient because of the large number of rules which renders the step-wise correction inoperative.

On the other hand, *pFDR*, *FDR* and *BS_FD* give moderately better results, what was expected. *BS_FD* appears to be the most stringent of the 3, especially on *Solar Flare II.* The reason is that the parameterization of *pFDR* and *FDR* ensures an average number of false discoveries equal to V_0, whereas *BS_FD* ensures that V_0 be exceeded only 0.05 of the time, which is quiet demanding. These three methods are efficient rule filters. *BS_FD* is the most complex, but is advantageously non-parametric (see next section for an example).

Thus, a filtering procedure based on controlling multiple risk eliminates that would otherwise be selected by a variety of measures. Logical rules whose consequent is very frequent (e.g. *Solar Flare II*) is an example of such measures. These attain a maximum under any measure that give a fixed maximum value to logical rules, though they present little interest and their p-values are non-significant. Conversely, computing p-values is independent of any sub-

Table 8. Predictive rules selected on CMC by applying *BS_FD* on LIFT and $GEII(2p_b)$

		LIFT		
		Selected	Not Selected	
$GEII_{\|2p_b}$	Selected	8	19	27
	Not Selected	6	11,042	11,048
		14	11,061	11,075

sequent ranking of the rules by descriptive measures that favour the more interesting rules, for example, asymmetric measures that favour rules with low frequency consequent.

Selecting Targeting Rules

Contrary to methods like FDR or $pFDR$, *BS_FD* does not require prior knowledge of the distribution of the measure under the null hypothesis. *BS_FD* can thus be applied to algebraically complex measures.

To illustrate this, let's turn our attention to targeting rules. These are rules for which knowing the antecedent multiplies by some constant λ the probability of observing the consequent. Here, we use $\lambda = 2$, which amounts to testing H_0: $\pi_{b/a} = 2\pi_b$. To assess this type of rule, 2 measures are used, $LIFT$ and $GEII(2p_b)$ (generalized entropic intensity index with parameter $2p_b$); the null distribution of *GEII* is not known. Under H_0, these measures are respectively 2 and 0.

The CMC base [6] was used for this experiment. First, using *Tanagra* [40] implementation of *Apriori*, $13,864$ rules with a support exceeding 0.05 were extracted. Among those, the $2,789$ rules for which $p_b > 0.5$ were removed. Of the $11,075$ remaining rules, *BS_FD* was applied on $LIFT$ and on $GEII(2p_b)$ by comparing the results to 2 and 0 respectively. Results are displayed in Table 8.

$LIFT$ and $GEII(2p_b)$ select respectively 14 and 27 rules of the $11,075$ extracted by *Apriori*. These rules ensure that B has twice as many chances of occurring if A is realized. The small number of rules allows expert examination. These are especially interesting in marketing and health sciences. In this latter case, the consequent is the occurrence of a disease, and the antecedents are possible factors of this disease. The proposed procedure detects factors that multiply notably the risk of disease.

Moreover, results show that $LIFT$ and $GEII(2p_b)$ do not select the same rules (only 8 are common). *BS_FD* applied to $LIFT$ naturally selects the rules with the highest measures. Thus, of the 224 rules with a $LIFT$ over 2, it retains those with a value above 2.479. Among those, *BS_FD* applied to $GEII(2p_b)$ does not select those rules with $p_a > 0.2$ and $p_b < 0.1$. Contrarily, it selects those with a low frequency antecedent. Using many measures allows different assessment of the interest of a given rule.

4 Conclusion and Perspectives

Means to search for association rules in databases is one of the principal contributions of data mining compared to traditional statistics. However, the usual extraction algorithms yields a very large number of not-all-interesting rules. On the other hand, these rules overfit the data [38], which makes them hard to generalize. This double problem calls for a double solution: a careful choice of the measure of interest and retaining those rules that are significant for the objective at hand. The authors have suggested a number of criteria to help the user to choose the most appropriate measure. To avoid overfitting, the significance of each rule must be tested raising the problem of controlling multiple risk and avoiding false discoveries. To this end, the authors suggest a bootstrap-based method, *BS_FD*; the proposed method controls the risk of exceeding a fixed number of false discoveries, accounting for the dependency among the rules, and allowing the test of several measures at once. *BS_FD* can be used for filtering rules where the antecedent increases the probability of the consequent (positive dependence), for filtering targeting rules, or filtering predictive rules. Experiments show the effectiveness and efficiency of the proposed strategy. An extension of this work to filtering discriminant rules in the context of genomics is being planned.

Acknowledgement. The authors gratefully acknowledge the referees for their helpful comments and Jean Dumais, Methodology Branch, Statistics Canada, for his help with the translation of this paper.

References

1. R. Agrawal, T. Imielinski, and A.N. Swami. Mining association rules between sets of items in large databases. In P. Buneman and S. Jajodia, editors, *ACM SIGMOD Int. Conf. on Management of Data*, pages 207–216, 1993.
2. R. Agrawal and R. Srikant. Fast algorithms for mining association rules. In J.B. Bocca, M. Jarke, and C. Zaniolo, editors, *Proceedings of the 20th Very Large Data Bases Conference*, pages 487–499. Morgan Kaufmann, 1994.
3. J. Azé and Y. Kodratoff. A study of the effect of noisy data in rule extraction systems. In *Sixteenth European Meeting on Cybernetics and Systems Research*, volume 2, pages 781–786, 2002.
4. Y. Benjamini and Y. Hochberg. Controlling the false discovery rate : a practical and powerful approach to multiple testing. *J. R. Stat. Soc., B*, 57:289–300, 1995.
5. Y. Benjamini and W. Liu. A step-down multiple-hypothesis procedure that controls the false discovery rate under independance. *J. Stat. Planng Inf.*, 82:163–170, 1999.
6. C.L. Blake and C.J. Merz. UCI repository of machine learning databases. http://www.ics.uci.edu/~mlearn/MLRepository.html, 1998.
7. J. Blanchard, F. Guillet, H. Briand, and R. Gras. Assessing the interestingness of rules with a probabilistic measure of deviation from equilibrium. In *ASMDA'05*, pages 191–200, 2005.

8. J. Blanchard, P. Kuntz, F. Guillet, and R. Gras. Mesure de la qualité des règles d'association par l'intensité d'implication entropique. *Mesures de Qualité pour la Fouille de Données*, (RNTI-E-1):33–45, 2004.
9. C. Borgelt and R. Kruse. Induction of association rules: APRIORI implementation. In *15th Conf. on Computational Statistics*, 2002.
10. S. Brin, R. Motwani, and C. Silverstein. Beyond market baskets: generalizing association rules to correlations. In *ACM SIGMOD/PODS'97*, pages 265–276, 1997.
11. S. Brin, R. Motwani, J.D. Ullman, and S. Tsur. Dynamic itemset counting and implication rules for market basket data. In Joan Peckham, editor, *ACM SIGMOD 1997 Int. Conf. on Management of Data*, pages 255–264, 1997.
12. B. Efron. Bootstrap methods: Another look at the jacknkife. *Annals of statistics*, 7:1–26, 1979.
13. J.-G. Ganascia. Deriving the learning bias from rule properties. *Machine intelligence*, 12:151–167, 1991.
14. Y. Ge, S. Dudoit, and T.P. Speed. Resampling-based multiple testing for microarray data analysis. Tech. rep. 663, Univ. of California, Berkeley, 2003.
15. E. Giné and J. Zinn. Bootstrapping general empirical measures. *Annals of probability*, 18:851–869, 1984.
16. R. Gras. *Contribution à l'étude expérimentale et à l'analyse de certaines acquisitions cognitives et de certains objectifs didactiques en mathématiques*. PhD thesis, Université de Rennes I, 1979.
17. R. Gras, R. Couturier, M. Bernadet, J. Blanchard, H. Briand, F. Guillet, P. Kuntz, R. Lehn, and P. Peter. Quelques critères pour une mesure de qualité de règles d'association - un exemple: l'intensité d'implication. *Mesures de Qualité pour la Fouille de Données*, (RNTI-E-1), 2004.
18. R. Gras, P. Kuntz, R. Couturier, and F. Guillet. Une version entropique de l'intensité d'implication pour les corpus volumineux. *EGC 2001*, 1(1-2):69–80, 2001.
19. R. Gras and A. Lahrer. L'implication statistique: une nouvelle méthode d'analyse des données. *Math. Inf. et Sc. Hum.*, 120:5–31, 1993.
20. F. Guillet. Mesure de la qualité des connaissances en ECD. Tutoriel 4e Conference Extraction et Gestion des Connaissances, EGC'04, 2004.
21. P. Hajek, I. Havel, and M. Chytil. The GUHA method of automatic hypotheses determination. *Computing*, (1):293–308, 1966.
22. P. Hajek and J. Rauch. Logics and statistics for association rules and beyond. Tutorial PKDD'99, 1999.
23. R.J. Hilderman and H.J. Hamilton. Evaluation of interestingness measures for ranking discovered knowledge. *LNCS*, 2035:247–259, 2001.
24. S. Holm. A simple sequentially rejective multiple test procedure. *J. Statistic.*, 6:65–70, 1979.
25. H. Jeffreys. Some test of significance treated by theory of probability. In *Proc. Of the Cambridge Phil. Soc.*, pages 203–222, 1935.
26. M. Kamber and R. Shingal. Evaluating the interestingness of characteristic rules. In *Proceedings of KDD'96*, pages 263–266, 1996.
27. Yves Kodratoff. Quelques contraintes symboliques sur le numérique en ECD et en ECT. Lecture Notes in Computer Science, 2000.
28. S. Lallich. Mesure et validation en extraction des connaissances à partir des données. Habilitation à Diriger des Recherches – Université Lyon 2, 2002.

29. S. Lallich, E. Prudhomme, and O. Teytaud. Contrôle du risque multiple en sélection de règles d'association significatives. *RNTI-E-2*, 2:305–316, 2004.
30. S. Lallich and O. Teytaud. Évaluation et validation de l'intérêt des règles d'association. *Mesures de Qualité pour la Fouille de Données*, (RNTI-E-1):193–217, 2004.
31. S. Lallich, B. Vaillant, and P. Lenca. Parametrised measures for the evaluation of association rule interestingness. In *Proc. of ASMDA'05*, pages 220–229, 2005.
32. P. Lenca, P. Meyer, B. Vaillant, P. Picouet, and S. Lallich. Évaluation et analyse multicritère des mesures de qualité des règles d'association. *Mesures de Qualité pour la Fouille de Données*, (RNTI-E-1):219–246, 2004.
33. I. Lerman. Comparing partitions, mathematical and statistical aspects. *Classification and Related Methods of Data Analysis*, pages 121–131, 1988.
34. I.C. Lerman and J. Azé. Indice probabiliste discriminant de vraisemblance du lien pour des données volumineuses. *Mesures de Qualité pour la Fouille de Données*, (RNTI-E-1):69–94, 2004.
35. I.C. Lerman, R. Gras, and H. Rostam. Elaboration d'un indice d'implication pour les données binaires, i et ii. *Mathématiques et Sciences Humaines*, (74, 75):5–35, 5–47, 1981.
36. J. Loevinger. A systemic approach to the construction and evaluation of tests of ability. *Psychological monographs*, 61(4), 1947.
37. H. Mannila, H. Toivonen, and A.I. Verkamo. Efficient algorithms for discovering association rules. In *Proc. AAAI'94 Workshop Knowledge Discovery in Databases (KDD'94)*, pages 181–192, 1994.
38. N. Meggido and R. Srikant. Discovering predictive association rules. *Knowledge Discovery and Data Mining (KDD-98)*, pages 274–278, 1998.
39. G. Piatetsky-Shapiro. Discovery, analysis and presentation of strong rules. In *Knowledge Discovery in Databases*, pages 229–248. AAAI/MIT Press, 1991.
40. R. Rakotomalala. Tanagra. http://eric.univ-lyon2.fr/~ricco/tanagra, 2003.
41. P. F. Russel and T. R. Rao. On habitat and association of species of anopheline larvae in south-eastern Madras. *J. Malar. Inst. India*, (3):153–178, 1940.
42. M. Sebag and M. Schoenauer. Generation of rules with certainty and confidence factors from incomplete and incoherent learning bases. In J. Boose, B. Gaines, and M. Linster, editors, *Proc. of EKAW'88*, pages 28-1 – 28–20. 1988.
43. J. D. Storey. A direct approach to false discovery rates. *J. R. Statisc. Soc., Series B*, 64:479–498, 2002.
44. P.-N. Tan, V. Kumar, and J. Srivastava. Selecting the right interestingness measure for association patterns. In *Eighth ACM SIGKDD Int. Conf. on KDD*, pages 32–41, 2002.
45. O. Teytaud and S. Lallich. Bornes uniformes en extraction de règles d'association. In *CAp'01*, pages 133–148, 2001.
46. B. Vaillant, P. Picouet, and P. Lenca. An extensible platform for rule quality measure benchmarking. In R. Bisdorff, editor, *HCP'2003*, pages 187–191, 2003.
47. A. Van Der Vaart and J.A. Wellner. *Weak Convergence and Empirical Processes.* Springer Series in Statistics. Springer-Verlag Publishers, 1996.
48. T. Zhang. Association rules. In *Proceedings of PAKDD 2000*, volume 1805 of *LNCS*, pages 245–256. Springer, 2000.

Comparing Classification Results between N-ary and Binary Problems

Mary Felkin

Laboratoire de Recherches en Informatique, Universite Paris Sud, Batiment 490, 91405 Orsay felkin@lri.fr

Summary. Many quality measures for rule discovery are binary measures, they are designed to rate binary rules (rules which separate the database examples into two categories eg. *"it is a bird"* vs. *"it is not a bird"*) and they cannot rate N-ary rules (rules which separate the database examples into N categories eg. *"it is a bird"* or *"it is an insect"* or *"it is a fish"*). Many quality measures for classification problems are also binary (they are meant to be applied when the class variable has exactly two possible, mutually exlusive, values).

This chapter gives the data-analyst a pratical tool enabling him or her to apply these quality measures to any rule or classification task, provided the outcome takes a known and finite number of possible values (fuzzy concepts are excluded). Its purpose is also to help the data-analyst during the delicate task of pre-processing before a classification experiment. When he or she is considering different formulations of the task at hand, and more precisely, the number of possible class values that the classification problem should have, a clear indication of the relative difficulties of the consequent problems will be welcome.

Key words: Classifier, rule discovery, quality measures, N-ary classification.

1 Introduction

This chapter uses the validation of classification problems (supervised machine learning) to explain and test our proposed method. As this book is also intended for non-specialists, a short and easy to understand explanation of classification is given is section 2. Once the principle of classification is understood, and once our formulas are understood, applying them to rule discovery will become trivial.

Formally, supervised machine learning is about constructing a classifier model to achieve the best possible approximation for an unknown function $f(x)$ given training examples of the form $< x_i, f(x_i) >$, where x_i is a list of constant length of attribute values and where $f(x)$ takes its possible values from a finite set. The elements of this finite set are called the possible class

Mary Felkin: *Comparing Classification Results between N-ary and Binary Problems*, Studies in Computational Intelligence (SCI) **43**, 277–301 (2007)
www.springerlink.com

values. We distinguish between the binary cases, where $f(x)$ has only two possible values, and N-ary cases, where $f(x)$ has N possible values, with $N > 2$.

In many real-world learning tasks the unknown function f is N-ary.[1] A problem which occurs regularly to people mining real datasets and which, to the best of our knowledge, has seldom been adressed is how to compare results between a binary problem and an N-ary problem. This issue is, however, comparable to the problem of evaluating the performance of a classifier on a database in which biases have been introduced on the underlying class distribution through a modification of the respective number of examples of each class: "a problem with multiple classes with equal prior probabilities can be expected to be harder than a problem with great differences between prior probabilities" [14]. This is intuitively obvious if we consider the extreme case where a large percentage of the examples belong to class A.[2] The trivial model which always predicts the majority class A will have the same large percentage of successes.

There are several possible reasons for wishing to compare the performance of a machine learning algorithm on a binary problem with its performance on an N-ary problem.

The earliest, and often the most time-consuming part of the data-analysis task, pre-processing, is crucial. It should be given full consideration. Our formulas can help at this stage by answering questions such as "If I consider that specimen belonging to group IIa and specimen belonging to group IIb are two different kinds, how much will it increase the difficulty of the problem? By how much can I expect this added difficulty to lower the resulting accuracy?".

Some classification algorithms deal poorly with problems where the number of possible class values is very high. So the user of a classification algorithm upon such a problem might wonder whether his poor results are due to the algorithm not being well adapted to his problem, or whether they are normal considering the number of possible class values involved.

---

[1] Prominent machine learning algorithms such as C4.5 [12] can construct decision trees which generally give very good approximations of f, in N-ary classification problems as well as in binary ones. Most support vector machines (SVM) are best suited to learning binary functions [13], [8]. All supervised learning algorithms can handle binary functions, but not all of them can handle N-ary functions.

[2] Speaking in terms of probabilities, if the class has two possible values occuring with equal frequencies, a random classifier model will obtain 50% accuracy. If 60% of the examples belong to class A and 40% belong to class B, a random model following the frequencies will answer "A" 60% of the time and "B" 40% of the time. When it answers "A" it will be right 60% of the time, and when it answer "B" it will be right 40% of the time. So, all in all, it will achieve an accuracy of 52%. If the imbalance reaches 90% of class "A" and 10% of class "B", the random classifier following the distribution frequencies will obtain 82% accuracy. (Note however that it would be doing worse than the classifier simply predincting the majority class, which would achieve 90% accuracy).

Some classification algorithms can only deal with binary problems. A researcher may wish to experiment with several classifiers on a 3-class-values dataset, and to compare their results, while one of these classifiers is only built to handle 2-class-values problems.

At the latest stage, knowing the relative difficulties may also help explain the relative performances of the induced models during these different experiments. A dataminer may wish to focus upon the attribute values characterising a particular class vs. the others. A user may find large trees difficult to read and so reduce the number of class values in order to reduce the size of the induced decision tree. Supposing he was using C4.5 on the iris dataset, he would notice that grouping the virginicas with the versicolors reduced the number of leaves from 5 to 2 and increased the accuracy from 96% to 99%, and he would then wonder which tree was "best".

Having such a measure would be convenient. The person presenting the results of several classification problems with different numbers of possible class values ends up saying something like "on the two-classes problems the average accuracy was 81%, while on the 5-classes problems it was only 75%. Of course, a 5-classes problem is intrisically more difficult than a two-classes problem". But if he cannot say on which dataset his classifier worked best while taking into account this increased difficulty, this part of his presentation is a bit unsatisfying.

Last but not least, though the basic performance measures, such as accuracy or recall[3] can be defined locally, class value by class value, most authors focus upon their global definition: one accuracy value and one recall value per classification experiment.

This chapter is about ways to precisely compare the performance of machine learning algorithms on binary vs. N-ary problems. Incidentally, this approach can also be used to compare classifiers results on N_1-ary vs. N_2-ary problems, with $N_1 \neq N_2$, because both can be compared via their two-class-values equivalence.

After a brief explanation of classification we will present some currently related research. We will then introduce the "Relative Classifier Information" measurement [14], and we will give a simple criterion, inspired by information theory, to compare accuracies. Then we will show how this method can be extended to transform an $N * N$ confusion matrix into a $2 * 2$ confusion matrix, and present a detailled example. In the "Experimental Results" section we will show how both our measurement and the "Relative Classifier Information" measurement [14] confirm each other, and that they are both efficient predictors.

[3] accuracy will be defined later, recall is outside the topic of this chapter

2 Brief explanations

These explanations are intended for readers unfamiliar with all the different fields of KDD. If you feel comfortable with terms such as "classification", please skip this section.

The explanation of classification is - by now traditionally - supported by the "play tennis" database (table 1):

Table 1. The "Play Tennis" database

Outlook	Temperature	Humidity	Windy	Play Tennis
sunny	85	85	false	Don't Play
sunny	80	90	true	Don't Play
overcast	83	78	false	Play
rain	70	96	false	Play
rain	68	80	false	Play
rain	65	70	true	Don't Play
overcast	64	65	true	Play
sunny	72	95	false	Don't Play
sunny	69	70	false	Play
rain	75	80	false	Play
sunny	75	70	true	Play
overcast	72	90	true	Play
overcast	81	75	false	Play
rain	71	80	true	Don't Play

This database has five attributes (or "variables"): Outlook, Temperature, Humidity, Windy and Play Tennis. The class attribute of the database (also called "the class" or "the class variable") is the attribute which value has to be predicted. The class of this database is "Play Tennis". Each example of the database (each row of the table) also has a "class". The class of an example is the value of its class attribute. (eg. the class of the first example of the play tennis database is "Don't Play").

In supervised machine learning the task is to learn how to predict the class attribute of the database. This means predicting whether these people will play tennis or not in all possible cases, including cases not mentioned in the database. This process is called classification.

Question: if the sky is overcast, the temperature equal to 85, the humidity equal to 81, and there is no wind, will these people play tennis? The algorithm doing the classification, called a classifier, might "notice" that these people always seem to play tennis when the sky is overcast. It would then answer "Play" to our question. It might also notice that the only time on record the temperature reached 85 they didn't play and answer "Don't Play"[4]. A rule discovery algorithm might "notice" exactly the same relationships: "Don't play if temperature equals 85" and "Play if outlook is overcast". So though they are different algorithms, with different purposes, a quality measure intended for rule discovery can be used to rate a classification result and vice-versa. Some measurments may not be very useful when applied to a kind of results they were never meant to rate, but this is another matter.

3 State of the art

As [16] showed in their error-correction approach, any N-classes classification problem can be transformed into several 2-classes problems. When the classes cannot be ordered this is usually done by classifying one class vs. the $N-1$ others, and repeating the process for each class. It must be noted that there are many other, dataset-dependent or goal-dependent, possibilities. When the classes can be ordered, this binarisation is usually done by classifying the top-most class vs. the others, then the two top-most classes together vs. the others, etc. However it is done, the respective performances of machine learning algorithms with respect to the binary problems are not accurate indicators of their performance on the N-ary problem.

Another issue concerns learning models from datasets with a large number of possible class values. A decision tree algorithm confronted with a nominal class variable which has 250 different values will either try to predict them all (and possibly run out of significant attributes long before completing this task) or will try to find the best way to group the class values into categories, an NP-complete problem in itself [16]. CART [3], for example, would attempt the NP-complete task of finding the best binary grouping of these class values.

Dealing with problems with a large or very large number of possible class values is an active field of research. [4] developed a technique that can be applied to a decision tree with any number of possible class values. Their proposed technique is equivalent to testing all 2^{n-1} possible splits of each categorical variable. It does not run out of memory because it uses a relational database instead of RAM memory to compute the model. Past approaches used for decision tree building were unable to deal with N-ary problems when N was larger than about 30 while insuring that the result was optimal with respect to the capacities of the chosen classifier. These past approaches included grouping similar values of the variable, not choosing a variable that had more than an acceptable maximum number of values for split tests ("Answer tree" of SPSS inspired by [6]), generating X random numbers, each number corresponding to a pair of binary partitions between 1 and 2^{n-1}, and testing them, among others.

[4] "Don't play if the temperature is 85" seems a worse solution, because this reasoning is only supported by one example while "Play if the sky is overcast" is supported by four examples.

When learning models from datasets with a large number of possible class values, being able to measure the relative efficiency of a classifier on datasets with different numbers of possible class values could help, for example, to change a brute force search into a best-first search.

Many formulas for evaluating the performance of a classifier can either be used directly on N-ary problems or have been adapted for use on such problems.

The famous Area Under the ROC Curve (AUC) [1], [17], [9], [5] is a metric which takes as input the complete table of the confusion matrix for a binary classification task. A confusion matrix separates the examples, not only according to the fact that they were or not misclassified, but also according to their original class and to their assigned class. Please refer to Sect. 6 for an illustrated example of a confusion matrix. For a binary problem it is a 2 *by* 2 matrix with:
$TP = True\ Positives$, the number of correctly classified positive examples.
$FP = False\ Positives$, the number of incorrectly classified negative examples (they have been falsely classified as being positive examples).
$FN = False\ Negatives$, the number of incorrectly classified positive examples (they have been falsely classified as being negative examples).
$TN = True\ Negatives$, the number of correctly classified negative examples

Bradley showed that the AUC was independent of class distribution [9]. It is a $2D$ measure with TPR = *True Positives Rate* and FPR = *False Positives Rate.*

$$TPR = \frac{TP}{TP + FN} \tag{1}$$

$$FPR = \frac{FP}{FP + TN} \tag{2}$$

Peter Flach has recently investigated 3-dimensional ROC curves, showing that scaling this metric upwards in the number of classes is far from trivial [10].

Kononenko and Bratko proposed a coefficient based on information theory [7]: The amount of information needed to correctly classify an example into class C, whose prior probability is $p(C)$, is defined as $-log_2(p(C))$. From this they derived the formula for the Information Score (IS):

$$IS = -log(\frac{TP + FN}{N}) + log(\frac{TP}{TP + FP}) \tag{3}$$

It would be impossible to quote here all metrics which, like this one, assume a binary problem. Kononenko and Bratko were however very concious of this restriction and provided a local, class by class, probabilistic[5] formula which

[5] Their argument in favour of their probabilistic approach deserves to be quoted here: "Suppose in 1950 a classifier was shown a list of one million names and asked "Who of these people will be the president of the U.S.A. in 1990? It looks

allows the IS to be calculated for N-ary problems [7]. They first argued then proved that a 44% accuracy on a classification problem with 22 possible class values (the primary tumor database) is much better than a 77% accuracy on a binary problem (the breast cancer database). So the IS is both an illustration of this restriction, and an illustration of an early attempt to lift it. It is also an early use of information theory for the purpose of constructing a measurement of classifier quality.

In a tutorial given in 2004, Benedikt Brors explains how confusion matrices can be compared between binary and N-ary classification results [2]. This method "Relative Classifier Information", was invented by [14] and we will present it in the next section for the sake of completeness and to allow comparisons with our our own formulas to be made.

4 Relative Classifier Information (RCI)

The basis of this method is comparable to the information gain calculations used by ID3 [11].

The input (or "prior") entropy is the entropy of the database before classification, Ent^{prior}. After classification, the entropy Ent^{out}_J of each group of examples of a given predicted class value J is calculated, and the output entropy Ent^{out} is the sum of the entropy of each group weighted according to the number of examples contained by that group (Fig. 1). The relative classifier information is the input entropy minus the output entropy.

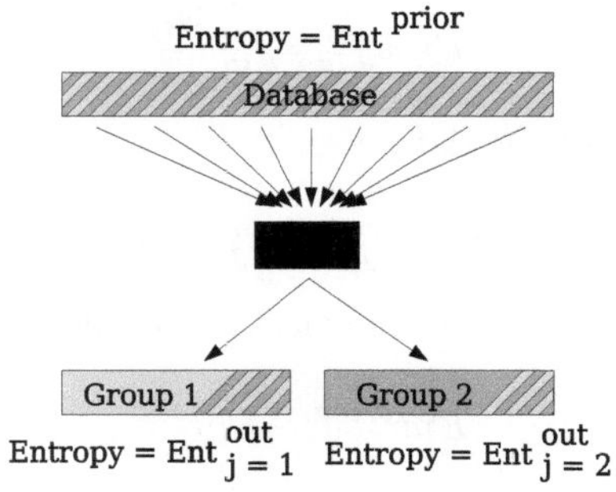

Fig. 1. From Ent^{prior} to Ent^{out}_J with a "black box" classifier

RCI was presented by [14]. A real-world application of this measure can be found in [15].

they are are all equally likely at the moment." So we have one million classes all with prior probabilities one in a million. Suppose the classifier in 1950 answered: $P(Bush) = 0.45$, $P(Dukakis) = 0.55$, $P(allothers) = 0$. Strictly speaking, this answer can be viewed as incorrect. However, normally it would be considered as highly insightful, informative and useful."

In all the following calculations, $0 * log(0)$ is replaced by 0 and $\frac{0}{0}$ is replaced by 0.
Let M be the confusion matrix[6] of a trained classifier.
Σ_{TOT} = The total number of examples
The frequencies: $Freq_{ij} = \frac{M_{ij}}{\Sigma_{TOT}}$
The prior class probabilities are the number of examples of class value I over the total number of examples:

$$Prob_I^{prior} = \frac{\Sigma_j M_{ij}}{\Sigma_{TOT}} \tag{4}$$

The prior uncertainty about the classes is the entropy of the database:

$$Ent^{prior} = \Sigma_i - Prob_i^{prior} log Prob_i^{prior} \tag{5}$$

Given that an example has been labelled J by the classifier, the probability that it belongs to class I is:

$$Prob_{I|J}^{out} = p_{ij} = \frac{M_{ij}}{\Sigma_i M_{ij}} \tag{6}$$

Once the confusion matrix is known, the entropy of class j is then:

$$Ent_J^{out} = \Sigma_i - p_{ij} log p_{ij} \tag{7}$$

The probability that the classifier assigns as a label the class value J to an example is:

$$Prob_J^{out} = \frac{\Sigma_i M_{ij}}{\Sigma_{TOT}} \tag{8}$$

And the total entropy of the database:

$$Ent^{out} = \Sigma_j (Prob_J^{out} * Ent_J^{out}) \tag{9}$$

The reduction in uncertainty by the classifier is then:

$$Ent^{prior} - Ent^{out} \tag{10}$$

This gives us two measurements of the efficiency of the classifier which are independent of the number of possible class values: Ent^{out}, the entropy after classification, and $Ent^{prior} - Ent^{out}$, the reduction in uncertainty produced by the classifier. As the formulas we will present deal with the end results of classification experiments, and not with the actual process, we will not take Ent^{prior} into account in the "Experimental Results" section. We will only compare the results of our formulas with those given by Ent^{out}.

[6] The element on the $i^t h$ row and in the $j^t h$ column of M is M_{ij}. Please refer to section 6 for a description of the meaning of such a matrix

5 Comparing Accuracies

5.1 Overview

We define a binary problem to be "equivalent" to an N-ary problem if the examples of the binary classification problem are the same examples as the examples of the N-ary classification problem, after the same pre-processing (the same attributes have been selected). For example, if the problem concerns the ternary iris dataset, grouping the virginicas with the versicolors reduces the number of possible class values from three to two and so transforms this ternary problem into an equivalent binary problem.
We are going to introduce a formula allowing to predict, for an N-ary problem on which a classifier algorithm has a known accuracy, what accuracy this classifier algorithm would have on an equivalent binary problem. Our formula reflects the fact that reducing the number of possible class values usually makes the problem easier (as explained in the introduction).
We will first state and justify four properties which such a formula should have. Next, we will explain how our formula is inspired by the classical Error-Correcting Code framework, [16], based on information theory. We will show that it can be obtained by construction for a specific set of equivalent binary problems. It should be noted that the binary problems of this set are theoretical constructions, unlikely to be obtainable from any given real-life database. We will then prove that our formula has the four desired properties for any equivalent binary problem. In the tests section, we will look at how well our formula actually predicts the accuracy of equivalent binary problems for equivalent binary problems **not** included in the hypothetical set used for the construction. Readers taking a practical interest in this formula can skip the justification and simply read the formula at the end of section 5.3.4.

5.2 Required properties

Let us first consider the requirements of a measure to compare accuracies between N-ary and binary classification problems. The accuracy is the number of correctly classified examples over the total number of examples.
N = Number of classes
Acc_N = Accuracy on the N possible class values problem
Acc_2 = The corresponding (average) accuracy for a binary sub-problem
Comparison function: $Cf(Acc_N, N) = Acc_2$

The best possible accuracy is, in all cases, 1. So, for all N:

$$Cf(1, N) = 1 \tag{11}$$

The worst possible accuracy is, in all cases, 0. So, for all N:

$$Cf(0, N) = 0 \tag{12}$$

If the classifier answered randomly, without taking into account the class distribution, the average accuracy on binary problems would be 1/2. On N-classes problems it would be $1/N$. So, for all N:

$$Cf(1/N, N) = 1/2 \tag{13}$$

Finally, by definition of the problem, for all accuracies $X \in [0, 1]$:

$$Cf(X, 2) = X \tag{14}$$

Many functions would fit these points, and one of them has a justification which nicely corresponds to information theory, well known to our field thanks to Shannon's entropy.

5.3 Theoretical Construction

Notation and Temporary Restriction

Let us first consider a classification problem P_N where the number of possible class values N is 2^k, and let the problem be divided into k binary sub-problems, $P_1, ..., P_k$. Let the combination of these k answers correspond to one and only one class value in the original problem P_N. The $N = 2^k$ restriction, with k an integer, is only there for the purpose of explaining the origin of our formulas and will be lifted afterwards. (This is possible because when $N > 2$ it is always equal to 2^k for some real number $k = log_2(N)$). After the classical Error-Correcting Code framework, [16], based on information theory, the P_N problem can be analysed as the conjunction of k independent sub-problems (figures 2 and 3).

Real life binarisation

It is very important to note here that we are neither assuming nor supposing that these k binary sub-problems of equivalent complexity could actually be constructed. We do not care. We are just interested in determining what the accuracy of our classifier on such binary sub-problems would be if the sub-problems could be constructed. These sub-problems are defined as "binary problems of a complexity equivalent to the complexity of the original N-ary problem", as such their sole purpose is to illustrate our definition of "equivalent complexity" and the construction of our formulas.
In real life, an N-ary classification problem can always be transformed into N binary classification sub-problems, by taking each possible class value in turn and predicting whether examples belong to this class or not. But an N-ary classification problem cannot always be transformed into $k = log_2(N)$ classification sub-problems. Once our formulas have been presented, we will test them on real-life databases, partitionning these databases, according to the real-life method mentioned above, into N sub-problems. We will not try

to subdivide the N-ary problems into $log_2(N)$ binary sub-problems. We will compare the results of the original problem with the results of the N sub-problems. We will not try to reconstruct a solution to the original problem from the solutions of the sub-problems and we will not consider issues such as "what should be done if two different sub-problem classification models each predict that example x belongs to the class they have been trained to recognise".[7]

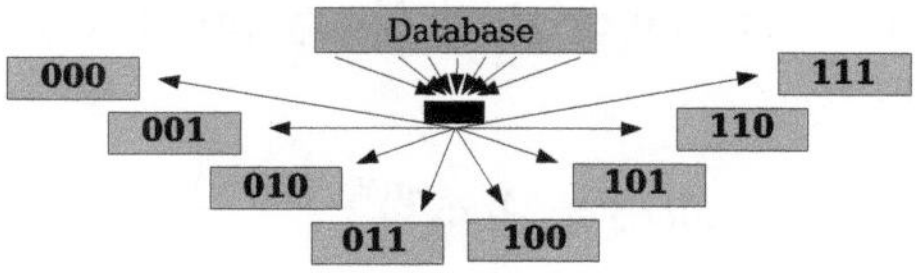

Fig. 2. A "black box" classifier sorting database examples into 8 groups of possible class values with binary labels

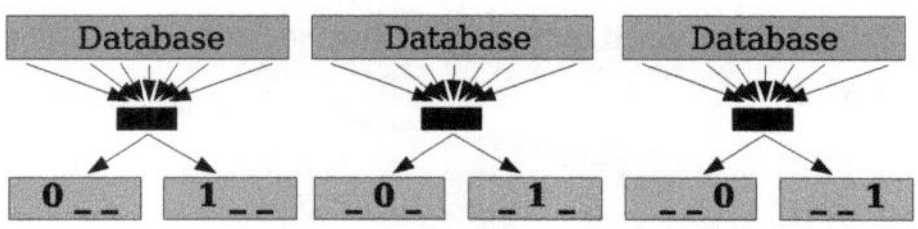

Fig. 3. 3 "black box" classifiers sorting the same database examples into $3*2$ groups of possible class values with binary labels

Construction

Fig. 2 shows a classifier partitioning a database into 8 groups corresponding to 8 possible class values. The possible class values can be given any convenient name, for example "bird", "insect", "fish", etc. Here each possible class value is a number, from 0 to 7, written in binary. To distinguish between 8 different possible class values by numbering them in binary, we need binary numbers of length $3 = log_2(8)$.

In Fig. 3, 3 binary classifiers each predict one digit of these binary numbers. By combining their answers the whole binary number is obtained. For example,

[7] 1 There are many, problem-dependent, reasons for the ease/difficulty of a classification task. The one in which we are interested here is the number of possible class values. So the number of sub-problems used for constructing our formula or for testing it is irrelevant to our case. The only important thing is that these sub-problems are all binary because our formula aims at predicting the accuracy of the classifier on a single, ideally equivalent, binary sub-problem.

if the first binary classifier predicts a zero, while the second and the third predict a one, we get the overall prediction 011 corresponding to the fourth possible class value (not the third because we are counting from zero).
Let Acc_2 be the accuracy on the $P_1, ..., P_k$ independent binary sub-problems. (As they are purely hypothetical, we can suppose that they all have the same accuracy). Expressing the accuracies as probabilities of success:

$$P(P_N) = P(P_1) * ... * P(P_k) \tag{15}$$

So the overall accuracy on the original N-ary problem is $Acc_N = Acc_2^k$.[8] As $k = log(N)/log(2)$:

$$Acc_2 = Acc_N^{log(2)/log(N)} \tag{16}$$

5.4 Properties

We can now see that this transformation is continuous and so, lifting our previous restriction, apply it to any $N > 2$ (figure 4).

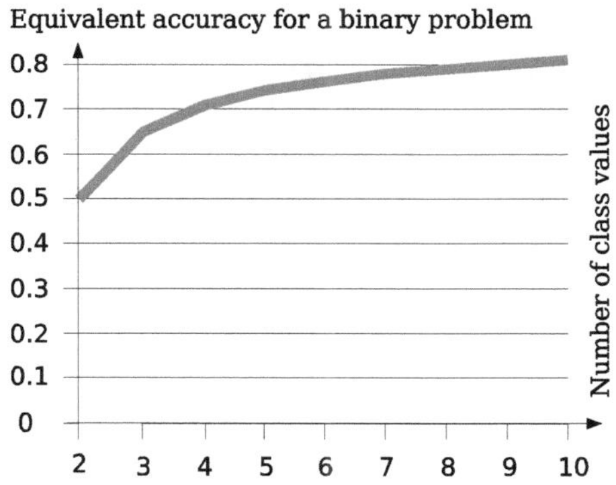

Fig. 4. What an accuracy of 0.5 means for classification problems when the number of classes increases

In Fig. 4 the points are joined together, for a better visualisation. They are joined together by straight segments because, though the mathematical function is indeed continuous, it makes no sense to speak about problems having a non-integer number of possible class values.
Our formula has the properties described above as being the requirements.

Property 1: $Cf(1, N) = 1$

[8] When the accuracies are not followed by a % sign, they are expressed as real numbers between 0 and 1.

$$Cf(1, N) = 1^{log(2)/log(N)} = 1 \tag{17}$$

Property 2: $Cf(0, N) = 0$

$$Cf(0, N) = 0^{log(2)/log(N)} = 0 \tag{18}$$

Property 3: $Cf(1/N, N) = 1/2$

$$Cf(1/N, N) = (1/N)^{log(2)/log(N)} \tag{19}$$

$$Cf(1/N, N) = e^{(ln(2)/ln(N))*ln(1/N)} \tag{20}$$

$$Cf(1/N, N) = e^{-(ln(2)/ln(N))*ln(N)} \tag{21}$$

$$Cf(1/N, N) = e^{-ln(2)} = 1/2 \tag{22}$$

Property 4: $Cf(X, 2) = X$

$$Cf(Acc_N, 2) = Acc_N^{log(2)/log(2)} = Acc_N \tag{23}$$

5.5 One formula is not enough

Beside accuracy, many other interesting performance measures have been proposed, summarising the information contained in the confusion matrix M, where M_{ij} indicates the number of examples of class I which have been classified as belonging to class J. A confusion matrix for an N-ary problem has N^2 numbers, N of which corresponding to properly classified examples.
No immediate correspondance can be found between an $N * N$ confusion matrix and a $2 * 2$ confusion matrix, because when the results are examined in detail, class by class, it makes no sense to regroup them: Which of the N classes should be considered as corresponding to which of the two classes? This means that all class-specific information will unavoidably be lost in the transformation of a $N * N$ confusion matrix into a $2 * 2$ confusion matrix.

6 Comparing Confusion Matrices

6.1 From an $N * N$ confusion matrix to a $2 * 2$ confusion matrix

Our proposed measurement goes a step further than the "Relative Classifier Information". It is not only independant of the number of possible class values, it can also be used to transform an $N * N$ confusion matrix into a $2 * 2$ confusion matrix. So it allows any measurement invented for evaluating the performance of a classifier on a binary problem to be used on an N-ary problem, provided class-specific information is not relevant to this measurement. It can also be used efficiently to predict the average results of the binarisation of a classification problem, as we will show in the "Experimental Results" section.

6.2 From a 3 * 3 confusion matrix to a 2 * 2 confusion matrix

Fig. 5 shows the transformations from a confusion matrix resulting from a 3 class-values classification problem into a 2 class-values confusion matrix, with class A of the ternary problem arbitrarily set to correspond to class P ("positive") of the binary problem.

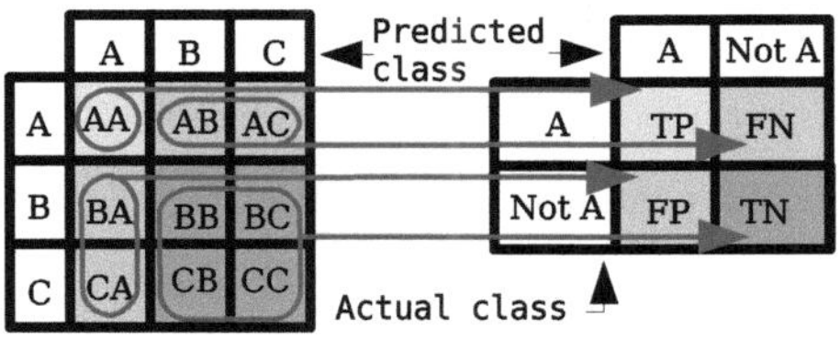

Fig. 5. Grouping values of a 3x3 confusion matrix

TP = number of correctly classified positive examples
FP = number of incorrectly classified negative examples
FN = number of incorrectly classified positive examples
TN = number of correctly classified negative examples
AA = number of correctly classified examples of class A
AB = number of examples of class A classified as belonging to class B
etc.

It should be noted that this picture shows a mapping (through the functions that will be defined) but not an equality. We are not saying $AA = TP$.

When considering class A vs. all others, examples of class B which have been classified as belonging to class C and vice-versa cannot be considered mistakes: they have "correctly" been classified as not belonging to class A.

On top of using our Cf formula:

$$Acc_2 = Acc_N^{log(2)/log(N)} \tag{24}$$

which is equivalent to:

$$\frac{TP + TN}{\Sigma_{TOT}} = (\frac{AA + BB + CC}{\Sigma_{TOT}})^{log(2)/log(N)} \tag{25}$$

we scale the contribution of class A towards Acc_N with the (hypothetical) contribution of class P towards Acc_2:

$$\frac{AA}{Acc_N} = \frac{TP}{Acc_2} \tag{26}$$

We also take into account that the relative number of examples belonging to class A must remain unchanged, as this is a descriptor of the database and not a classification result:

$$\frac{AA + AB + AC}{\Sigma_{TOT}} = \frac{TP + FN}{\Sigma_{TOT}} \tag{27}$$

The remaining formula which we use to complete our calculations is just the trivial fact that:

$$TP + FN + FP + TN = \Sigma_{TOT} \tag{28}$$

These last equations give us a system of four equations for four variables: TP, FP, FN and TN. We solve this system then take, one by one, all class values of the original problem and set them to correspond to class value P. We average the 3 results and simplify the resulting formulas. As this subsection is only about explaining the method we used to obtain our formulas, there is no need to pursue these calculations here. Instead, we will now go to the general case. The results of the equivalents of these equations for N possible class values are given in the next subsection.

6.3 Generalisation

Let Σ_{Xii} = The sum of all the correctly classified examples.[9]
Σ_{TOT} = The total number of examples

$$TP = \frac{Acc_2.\Sigma_{Xii}}{N.Acc_N} \tag{29}$$

$$FN = \frac{\Sigma_{Xii}.(1 - Acc_2)}{N.Acc_N} \tag{30}$$

$$FP = \Sigma_{TOT} - \frac{\Sigma_{Xii}(N.Acc_2 - Acc_2 + 1)}{N.Acc_N} \tag{31}$$

$$TN = \frac{\Sigma_{Xii}.Acc_2.(N - 1)}{N.Acc_N} \tag{32}$$

Transforming the results of an N-ary classification problem into the results of a hypothetical corresponding binary classification problem might be a useful or a useless move. As you will have seen in this book, quality measures for classification results are numerous. Some are independant of the number of possible class values, some are not. Some can be applied in a straightforward way to N-ary classification results, some cannot. When there is a straightforward way to apply your chosen quality/interestingness measure to your N-ary classification results, and if you are not comparing results across problems with different numbers of possible class values, use this straightforward way. It will be both simpler and more accurate to apply the formulas as they are meant to be applied than to use the confusion matrix transformation described in this chapter.

[9] in the $3 * 3$ example Σ_{Xii} was AA + BB + CC

7 Example: the iris dataset

The iris dataset is part of the UCI Machine Learning Repository. It is a set describing 50 setosa irises, 50 virginica irises and 50 versicolor irises. The task is to learn to recognise which iris is of which kind according to its description.

7.1 C4.5 and Naives Bayes results

The classifier algorithms used for this example are used with the default settings of the Weka platform and using 10 folds cross validation. With C4.5 (J48) we obtained the confusion matrix in (table 2).

Table 2. Original ternary confusion matrix obtained with C4.5

	Setosa	*Versicolor*	*Virginica*
Setosa	49	1	0
Versicolor	0	47	3
Virginica	0	2	48

As the accuracy on the ternary problem is 0.96, we predict an accuracy of $0.96^{\,log(2)/log(3)} = 0.97$ on equivalent binary problems. On this problem our measurement is too optimistic (we will see later that on some other databases it is too pessimistic). The real value, when each class is predicted in turn vs. the others and the accuracies are averaged, is 0.95. It is more difficult for C4.5 to accurately predict iris categories individually than it is to predict them together!

Fig. 5 detailed how one binary confusion matrix was obtained from a ternary confusion matrix. Fig. 6 shows this process being repeated for each of the 3 class values of the iris dataset, and the 3 resulting binary confusion matrices being averaged to build the final one.

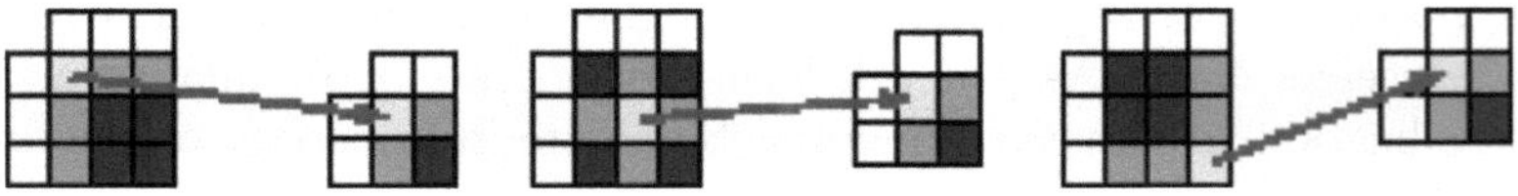

Fig. 6. Calculating the 3 intermediate confusion matrices

The tables 3, 4 and 5 show the values of the 3 intermediate binary confusion matrices (predicted values left and real values right). After each class value in turn has been taken as the "positive" class value, the results are averaged in table 6 according to the method illustrated by (Fig. 6).

The Ent^{out} value, calculated as is often done with the number of possible class values taken as the base of the logarithm, is 0.15 for the original confusion

Table 3. Predicting setosa (Yes/No)

	Setosa	*Not Set.*
Setosa	49.74 / 49	0.26 / 1
Not Setosa	3.56 / 0	96.44 / 100

Table 4. Predicting versicolor (Yes/No)

	Versicolor	*Not Vers.*
Versicolor	47.71 / 44	2.29 / 6
Not Versicolor	1.53 / 6	98.47 / 94

Table 5. Predicting virginica (Yes/No)

	Virginica	*Not Virg.*
Virginica	48.73 / 47	1.27 / 3
Not Virginica	2.54 / 7	97.46 / 93

Table 6. Predicted confusion matrix

Average	*P*	*N*
P	48.73 / 46.67	1.27 / 3.33
N	2.54 / 4.33	97.46 / 95.67

matrix and 0.16 for our estimated binary confusion matrix. It rises to 0.28 for the real average binary confusion matrix.

The same experiment with Naives Bayes reveals one of the weaknesses of our measurement: When all examples of a particular class value, here the setosas, have been correctly classified in the N-ary experiment, the predicted error for this class value in the binary experiment will become negative(table 7).

Table 7. Original ternary confusion matrix obtained with Naive Bayes

	Setosa	*Versicolor*	*Virginica*
Setosa	50	0	0
Versicolor	0	48	2
Virginica	0	4	46

The accuracy on the ternary problem is 0.96, the predicted accuracy on the binary problem is 0.97 and the actual average accuracy on the binary problems is 0.95. This is again too optimistic. The negative error not withstanding, the correlation between predictions and actual results is still acceptable, though it was slightly better with C4.5 than with Naive Bayes: 0.997 vs. 0.993(tables 8–11).

Table 8. Predicting setosa (Yes/No)

	Setosa	*Not Set.*
Setosa	50.76 / 50	-0.76 / 0
Not Setosa	4.57 / 0	95.43 / 100

Table 9. Predicting versicolor (Yes/No)

	Versicolor	*Not Vers.*
Versicolor	48.73 / 43	1.27 / 7
Not Versicolor	2.54 / 4	97.46 / 96

Table 10. Predicting virginica (Yes/No)

	Virginica	*Not Virg.*
Virginica	46.37 / 48	3.3 / 2
Not Virginica	0.51 / 10	99.49 / 90

Table 11. Predicted confusion matrix

Average	*P*	*N*
P	48.73 / 47	1.27 / 3
N	2.54 / 4.67	97.46 / 95.33

The Ent^{out} value on the actual averaged binary confusion matrix is again 0.28 vs. 0.15 on the original ternary confusion matrix. So [14] measurement is also too optimistic in this case.

7.2 Discussion of this example

A possible explanation for the fact that it is "easier" to solve the ternary iris classification problem than, on average, the three binary iris classification problems could be thought to reside in the fact that the ternary problem is balanced in the number of examples: there are 50 iris-setosas, 50 iris-versicolor and 50 iris-virginicas described. The binary problems are therefore unbalanced by a two to one ratio. However, this problem is recurrent and would become worse as the number of possible class values increases. The experimental results (next section) show us that this is not happening. The reason why our Cf formula and its associated formulas are poor predictors for the iris dataset is dataset dependent. As a support for a didactic step-by-step demonstration of our testing procedure, one dataset is just as good as any other. Choosing a failure simply makes this demonstration slightly more interesting, because when an experiment performs as expected we just think "good" whereas when it fails we wonder why.

8 Experimental Results

As our transformation is intended to be useful to machine learner practitioners, we chose to test it on real-world databases, selecting these whose number of possible class values range between 3 and 19. We binarised some of the real world databases of the UCI repository by selecting each class value in turn and using C4.5 and Naive Bayes to build a classification model separating this class from all the others taken together.

8.1 Accuracies

Acc_N is the accuracy on the N-ary problem, Acc_2R the real average accuracy on the binarised problem, and Acc_2P the predicted binary accuracy calculated to correspond to our hypothetical binary problem of equivalent difficulty.

The correlation between the predicted binary accuracy and the real binary accuracy is 0.97 with C4.5 and 0.95 with Naive Bayes (table 12), so the behaviour of C4.5 is slightly more predictable with respect to the binarisation of these datasets than the behaviour of Naive Bayes. We can also note that in the real world the average accuracy on the binary problems can be smaller than the accuracy on the N-ary problem. Our formula will never predict a smaller value so its predictions will be far of target in these cases.

Table 12. Accuracies: C4.5 (left) Naive Bayes (right)

Database	N	Acc_N	Acc_2R	Acc_2P	Acc_N	Acc_2R	Acc_2P
1: iris	3	0.96	0.95	0.97	0.96	0.94	0.97
2: hayes-roth	3	0.81	0.81	0.88	0.77	0.73	0.84
3: lenses	3	0.79	0.85	0.86	0.75	0.86	0.83
4: lung-cancer	3	0.5	0.63	0.65	0.5	0.69	0.65
5: new-thyroid	3	0.92	0.95	0.95	0.97	0.96	0.98
6: wine	3	0.97	0.97	0.98	0.97	0.97	0.98
7: ann-thyroid	3	1	1	1	0.96	0.96	0.97
8: allrep	4	0.99	1	1	0.94	0.97	0.97
9: allhyper	4	0.99	0.99	0.99	0.96	0.98	0.98
10: allhypo	4	0.99	1	1	0.95	0.97	0.98
11: soybean-small	4	0.98	0.99	0.99	0.98	0.99	0.99
12: anneal	5	0.92	0.97	0.97	0.79	0.9	0.9
13: glass	6	0.66	0.9	0.85	0.5	0.75	0.76
14: zoo	7	0.93	0.98	0.97	0.96	0.98	0.99
15: soybean-large	19	0.85	0.99	0.96	0.92	0.95	0.98

Table 13. Properly classified examples: C4.5 (left) Naive Bayes (right)

DB	N	TP_R	TP_P	TN_R	TN_P	TP_R	TP_P	TN_R	TN_P
1	3	46.66	48.72	95.66	97.45	47	48.72	94.66	96.80
2	3	32	38.54	74.33	77.08	29	37.16	68	74.32
3	3	6	6.90	14.33	13.80	5.66	6.67	15	13.34
4	3	3.33	6.88	16.66	13.77	5.33	6.88	16.66	13.77
5	3	67.66	68.03	137.33	136.07	70	70.18	137.33	140.37
6	3	54	56.99	115.66	113.98	56.33	58.06	116	116.12
7	3	1252.33	1255.01	2511.33	2510.03	1186.33	1221.49	2449	2442.98
8	4	695.25	697.49	2094.25	2092.48	665.75	678.68	2042.25	2036.02
9	4	690.25	695.10	2090.75	2085.32	666.75	685.34	2067.5	2056.03
10	4	694.75	697.87	2097	2093.62	658	682.53	2055	2047.59
11	4	11.87	11.62	35	34.87	11.53	11.62	35	34.87
12	5	145.87	154.22	627.6	616.90	133.53	144.05	586	576.21
13	6	25	30.35	168.5	151.75	30.33	27.17	130	135.89
14	7	12.85	14.06	85.71	84.38	13.42	14.22	85.85	85.33
15	19	14.10	15.53	289	279.68	15.68	15.85	275.26	284.90

8.2 Confusion matrice results

When looking at the details of the predicted confusion matrices, we also note a strong correlation between the number of TP and TN (True Positives and True Negatives) examples predicted and the numbers of TP and TN found by averaging the confusion matrices resulting from the real datamining experiment (table 13).

The correlation for TP is 1 with C4.5 and 1 with Naive Bayes (all correlations have been rounded up to two decimals). The correlation for TN is 1 with C4.5 and 1 with Naive Bayes (table 13).

Table 14. Badly classified examples: C4.5 (left) Naive Bayes (right)

DB	N	FP_R	FP_P	FN_R	FN_P	FP_R	FP_P	FN_R	FN_P
1	3	4.33	2.54	3.33	1.27	4.66	1.87	3	1.04
2	3	13.66	10.91	12	5.45	20	13.67	15	6.83
3	3	1.66	2.19	2	1.09	1	2.65	2.33	1.32
4	3	4.66	7.55	7.33	3.77	4.66	7.55	5.33	3.77
5	3	6	7.25	4	3.62	6	2.96	1.66	1.48
6	3	3	4.68	5.33	2.34	2.66	2.53	3	1.26
7	3	3.33	4.62	5	2.31	65.66	71.68	71	35.84
8	4	5.75	7.51	4.75	2.50	57.75	63.97	34.25	21.32
9	4	9.25	14.67	9.5	4.89	32.75	43.96	33.25	14.65
10	4	3	6.38	5.25	2.12	45	52.40	42	17.46
11	4	0.25	0.37	0.25	0.12	0.25	0.37	0.25	0.12
12	5	10.25	21.49	14	5.37	52.25	62.18	25.8	15.54
13	6	9.83	26.58	10.66	5.31	48.33	42.43	5.33	8.48
14	7	0.85	2.18	1.57	0.36	0.71	1.23	1	0.20
15	19	1.84	11.15	2.05	0.61	15.15	5.09	0.47	0.28

The correlation for FP is 0.74 with C4.5 and 0.98 with Naive Bayes. The correlation for FN is 0.95 with C4.5 and 0.98 with Naive Bayes (table 14).

The number of examples concerned by the predictions of failure are smaller. The correlations show these are worse approximations when C4.5 is used than when Naive Bayes is used, which was an unexpected result (table 14).

8.3 Ent^{out}

In table 15 Ent_R^{out} is the entropy of the averaged binary real confusion matrix, Ent_P^{out} is the entropy of the predicted binary confusion matrix, and Ent_N^{out} is the entropy of the original $N * N$ confusion matrix.

When C4.5 is used, the correlations of Acc_N with Ent_R^{out}, Ent_P^{out} and Ent_N^{out} are respectively -0.85, -0.91 and -0.96. The correlation between Ent_R^{out} and Ent_P^{out} is 0.98, that between Ent_R^{out} and Ent_N^{out} is 0.91 and that between Ent_P^{out} and Ent_N^{out} is 0.97 (table 15).

When Naive Bayes is used, the correlations of Acc_N with Ent_R^{out}, Ent_P^{out} and Ent_N^{out} are respectively -0.83, -0.91 and -0.9. The correlation between Ent_R^{out} and Ent_P^{out} is 0.97, that between Ent_R^{out} and Ent_N^{out} is 0.93 and that between Ent_P^{out} and Ent_N^{out} is 0.97 (table 15).

8.4 Unbalanced datasets

The underlying class distribution, that is the relative number of examples of each class, can have an effect upon some predictors. When there are many more examples of one class than there are examples of any another class, some classifier algorithms perform poorly. For this reason, we chose to do

Table 15. Ent^{out}: C4.5 (left) Naive Bayes (right)

DB	*N*	Acc_N	Ent_R^{out}	Ent_P^{out}	Ent_N^{out}	Acc_N	Ent_R^{out}	Ent_P^{out}	Ent_N^{out}
1	3	0.96	0.28	0.16	0	0.96	0.28	0.13	0
2	3	0.81	0.68	0.50	0.39	0.76	0.79	0.57	0.50
3	3	0.79	0.60	0.53	0.42	0.75	0.58	0.60	0.55
4	3	0.79	0.91	0.86	0.86	0.75	0.86	0.86	0.69
5	3	0.92	0.26	0.27	0.27	0.96	0.20	0.13	0.14
6	3	0.93	0.27	0.22	0.23	0.96	0.20	0.14	0.12
7	3	0.99	0.02	0.01	0.01	0.95	0.22	0.17	0.16
8	4	0.99	0.03	0.03	0.02	0.94	0.19	0.17	0.08
9	4	0.98	0.05	0.05	0.04	0.95	0.15	0.10	0.04
10	4	0.99	0.03	0.03	0.02	0.95	0.19	0.14	0.15
11	4	0.97	0.08	0.07	0.05	0.97	0.08	0.07	0.05
12	5	0.92	0.19	0.17	0.19	0.78	0.39	0.35	0.20
13	6	0.65	0.41	0.42	0.50	0.49	0.50	0.53	0.52
14	7	0.93	0.15	0.12	0.09	0.96	0.11	0.08	0.06
15	19	0.84	0.08	0.10	0.15	0.92	0.11	0.06	0.07

the following experiment with C4.5 [12], which is known to be robust in the presence of unbalanced datasets.

We generated concepts which were CNF formulas.[10] The boolean variables were randomly picked among 10 possible choices and their truth values in the CNF were randomly assigned. Each disjunctive clause contained 5 literals and each CNF was a conjunction of 5 clauses. No checks were made to ensure that the same variable was not chosen more than once in a clause. Under such conditions, there are on average between 4 and 5 time as many positive examples as there are negative examples (when all possible examples are represented). We increased the imbalance by using two such concepts for each database. The first concept separated the examples of class 0 (the negatives) from the examples of class 1 (the positives). The second concept separated the examples previously of class 0 into examples still of class 0 (the negatives according to the second concept) and examples of class 2 (the positives according to the second concept). We obtained 47 very imbalanced databases. Databases containing no examples of class 0 (always the minority class) were eliminated. The training sets were built by randomly selecting 341 examples of each class (some were chosen several times) and the training set contained the 1024 examples generated by all possible combinations of the ten boolean variables.

We binarised the datasets by grouping together two class values each time and we averaged the accuracy of C4.5 upon the three binarised database. The error shown on Fig. 7 is the binary accuracy we predicted minus the

[10] A CNF formula, meaning a boolean function in Conjunctive Normal Form, is a conjunction of disjunction of literals. A literal is a variable which has one of two possible values: 0 or 1. For example, if A, B, C and D are boolean literals, then the logical formula (A OR B) AND (C OR (NOT D)) is a CNF.

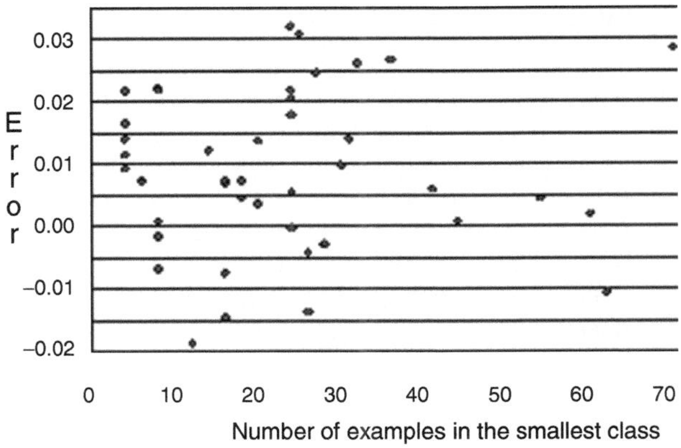

Fig. 7. The error of the predicted binary accuracy for imbalanced datasets as the number of examples in the minority class increases.

averaged binary accuracy (which is why there are "negative" error values); 0.01 corresponds to an error of 1%.

8.5 Summary of Experimental Results

The entropy of the classified database measurement [14] and our predicted 2 by 2 confusion matrix strongly agree. These are two measurements which can be used to compare results between an N-ary and a binary classification problem, because they are independent of the number of possible class values. The two methods confirm each other in the sense that the correlations are better between the entropy of the original, Ent_N^{out}, and the entropy of the predicted confusion matrix, Ent_P^{out} than between Ent_N^{out} and the entropy of the averaged real confusion matrix Ent_R^{out}. The underlying class distribution can be imbalanced without affecting the results.

9 Discussion and Limitations

9.1 Limitations

Our experiments showed that, although our predicted values globally matched the values obtained by averaging the N real binary classification results, the variance whithin problems was quite high. It also showed that when the classification algorithm performed "too well", for example on the contact lenses database, the FP and FN values of the predicted confusion matrix could become negative for a given class value taken as "the positives". Eventually these variations mostly cancel out when all the class values in turn have been taken as "the positives", so though the individual results of binary classification experiments may be quite different from the predicted result, the average result is much closer to the predictions, for both the accuracy and for the values of the confusion matrix.

9.2 Discussion

The purpose of our proposed transformation is not to measure the performance of a classifier taken on its own, but to allow comparisons to be made between classifier performances on different splits of a database. So the relative progression of Acc_N vs. the predicted Acc_2 has to be regular in order to insure scalability, reversibility, and so on. Neither the chi square measurement nor any of the derived measurements can give us this regularity. Fig. 8 shows the increase of the accuracy of the theoretical binary problem as the corresponding accuracy of the N-ary problem increases linearly (left) and the cumulative probability density function of the chi square statistic for a binary and a ternary classification problem, for values of chi square between 0 and 8 (right).

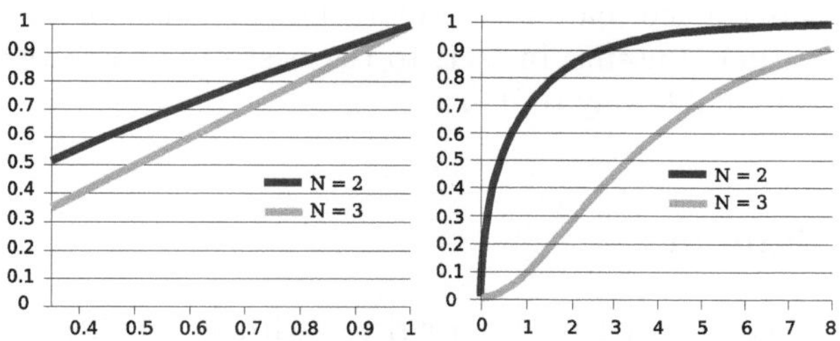

Fig. 8. Comparing straightforwardnesses

In the same way, binomial and multinomial probabilities are not meant to be used together to compare results across problems which differ in their number of possible outcomes. Because the multinomial distribution has, in contrast to a binomial distribution, more than two possible outcomes, it becomes necessary to treat the "expected value" (mean) of the distribution as a vector quantity, rather than a single scalar value. So going this way would complicate the problem before (or instead of) solving it.

When considering the results of an N-ary classification problem, the accuracy of the corresponding binary problem might be of some interest on its own. But the raw corresponding binary confusion matrix is much less informative than the N-ary confusion matrix, as all class-specific information has been lost. Consider the two binary confusion matrices in table 16 (where "C. Pos" stands for "Classified as Positive" and "C. Neg" for "Classified as Negative"): If they show the results of two binary experiments they could be ordered throught the use of problem-dependant considerations such as "False positives represent a more serious problem than false negatives, so Problem 1 gave a better result". If they are derived from the results of N-ary classification problems they cannot even be ordered.

Table 16. Example of comparison difficulties

Problem 1	*C. Pos.*	*C. Neg.*	*Problem 2*	*C. Pos.*	*C. Neg.*
Positive	50%	15%	*Positive*	55%	10%
Negative	15%	20%	*Negative*	20%	15%

The TP, FP, FN and TN of a hypothetical binary problem corresponding to an N-ary problem are building blocks enabling to:

- Use result quality measures which are not independant of the number of possible class values to calculate a fair comparison between classification problem results with a different number of possible class values (N-ary problems with different values for N can be compared via their binary equivalence).

- Use result quality measures which cannot be applied in a straightforward way to N-ary classification results on the binary equivalent of such a problem.

By itself, this transformation is useless, but in either of these two cases, this transformation is useful. In fact, to the best of our knowledge, it is the only generally appliable way to do it.

Acknowledgments

This chapter would never have been written without the help and encouragments received from Michele Sebag and Yves Kodratoff.

References

1. Hanley J. A. and McNeil B. J. The meaning and use of the area under a receiver operating characteristic (roc) curve. *Radiology*, 143:29–36, 1982.
2. Brors B. and Warnat P. Comparisons of machine learning algorithms on different microarray data sets. *A tutorial of the Computational Oncology Group, Div. Theoretical Bioinformatics, German Cancer Research Center*, 2004.
3. Olshen R. Breiman L., Friedman J. and Stone C. Classification and regression trees. *Wadsworth International Group*, 1984.
4. Leite E. and Harper P. Scaling regression trees: Reducing the np-complete problem for binary grouping of regression tree splits to complexity of order n. Southampton University, 2005.
5. Hernandez-Orallo J. Ferri C., Flach P. Learning decision trees using the area under the roc curve. *Proceedings of the Nineteenth International Conference on Machine Learning (ICML 2002)*, pages 139–146, 2002.

6. Kass G. An exploratory technique for investigating large quantities of categorical data. *Applied Statistics*, 29:119–127, 1980.
7. Kononenko I. and Bratko I. Information-based evaluation criterion for classifier's performance. *Machine Learning*, 6:67–80, 1991.
8. Suykens J. A. K. and Vanderwalle J. Least square support vector machine classifiers. *Neural Processing Letters*, 9:293–300, 1999.
9. Bradley A. P. The use of the area under the roc curve in the evaluation of machine learning algorithms. *Pattern Recognition*, 30:1145–1159, 1997.
10. Flach P. The geometry of roc space: Understanding machine learning metrics through roc isometrics. *Proceedings of the twentieth International Conference on Machine Learning (ICML 2003)*, pages 194–201, 2003.
11. Quinlan R. Induction of decision trees. *Machine Learning*, 1:81–106, 1986.
12. Quinlan R. *C4.5: Programs for Machine Learning.* Morgan Kaufmann, 1993.
13. Gammerman A. Saunders C. and Vovk V. Ridge regression learning algorithm in dual variables. *Proceedings of the fifteenth International Conferencence on Machine Learning (ICML 1998)*, pages 515–521, 1998.
14. Bhattacharya P. Sindhwani V. and Rakshit S. Information theoretic feature crediting in multiclass support vector machines. *Proceedings of the first SIAM International Conference on Data Mining*, 2001.
15. Tsamardinos I. Hardin D. Statnikov A., Aliferis C. F. and Levy S. A comprehensive evaluation of multicategory classification methods for microarray gene expression cancer diagnosis. *Bioinformatics*, 25:631–643, 2005.
16. Diettrich T. and Bakiri G. Solving multiclass learning problems via error-correcting output codes. *Journal of Artificial Intelligence Research*, 2:263–286, 1995.
17. M. H. Zweig and G. Campbell. Receiver operating characteristic (roc) plots. *Clinical Chemistry*, 29:561–577, 1993.

A

About the Authors

Jérôme Azé is an Assistant Professor in the Bioinformatics group at the LRI (Orsay, France). He hold a Ph.D. in Computer Science in 2003 at the University of Paris-Sud Orsay, France. The research subject of his Ph.D. concerns association rules extraction and quality measures in data mining. Since 2003, he is working in Data Mining applied to specific biological problems. These, consist of protein-protein interaction on one hand and learning functional annotation rules applied to bacterial genome on the other hand.

Laure Berti-Équille is currently a permanent Associate Professor at the Computer Science Department (IFSIC) of the University of Rennes (France). Her research interests at IRISA lab (INRIA-Rennes, France) are multi-source data quality, quality-aware data integration and query processing, data cleaning techniques, recommender system, and quality-aware data mining. She published over thirty papers on international conferences, workshops, and refereed journals devoted to database and information systems. She co-organized the second edition of the ACM workshop on Information Quality in Information Systems (IQIS) in conjunction with ACM SIGMOD/PODS conference in Baltimore in 2005. She has also initiated and co-organized the first editions of the French workshop entitled "Data and Knowledge Quality (DKQ)" in Paris (DKQ'05) and in Villeneuve d'Ascq (DKQ'06) in conjunction with the French conference EGC (Extraction et Gestion des Connaissances).

Julien Blanchard earned his Ph.D. in 2005 from Nantes University (France) and is currently an assistant professor at Polytech'Nantes[1]. He is the author of a book chapter and 7 journal and international conference papers in the areas of visualization and interestingness measures for data mining.

Henri Briand is a professor in computer science at Polytech'Nantes[1]. He earned his Ph.D. in 1983 from Paul Sabatier University of Toulouse, and has over 100 publications in data mining and database systems. He was at the head of the computer engineering departement of Polytech'Nantes, and was in charge of a research team in data mining. He is responsible of the organization of the data mining master in university of Nantes.

Peter Christen is a Lecturer and Researcher at the Department of Computer Science at the Australian National University in Canberra. His main research interests are in data mining, specifically data linkage and data-preprocessing. Other research interests include parallel and high-performance computing aspects within data mining. He received his Diploma in Computer Science Engineering from the ETH Zürich in 1995 and his PhD in Computer Science from the University of Basel (Switzerland) in 1999.

Jean Diatta is an Assistant Professor in Computer Science at the Université de la Réunion, France. He received his Ph.D (1996) in Applied Mathematics from the Université de Provence, France, and his accreditation to supervise research (2003) in Computer Science from the Université de La Réunion, France. His research interests include numerical clustering, conceptual clustering, and data mining.

Béatrice Duval is an Assistant Professor in Computer Science at the University of Angers in France. She received her Ph.D. Degree in Computer Science in 1991 from the University of Paris XI, France. Her research interests include Machine learning, specially for non monotonic formalisms, and Data Mining. In the field of Bioinformatics, she studies the problem of gene selection for classification of microarray data.

Mary Felkin is a PhD student at the Department of Computer Science in the University of Orsay, under the supervision of Yves Kodratoff. Her area of research is robotics and artificial intelligence. She has a BSc in Computer Science from the University of Bath, an MSc in Machine Learning from the University of Bristol and a DEA in Data Mining from the University of Lyon2.

Liqiang Geng is currently working as a software developer at ADXSTUDIO, Inc. He received a Ph.D. in Computer Science in 2005 from the University of Regina in Regina, Canada. He has authored papers for book chapters, journals, and conference proceedings in the areas of data mining, data cubes, expert systems, and case-based reasoning systems.

Karl Goiser is a PhD student at the Department of Computer Science at the Australian National University in Canberra. His area of research is in data linkage within data mining. He has previously worked as a social researcher and software engineer. He has a bachelor of computing with honours from the University of Tasmania.

Régis Gras is an Emeritas professor at Polytech'Nantes[1] and he is member of the "KnOwledge and Decision" team (KOD) in the Nantes-Atlantic Laboratory of Computer Sciences (LINA CNRS 2729). He has made his first researches on Stockastic dynamic programming. Then, while teaching to professors of mathematics, he was interested in didactics of mathematics. In this context, he has designed a set of methods gathered in his "Statistical Analysis Implicative" original approach. Since, he continues to developp and extend this approach to data mining issues.

Fabrice Guillet is an Assistant Professor in Computer Science at Polytech'Nantes[1], and he is member of the "KnOwledge and Decision" team (KOD) in the Nantes-Atlantic Laboratory of Computer Sciences (LINA CNRS 2729) since 1997. He hold a PhD in Computer Sciences in 1995 at the Ecole Nationale Supérieure des Télécommunications de Bretagne. He is a foundator member of the "Knowledge Extraction and Management" French-speaking association of research[2], and he is involved also in the steering committee of the annual EGC French-speaking conference since 2001. His research interests include knowledge quality and knowledge visualization in the frameworks of Data Mining and Knowledge Management.

Howard J. Hamilton received a B. Sc. degree with High Honours and an M. Sc. degree in Computational Science from the University of Saskatchewan in Saskatoon, Canada and a Ph. D. degree from Simon Fraser University in Burnaby, Canada. He has been a professor at the University of Regina since 1991. Since 1999, he has also directed the Laboratory for Computational Discovery. He is a member of the Canadian Society for Computational Studies of Intelligence, the American Association for Artificial Intelligence, the Association for Computing Machinery, and the IEEE Computer Society. His research interests include knowledge discovery, data mining, machine learning, temporal representation and reasoning, and computer animation.

Robert J. Hilderman received his B.A. degree in Mathematics and Computer Science from Concordia College at Moorhead, Minnesota in 1980. He worked as a consultant in the software development industry from 1980 to 1992. He received his M.Sc. and Ph.D. degrees in Computer Science from the University of Regina at Regina, Saskatchewan in 1995 and 2000, respectively, and is currently employed there as an Associate Professor. His research interests include knowledge discovery and data mining, parallel and distributed computing, and software engineering. He has authored papers for refereed journals and conference proceedings in the areas of knowledge discovery and data mining, data visualization, parallel and distributed

[1] Polytechnic post-graduate School of Nantes University, Nantes, France
[2] Extraction et Gestion des Connaissances (EGC) `http://www.polytech.univ-nantes.fr/associationEGC`

algorithms, and protocol verification and validation. He co-authored a book entitled "Knowledge Discovery and Measures of Interest" published by Kluwer Academic Publishers in 2001.

Xuan-Hiep Huynh works at the Department of Information System, College of Information Technology, Can Tho University (CIT-CTU), Vietnam since 1996. He is currently a PhD student at the Nantes-Atlantic Laboratory of Computer Sciences (LINA CNRS 2729), Polytech'Nantes[1], France. He received his Diploma in Computer Science Engineering from Can Tho University in 1996, and a Master in Computer Science from Institut de la Francophonie pour l'Informatique (Ha No, Vietnam) in 1998. Since 1999 he is a Lecturer at CIT-CTU and served as a vice-director of the Department of Information System (CIT-CTU) from 2002 to 2004. His main research interests are in data mining, specifically measures of interestingness. Other research interests include data analysis, artificial intelligence, and fuzzy logic.

Pascale Kuntz is Professor in Computer Science at Polytechn'Nantes[1] since 2002. She is at the head of the KOD (KnOwledge and Decision) team at the Nantes-Atlantic Laboratory of Computer Sciences (LINA CNRS 2729). From 1992 to 1998 she was Assistant Professor at the Ecole Nationale Supérieure des Télécommunications de Bretagne; and from 1998 to 2001 at the Polytech'Nantes[1]. She hold a PhD in Applied Mathematics in 1992 at the Ecole des Hautes Etudes en Sciences Sociales, Paris. She is at the editorial board of Mathématiques and Sciences Humaines and the Revue dIntelligence Artificielle. Until 2003 she was the editor of the bulletin of the French-speaking Classification Society. Her main research of interest are classification, data mining and meta-heuristics.

Stéphane Lallich is currently Professor in Computer Sciences at the University of Lyon with ERIC laboratory. His research interests are related to knowledge discovery in databases, especially the statistical aspects of data mining.

Philippe Lenca is currently Associate Professor in Computer Sciences at the ENST Bretagne, a French graduate engineering school. ENST Bretagne is a member of the Group of Telecommunications Schools. His research interests are related to knowledge discovery in databases and decision aiding, especially the selection of rules.

Israël-César Lerman is Professor Emeritus at the University of Rennes1 and researcher at the IRISA computing science institute in the Symbiose project (Bioinformatics). His research domain is data classification (foundations, methods, algorithms and real applications in many fields). Several methods have been built, among them the Likelihood of the Linkage Analysis (LLA) hierarchical classification method. His most important contribution adresses the problem of probabilistic comparison between complex structures in data analysis and in data mining (association coefficients between combinatorial and relational attributes, similarity indices between complex objects, similarity indices between clusters, statistical tools for forming class explanation, ...). Other facets of his work concern computational complexity of clustering algorithms, pattern recognition, decision trees, satisfiability problem, ... He began his research at the Maison des Sciences de l'Homme (Paris) in 1966. Since 1973 he is Professor at the University of Rennes 1. He hold a PhD in Mathematical Statistics in 1966 at the University of Paris 6. The author recieved the diploma of "Docteur ès Sciences Mathématiques" in 1971 at the University of Paris 6. For many years he has been at the editorial board of several journals: RAIRO-Operations Research (EDP Sciences), Applied Stochastic Models and Data Analysis (Wiley), Mathématiques et Sciences Humaines (Mathematics and Social Sciences) (EHESS, Paris), La Revue de Modulad (INRIA). He wrote two books (1970, 1981) and more than one hundred papers, partly in french and partly in english. His second book "Classification et Analyse Ordinale des Données" (Dunod, 1981) has been edited in a CD devoted to the out of print classics in Classification by the Classification Society of North America (2005).

Patrick Meyer currently works at the University of Luxembourg, where he finishes his Ph.D. thesis. He received a Master's degree in Mathematics in 2003 at the Faculté Polytechnique of Mons in Belgium. His main research interests are in the domain of multiple criteria decision aiding. In the past he has worked on projects involving financial portfolio management and analysis of large amounts of financial data from stocks. He has recently contributed to the development of the R package Kappalab, a toolbox for capacity and integral manipulation on a finite setting.

Rajesh Natarajan currently works as Assistant Manager-Projects at Cognizant Technology Solutions, Chennai, India. He is a Fellow of the Indian Institute of Management Bangalore (IIMB) and has served as Assistant Professor, IT and Systems Group at the Indian Institute of Management Lucknow, India for two years. He has also worked in Reliance Industries Limited as Assistant Manager-Instrumentation for about two years. He has published over ten articles in various refereed international journals such as Fuzzy Optimization and Decision Making and international conferences like IEEE ICDM 2004, ACM SAC 2005 and others. He has served in the Program committee of the Data Mining Track of ACM SAC 2006. His research interests include Artificial Intelligence, Systems Analysis and Design, Data Modeling, Data Mining and Knowledge Discovery, Applications of Soft Computing Methods and Knowledge Management.

Terry Peckham received his M.Sc. degree in Computer Science from the University of Regina at Regina, Saskatchewan in 2005. He is currently employed as an Instructor with the Saskatchewan Institute of Applied Science and Technology in the Computer Systems Technology program. His research interests include knowledge discovery and data mining, human computer interaction, and 3-D visualization of large scale real-time streaming data.

Elie Prudhomme is currently PhD Student in Computer Science at the ERIC laboratory, University of Lyon, France. His main research interests concern High-Dimensional data analysis with focus on features selection and data representation.

Henri Ralambondrainy is a Professor of Computer Science at the Université de la Réunion, France. He received his Ph.D (1986) from the Université de Paris Dauphine, France. His research interests include numerical clustering, conceptual clustering, classification, statistical implication in data mining.

Ansaf Salleb received her Engineer degree in Computer Science in 1996 from the University of Science and Technology (USTHB), Algeria. She earned the M.Sc. and Ph.D. degrees in Computer Science in 1999 and 2003 respectively, from the University of Orleans (France). From 2002 to 2004, Ansaf worked as an assistant professor at the University of Orleans and as a postdoctoral Fellow between 2004 and 2005 at the French national institute of computer science and control (INRIA), Rennes (France). She is currently an associate research scientist at the Center for Computational Learning Systems, Columbia University, where she is involved in several projects in Machine Learning and Data Mining.

B. Shekar is Professor of Quantitative Methods and Information Systems at the Indian Institute of Management Bangalore (IIMB), India. He has over fifteen years of rich academic experience. Prior to completing his PhD from the Indian Institute of Science, Bangalore in 1989, he worked in various capacities in the industry for over ten years. He has published over 35 articles in refereed international conferences and journals including Decision Support systems, Fuzzy sets and systems, Fuzzy optimization and Decision Making, Pattern Recognition and others. He has served in the Program Committee of various international conferences and has been actively involved in reviewing articles for international journals like IEEE SMC and others. He is the recipient of The Government of India Merit Scholarship (1969-74) and has been listed in Marquis "Who's Who in the World 2000" and Marquis "Who's Who in Science and Engineering, 2005-2006." His research interests include Knowledge engineering and management, Decision support systems, Fuzzy sets and logic for applications, Qualitative reasoning, Data Modeling and Data Mining.

Olivier Teytaud is research fellow in the Tao-team (Inria-Futurs, Lri, université Paris-Sud, UMR Cnrs 8623, France). He works in various areas of artificial intelligence, especially at the intersection of optimization and statistics.

André Totohasina is an Assistant Professor in mathematics at the Université d'Antsiranana, Madagascar. He received his M.Sc.(1989) from Université Louis Pasteur, France, and Ph.D (1992) from the Université de Rennes I, France. His research interests include statistical implication in data mining, its application in mathematics education research, mathematics education, teacher training in mathematics.

Benoît Vaillant is a PhD Student of the ENST Bretagne engineering school, from which he graduated in 2002. He is working as an assistant professor at the Statistical and Computer Sciences department of the Institute of Technology of Vannes. His research interests are in the areas of data mining and the assessment of the quality of automatically extracted knowledge.

Christel Vrain is professor at the university of Orlans, France. She works in the field of Machine Learning and Knowledge Discovery in Databases. Currently, she is leading the Constraint and Learning research group in LIFO (*Laboratoire d'Informatique Fondamentale d'Orlans*) Her main research topics are Inductive Logic Programming and Relational Learning and with her research group, she has addressed several Data Mining tasks: mining classification rules, characterization rules, association rules. She has worked on several applications, as for instance mining Geographic Information systems in collaboration with BRGM (a French institute on Earth Sciences), or text mining for building ontologies.

Index

MIX
Papier aus verantwortungsvollen Quellen
Paper from responsible sources
FSC® C105338
FSC
www.fsc.org

If you have any concerns about our products,
you can contact us on
ProductSafety@springernature.com

In case Publisher is established outside the EU,
the EU authorized representative is:
Springer Nature Customer Service Center GmbH
Europaplatz 3, 69115 Heidelberg, Germany

Printed by Libri Plureos GmbH
in Hamburg, Germany